CONSTITUTIVE AND CENTRIFUGE MODELLING: TWO EXTREMES

PROCEEDINGS OF THE WORKSHOP ON CONSTITUTIVE AND CENTRIFUGE MODELLING: TWO EXTREMES, MONTE VERITÀ, SWITZERLAND, 8–13 JULY 2001

Constitutive and Centrifuge Modelling: Two Extremes

Edited by

Sarah Springman
Institute for Geotechnical Engineering, Swiss Federal Institute of Technology, Zurich, Switzerland

A.A. BALKEMA PUBLISHERS   LISSE / ABINGDON / EXTON (PA) / TOKYO

Published by: A.A. Balkema, a member of Swets & Zeitlinger Publishers
        www.balkema.nl and www.szp.swets.nl

ISBN 90 5809 361 1

Printed in The Netherlands

Constitutive and Centrifuge Modelling: Two Extremes, Springman (ed.)
© 2002 Swets & Zeitlinger, Lisse, ISBN 90 5809 361 1

Table of contents

Problems governed by deformation

Problems governed by interfaces

Closure

Foreword: Modelling…& truth…at Monte Verità

S.M. Springman
Institute for Geotechnical Engineering, Swiss Federal Institute of Technology, Zurich, Switzerland

ABSTRACT: This book has arisen from a causative chain of events, stimulated largely by the outstanding challenges offered to three of the organisers of the associated workshop by researching and teaching within the Soil Mechanics Group at Cambridge University. This was followed by appointment of the author to a chair at the Swiss Federal Institute of Technology in Zurich in January 1997, where subsequent research themes have included constitutive modelling of typical Swiss soils together with the development of a geotechnical centrifuge facility.

1 INTRODUCTION

An international workshop was planned for the summer of 2001 to celebrate on a number of levels. Primarily, the purchase, manufacture and commissioning of the ETHZ Geotechnical Drum Centrifuge for research and teaching, one of the two largest 'geotech drums' in the world, was to be signalled by a public inauguration and followed by a stimulating and related scientific event. A parallel goal was to recognise the contribution of Professor Andrew Schofield FRS (Fig. 1) to geomechanics, as an informal Festschrift. Andrew had inspired three of the organisers, initially as a teacher and later as a colleague. His early work in constitutive modelling, followed by his crusade for world-wide acceptance of centrifuge modelling as an effective, and sometimes the only possible, design tool, has been unsurpassed.

Figure 1. Professor Andrew Schofield FRS (painting by Margaret Schofield, photograph by Sarah Springman)

2 PLANNING THE WORKSHOP

The two themes of constitutive and centrifuge modelling were adopted in the present format following brainstorming between Professor Malcolm Bolton (Cambridge University), Professor David Muir Wood (formerly a lecturer at Cambridge University and now Head of the Engineering Department at Bristol University) and the author (see Fig. 2), who is most grateful to them for their stimulating ideas as well as the discussion arising from these! Dr Lisa Martinenghi of the Swiss Italian University joined the organising team once the location of the event had been settled in the Italian speaking canton of Ticino. She organised the local events, including a public lecture on the leaning tower of Pisa and the excursions to the Verzasca valley and the Cardada, of which photographic evidence can be seen later in this foreword.

Workshop participants included constitutive modellers and centrifuge modellers from 5 continents, although the majority came from Europe. Key lecturers were invited alongside other senior and younger researchers, all of whom were prepared to challenge established orthodoxies and to work across the physical - theoretical - practical modelling divide.

Figure 2. Malcolm Bolton (left) and David Muir Wood (right) (drawn by Margaret Schofield)

3 MONTE VERITÀ & THE MEANING OF TRUTH AND KNOWLEDGE

This international workshop took place in July 2001 at the Centro Stefano Franscini (CSF) on the Monte Verità (Figs 3a, b, 4), the Hill of Truth, just overlooking Ascona (Fig. 3c). Not only is the CSF a world-renowned conference centre, set in idyllic surroundings overlooking Lago Maggiore, it also has had an extremely colourful past in the search for truth (Riess 1964, Landmann 2000). At the turn of the 19th century, a group with variously bohemian, naturistic, theosophistic, communistic even anarchistic tendencies settled there in order to escape the tedium of a European society, shortly to be riven by the 1st World War. A new way of life was promoted as counterpoint to capitalism and commercialism, in which free love, vegetarianism and many forms of artistic expression were practised and pursued on the Monte Verità, which also possesses an extremely high magnetic attraction based on the high radium content locally in the area! Colonies of artists of various genre (e.g. Dadaists, Bauhaus school, Swiss expressionists) inhabited Ascona for significant periods over the last century. A centre was founded by Rudolf von Laban in 1910 for new forms of dancing, as an art, as a ritual and as a religion. Over the ensuing decades, theatre, poetry, psychoanalysis, video and cinematographic art have rounded out some of the activities that have taken place on the Mountain of Truth. Isidora Duncan, Carl Gustav Jung, Herman Hesse, Paul Klee and Michael Balint are counted among the many famous visitors. The Museum at the Casa Anatta (Fig. 3b) contains a permanent and impressive exhibition of this exciting past.

(a) Hotel: Bauhaus style (drawn by Margaret Schofield)

(b) Museum: Casa Anatta (left) and Casa Giovanna (right) (drawn by Margaret Schofield)

(c) Lake frontage, Lake Maggiore: Ascona (drawn by Margaret Schofield)

Figure 3. Buildings and scenes from the Centro Stefano Franscini, Monte Verità and Ascona

Figure 4. Monte Verità from Cardada, with Ascona to the left and adjacent to Lake Maggiore (IGT)

Figure 5. Concert by Jean and Laurence Sulem (photography by Ivo Herle)

It is fitting that the workshop should have been privileged to have had both an artist in residence, Margaret Schofield, and musicians in residence, Jean and Laurence Sulem. Margaret has contributed the beautiful frontispiece of the Casa Giovanna at the entrance to Monte Verità as well as a number of sketches of notable places and events, including those on the preceeding page. Jean and Laurence gave us a wonderful evening recital on the viola and piano (Fig. 5), reflecting Jean's dual role as a professor of geomechanics at CERMES and of viola at the Conservatoire National Supérieur de Musique de Paris! I am grateful to them both for helping us to appreciate the artistic heritage of Monte Verità.

Bertrand Russell (1872-1970) was himself extremely active parallel to the most exciting periods in the history of Monte Verità (Fig. 4). He has not been reported to have visited Monte Verità, although he was moved to thoughts on problems in philosophy in the early 20th century (Russell 1912, reprinted 1998). He pointed out that *error* can creep into our knowledge of truth and he referred to a *dualism*. We may believe what is false as well as what is true since people hold incompatible and very strong opinions on many subjects, and so some of these beliefs must be erroneous. So in considering what is truth, he observed that there were three requisites to be fulfilled by the theory being postulated. Firstly, the theory of truth must be able to permit the opposite, falsehood, to emerge. Secondly, a world without truth and falsehood would be a world of mere matter because 'things' cannot be untrue, whereas beliefs and statements can be. Finally, truth and falsehood are properties dependent upon the relationship of the belief to, for example, a known past series of events and not to the internal quality of the belief itself.

Russell discussed the requirements for truth, examining whether *coherency* would be sufficient, in that falsehood would be the failure of the body of our beliefs to be coherent, leading to the conclusion that there can be only one coherent body of belief. But in science, two or more hypotheses may be coherent (we may choose in our context to think of continua and of discrete systems...) and account for all the known facts on the subject - trying to find facts to expunge all the hypotheses except for one, which would be the truth, may not succeed! Russell came down in favour of the need for beliefs to have correspondence with *fact*, which might be determined by the constituents of a judgement in respect of an order of, for example, these constituents, e.g., between a subject and an object. He cited a complex unity - when belief is true - as corresponding to a certain associated complex. Certain truths might also be self evident, ensuring infallibility and constituting *knowledge* of the fact. Perhaps this knowledge could be described, as Russell did, as derivative knowledge, with a degree of self-evidence, rather than intuitive knowledge - which can be dangerous!

To summarise Russell's statements in the light of our deliberations at Monte Verità, coherence cannot be a definition of truth but may be used as a criterion, and a greater part of what would pass as knowledge, is more or less *probable opinion*. A body of individually probable opinions - if mutually coherent - become more probable than one alone and this is how scientific hypotheses acquire probability. Order and coherence lead to probable opinion delivering near certainty of truth, but the organisation of probable opinions never transforms into indubitable knowledge alone and what we firmly believe, if it is not true, is called error!

4 CONSTITUTIVE AND CENTRIFUGE MODELLING: TWO EXTREMES

Armed with these thoughts, knowledge of the tradition at Monte Verità and historical and mythological examples from many different cultures (that wondrous revelations of truth often take place on top of a mountain: Mount Sinai, Tabor, Olympus, Parnassus and Fujiyama have been cited by some authors, e.g. Fraccaroli 2001), truth (Veritas) together with anarchy of thought, became watchwords for the conference, which, as its title acknowledges, is challenging two extremes of modelling. Refreshing discussions delved into the current state of the art as well as what could not be modelled effectively at present. Typical considerations identified by the organisers were:

- What are the constitutive modelling needs that are thrown up by the observations made in physical models?
- What are the physical tests that can be used to choose between competing constitutive modelling strategies?
- Why do constitutive models, which are clearly inadequate in reproducing phenomena observed either in element or full-scale tests, appear to provide satisfactory predictions of physical geotechnical system performance?

In short the questions related to *'testing, modelling and reality'*. Special lectures were given to set the scene:

- constitutive modelling: alternative frameworks – from continuum to discrete i.e. *'modelling & reality'* (David Muir Wood, Malcolm Bolton)
- centrifuge technology: key developments, philosophy of specific 'tests' or generic 'models' i.e. *'testing, modelling & reality'* (Jan Laue, Fook Hou Lee)
- application of modelling to design:
  - Tower of Pisa: back to the future (public lecture, including the role of physical modelling in solving Pisa's problems) (Carlo Viggiani, Fig. 6)
  - The impact of centrifuge and constitutive modelling on offshore foundation design. (Chris Martin)

Figure 6. Public lecture: Carlo Viggiani (drawn by Margaret Schofield)

Figure 7. Protagonists in the debate (photography by Ivo Herle)

A debate was also held to explore whether centrifuge tests are a sufficient means of solution to geotechnical design problems (Pro: Andrzej Niemunis & Bill Craig, Con: Laurent Vulliet & Diethard König, see left to right, Fig. 7). The outcome was, perhaps, predictable. The criteria "sufficient" led to a slightly uneven playing field in favour of the opponents, who duly won by a handsome margin.

The main focus then transferred to the impact of constitutive modelling and centrifuge modelling on a series of problems that must be confronted regularly and dealt with in geotechnical engineering construction:

- problems governed by FAILURE: e.g. landslides, liquefaction, jack-up platform, pile installation (Theme lecturers: Alessandro Gajo, Ryan Phillips)
- problems governed by DEFORMATION: e.g. retaining walls, foundations, tunnels (Theme lecturers: Ivo Herle, Jitendra Sharma (Fig. 8))
- problems governed by INTERFACES: e.g. piles, reinforcement (Theme lecturers: Jean Sulem, Jacques Garnier)

Each 'theme' was introduced from the perspective of both constitutive and centrifuge modelling to identify the problems and needs, set key challenges for the community at Monte Verità, and to provide the focus for the discussion sessions, which followed. These discussion sessions were kicked off by very short presentations from several young researchers, who stimulated discussion by presenting some of their work and posing questions relating to their research, to be answered from the floor, where possible. These postdoctoral or doctoral researchers (mainly from ETHZ, Bristol, Cambridge, Oxford, University of Western Australia, T.U. Darmstadt), were also invited to comment on each of these three problem series as 'Young reporters'. They were charged with reviewing the progress over the week and reporting back at the end of the workshop, in terms of what they found to be most noteworthy from the proceedings and where they viewed fruitful future research directions to lie.

Figure 8. Jitendra Sharma inspecting a 'retaining wall': the Verzasca Dam and Power Station (Ivo Herle)

Figure 9. The participants (IGT)

Discussions were extremely lively: references to anarchy abounded and there were some heated debates about the way forward and what precisely constituted truth in relation to geotechnical modelling, be it of a constitutive or a physical nature. It was clear that the rival hypotheses still retained sufficient coherence to be able to be considered as a *probable opinion* if not the *certain truth*. Much was made of the exciting work to be done in the future. Perhaps this group (Fig. 9) should meet back on the Monte Verità in a decade, to revisit the past and swap notes on the development of knowledge and truth?

5 A FAMILY DIGRESSION

Coincidence often offers a congruent explanation for happenstance, and the editor of these Proceedings begs leave for a small digression. A certain Dr Eduard Freiherr von der Heydt, originally from Wuppertal in Germany, purchased the commune at Monte Verità in 1926. He had acted as a banker to the Kaiser Wilhelm II before, during and after the war and was a celebrated collector of art. Upon his death in 1964, the Canton of Ticino inherited a far more substantial development than he had purchased, including the present hotel and the original meeting and recital rooms. These remained underused for some time until the canton offered to operate the centre as a Foundation, together with the Ascona Town Council and the two Swiss Federal Institutes of Technology in Zurich and Lausanne, so that world class scientific events could be held there on a regular basis.

The family tree (Fig. 10) traces the relationship of the Freiherr to his great great grandfather, Daniel Heinrich von der Heydt (1767-1832), who was himself of humble origins, but was fortunate both to marry an heiress to a Wuppertaler bank, Wilhelmina Kersten and to have talent for financial dealings! He became principal of the newly formed von der Heydt-Kersten bank and sired a large family. Leaving the bank primarily in the hands of his second son, Daniel, his eldest son August (1801-1874) is reported to have 'dabbled' in politics, however this was clearly successful because it culminated in him becoming finance minister for Bismarck during the war

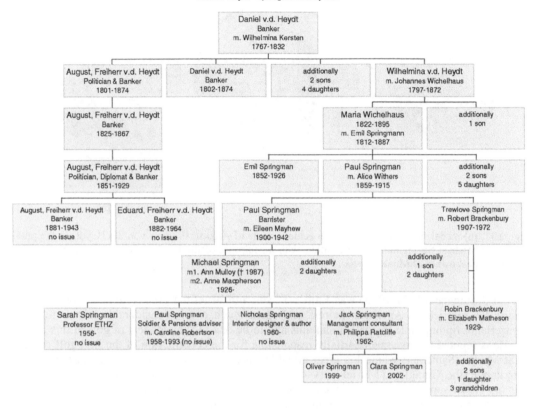

Figure 10. Elements from the von der Heydt-Springman family tree

against Austria in 1866. The medal in Figure 11a recognises his contribution posthumously: 'Er schaffte die Mittel für den Krieg 1866' – he raised the funds for the war of 1866. He was elevated to becoming a Freiherr in 1863 for his service to Prussia, more or less the equivalent of a baronet within the British aristocracy. His younger great grandson was none other than the Eduard (1882-1964) who settled at Monte Verità, and who was himself a significant patron and collector of contemporary, oriental and primitive art (Riess 1964). The von der Heydt Museum in Wuppertal and the Rietberg Museum in Zurich are two collections that have benefited greatly from his generosity and artistic acuity. Indeed, the year 2001 was celebrated by the von der Heydt Museum as the 150th Anniversary of the birth of Eduard's father, during which there were exhibitions of artistic contributions from the family, as well as some aspects of the family history.

Following the female line, one may note the eldest of Daniel's children, who survived the childhood years, Wilhelmina (1797-1872), married a Pastor Wichelhaus. Their daughter, Maria (1822-1895), married a burgher of Wuppertal who left Germany to trade in cotton in Liverpool alongside his brother in law E. Busch. His name was Emil Springmann (1812-1887) and he was the founder of the Cotton & Commodity Broking Firm, Emil Springman & Co. N.B., this 'English' branch of the Springmann family lost one of the 'n's during the mid 19th Century when the next generation are recorded with the surname Springman.

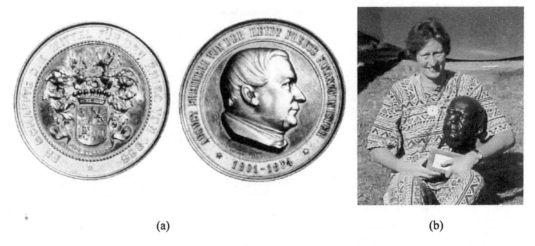

(a)   (b)

Figure 11. (a) medal struck to recognise the achievements of August, Freiherr von der Heydt (in the possession of Michael Springman) (b) the author with a sculpture of the head of Eduard, Freiherr von der Heydt, the past owner of Monte Verità

Paul Springman (1859-1915) was Emil's 3rd son and benefited greatly from his father's industry. He married an American heiress and spent much of his considerable leisure time in Switzerland, finding time to scale mountains and to be one of the pioneers of golf in this country. He is reputed to have been partially responsible for the construction of the first golf course in St Moritz (Samedan) where his great granddaughter now carries out research in the mountains above (the Muragl and Murtel Corvatsch rock glaciers)! Paul Eyre Springman (1900-1942) was his eldest son, and the editor's grandfather, serving as a barrister (and in the Royal Air Force) prior to being killed in the second world war. The editor is grateful to her second cousin, Robin Brackenbury (q.v.) as well as to her father, Michael Springman (q.v.), for details of the von der Heydt-Springman family history, and for permission to print the pictures in Figure 11a.

So not only has the requirement to learn and teach in German been somewhat of a return to a family past, the election to a chair at the ETHZ and the selection of Monte Verità as the location of the workshop is also a happy coincidence for the author. Figure 11b shows that family resemblance is not greatly in evidence, perhaps a result of 5 generations of cross breeding with other Irish, English and American ancestors.

ACKNOWLEDGEMENTS

The Centro Stefano Franscini is to be thanked for hosting a thoroughly fruitful workshop. Claudia Lafranchi, Mr and Mrs Stevenoni-Albertini and Ms Balli (in Ascona) and Brigitta Pichler (in Zurich) are to be thanked for their patience and ever ready offers of assistance. The CSF, led by Prof. Hannes Flühler, were also generous in granting money towards the lecturers' travelling costs, scholarships for most of the younger researchers and in underwriting some of the daily costs via financial contributions from both Swiss Federal Institutes of Technology and the Swiss National Science Foundation. The 'Scientific Co-operation between Eastern Europe and Switzerland' SCOPES 2000-2003 programme, also run by the Swiss National Science Foundation, assisted by funding the majority of the travelling and attendance costs for one lecturer (Dr. Ivo Herle) and one discussion leader (Dr. Jan Boháč) from the Czech Republic.

This workshop would not have taken off without the calm and efficient organisation of my multilingual secretary, Therese Frei. This was her last major challenge in the Institute prior to taking her well-earnt retirement, and this book stands as a testament to 23 years of loyal service. Gabriela Laios has kindly taken up the baton and has been a willing and enthusiastic helper during the final editing of the papers. I am also grateful to Kalman Kovári for introducing me to the works of Bertrand Russell, to Ivo Herle for permission to use his photographs and to Mengia Amberg, who assisted with some of the figures. Dr, now Prof., Jitendra Sharma was ably aided by Thomas Weber and Ravikiran Chikatamarla in providing technical support during the workshop, including recording the discussion sessions and ensuring that the various forms of electronic assistance functioned smoothly. Dr. Jan Laue, Lukas Arenson and Philippe Nater were willing guides and drivers for our visitors. Thomas Weber faithfully reproduced the recordings of the discussion sessions. He has also been of tremendous assistance in dealing with the complications associated with editing large files.

REFERENCES

Fraccaroli, M. 2001. Website to Centro Stefano Franscini. www.csf.ethz.ch.
Landmann, R. 2000. Ascona - Monte Verità: Auf der Suche nach dem Paradies. Reprinted. Frauenfeld: Huber.
Riess, C. 1964. Ascona. Die Geschichte des seltsamsten Dorfes der Welt. Zürich: Europa.
Russell, B. 1912, reprinted 1998. The problems of philosophy. Oxford: Oxford University Press.

Special lectures & events

Constitutive and Centrifuge Modelling: Two Extremes, Springman (ed.)
© 2002 Swets & Zeitlinger, Lisse, ISBN 90 5809 361 1

The Tower of Pisa is back to the future

Carlo Viggiani
University of Napoli Federico II, Naples, Italy

ABSTRACT: The increasingly hazardous inclination of the Leaning Tower of Pisa has been the subject of intense investigations in recent years. Past history has shown that any attempt to reduce the inclination has met with a contrary response from the Tower. On this occasion, however, it has been possible to "right" the Tower by half a degree. The inclination has been reduced to 5°, a value equivalent to that from about 160 years ago.

1 INTRODUCTION

A cross section of the Leaning Tower of Pisa is reported in Figure 1. It is nearly 60 m high and the foundation is 19.6 m in diameter; the weight is 141.8 MN. In the early 90's, the foundation was inclined southwards at about 5.4° to the horizontal. The average inclination of the axis of the tower to the vertical is somewhat less, due to its slight curvature resulting from corrections that were made by masons during the construction, to counteract the inclination already occurring. The seventh cornice overhangs the first one by about 4.1 m.

Figure 2 shows the ground profile underlying the tower. It consists of three distinct horizons. Horizon A is about 10 m thick and primarily consists of estuarine deposits, laid down under tidal conditions. As a consequence, the soil consists of rather variable interbedded sandy and clayey silts. At the bottom of horizon A, there is a 2 m thick medium dense fine sand layer. Based on sample descriptions and piezocone tests, the materials to the south of the tower appear to be more silty and clayey than to the north and the sand layer is locally thinner. This is believed to be at the origin of the southward inclination of the tower.

Horizon B consists primarily of marine clay which extends to a depth of about 40 m. It is subdivided into four distinct layers. The upper layer is soft sensitive clay locally known as the Pancone, underlain by an intermediate layer of stiffer clay, which in turn overlies a sand layer (the intermediate sand). The bottom layer of horizon B is normally consolidated clay known as the lower clay. Horizon B is very uniform laterally in the vicinity of the tower. Horizon C is dense sand (the lower sand) which extends to considerable depth.

The water table in horizon A is between 1 m and 2 m below the ground surface, depending on the rain intensity and duration. Pumping from the lower sand has resulted in downward seepage from horizon A with a pore pressure distribution below hydrostatic through horizons B and C.

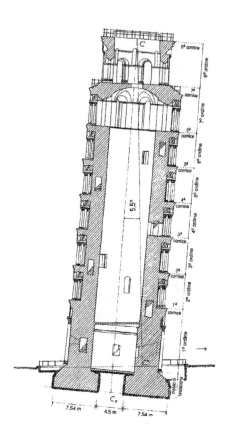

Figure 1. Cross section of the Leaning Tower of Pisa

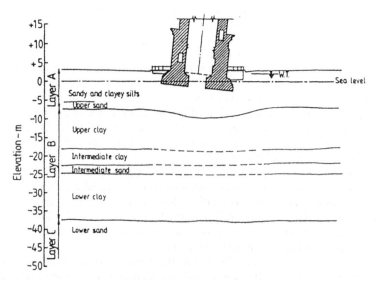

Figure 2. The subsoil of the leaning tower

4

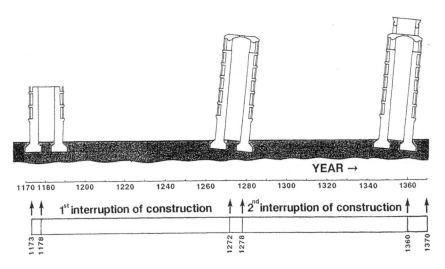

YEAR →

| 1170 1180 | 1200 | 1220 | 1240 | 1260 | 1280 | 1300 | 1320 | 1340 | 1360 |

1st interruption of construction   2nd interruption of construction

1173 1178   1272 1278   1360 1370

Figure 3. History of the construction

The many borings beneath and around the tower show that the surface of the Pancone clay is dished beneath the tower, from which it can be deduced that the average settlement of the monument is approximately 3 m. Fuller details about the tower and its subsoil, including a wide list of references, are reported by Burland et al. (1999).

Work on the tower began in 1173 (Fig. 3). Construction had progressed to half the 4th storey by 1178, when the work was interrupted. The reason for stoppage is not known, but had it continued much further, the foundations would have experienced an undrained bearing capacity-failure. The work recommenced in 1272, after a pause of nearly 100 years, by which time the strength of the ground had increased due to consolidation under the weight of the structure. By about 1278, construction had reached the 7th cornice when work stopped again. Once again, there can be no doubt that, had work continued, the tower would have fallen over. In about 1360 work on the bell chamber was commenced and was completed in about 1370, two centuries after the start of the work.

It is known that the tower must have been tilting to the south when work on the bell chamber began, as it is noticeably more vertical than the remainder of the tower. Indeed on the north side there are four steps from the seventh cornice up to the floor of the bell chamber, while on the south side there are six steps (Fig. 4).

Another important detail of the history of the tower is that in 1838 a walkway was excavated around the foundation. This is known as the *catino*, and its purpose was to expose the column plinths and foundation steps for all to see, as was originally intended. The operation resulted in an inrush of water on the south side, since here the excavation is below the water table.

A reliable clue to the history of the tilt lies in the adjustments made to the masonry layers during construction and in the resulting shape of the axis of the tower. This is depicted in Figure 5; based on this shape and a hypothesis on the manner in which the masons corrected for the progressive lean of the tower, the history of inclination of the foundation of the tower may be deduced (as reported in Fig. 6). During the first phase of construction to just above the third cornice (1173 to 1178), the tower inclined slightly to the north. The northward inclination increased slightly during the rest period of nearly 100 years to about 0.2°. When construction recommenced in about 1272, the tower began to move towards the south and accelerated shortly before construction reached the seventh cornice in about 1278, when work again ceased, at which stage the inclination was about 0.6° towards the south.

5

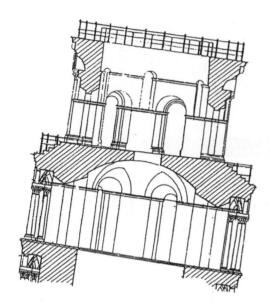

Figure 4. Detail of the bell chamber

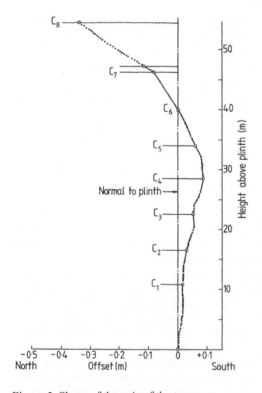

Figure 5. Shape of the axis of the tower

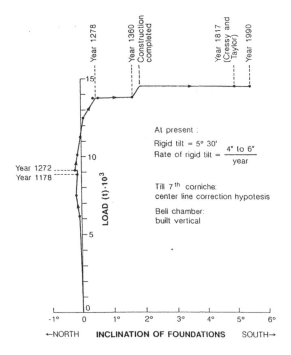

Figure 6. History of inclination

During the next 90 years, the inclination increased to about 1.6°. After the completion of the bell chamber in about 1370, the inclination of the tower increased significantly. In 1817, when Cresy and Taylor made the first recorded measurement with a plumb line, the inclination of the tower was about 4.9°. The excavation of the *catino* in 1834 appears to have caused an increase of inclination of approximately 0.5° and the inclination of the foundation in the early 1990s was about 5.5°.

In 1990, the Italian Government appointed an International Committee to safeguard and stabilise the Tower. It was conceived as a multidisciplinary body composed of: experts in arts, restoration and materials; structural and geotechnical engineers.

The Committee has developed a detailed understanding of the history of the inclination of the tower, and in particular of the movements it has experienced in the last century. These have been observed by a very comprehensive monitoring system, installed on the tower since the beginning of the 20th century and progressively enriched by a greater variety and quantity of measuring devices. The behaviour of the tower clearly indicates that it is affected by leaning instability, a phenomenon controlled by the stiffness of the soil rather than by its strength. (Gorbunov Possadov & Serebriany 1961, Habib & Puyo 1970, Schultze 1973, Hambly 1985, Lancellotta 1993, Desideri & Viggiani 1994, Desideri et al. 1997, Potts & Burland 2000).

The present paper reports the various physical and numerical models developed to investigate the stability of the tower and to design the final stabilisation measures.

2 LEANING INSTABILITY

To demonstrate leaning instability, the simple conceptual model of an inverted pendulum may be used. It is a rigid vertical pole (Fig. 7) of length h, with a concentrated mass of weight W at the

7

top and hinged at the base to a constraint that reacts to a rotation α with a stabilising moment $M_s = M_s(\alpha)$. On the other hand, a rotation α induces an overturning moment $M_o = W h \sin\alpha$.

In the vertical position ($\alpha = 0°$), the system is in equilibrium. Let us imagine that a rotation α occurs. If $M_s > M_o$, the equilibrium is stable; the system returns to the vertical configuration. If $M_s < M_o$, the equilibrium is unstable; the system collapses. If $M_s = M_o$, the equilibrium is neutral; the system stays in the displaced configuration. The stability of the equilibrium may be characterised by a safety factor $FS = M_s/M_o$.

Edmonds (1993) performed a number of small scale physical tests on a model tower resting on a bed of fine sand. A sketch of the experimental setup is reported in Figure 8. A model tower with a diameter of 102 mm was placed on top of a very loose fine sand bed, and loaded through a hanger at a height of 126 mm over the base. The ratio 126/102 is approximately equal to that of the height of the centre of gravity of the tower of Pisa to the diameter of its foundation.

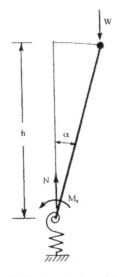

Figure 7. The inverted pendulum

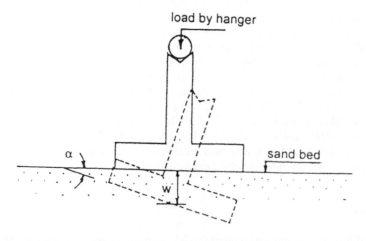

Figure 8. Set up of the experiments by Edmonds (1993)

8

Loading the model tower produced a settlement and a rotation α. A total of 8 load tests were carried out. Failure in all cases was by toppling, with the lowest edge of the model tower's base sinking into the sand as the tower rotated.

The individual plots of α varying with load give somewhat variable results, but when combined into one plot, as in Figure 9, an envelope of results emerges. The envelope shows a pronounced change in curvature at a load of 160 to 165 N, where the inclination averages 0.09 (α = 5°).

Potts & Burland (2000) investigated the differences between leaning instability and the usual bearing capacity failure, using finite element analysis. Figure 10 shows the plane strain problem of a tower resting on a uniform deposit of undrained clay. The clay is modelled as a linear elastic-perfectly plastic Tresca material with an undrained strength s_u = 80 kPa. The tower was given an initial tilt of 0.5°; the self weight was then increased in a large displacement finite element analysis.

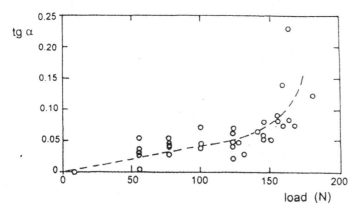

Figure 9. Results of the experiments by Edmonds (1993)

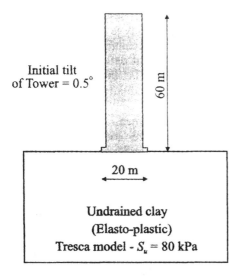

Figure 10. Plane strain problem of a tower resting on a uniform deposit of undrained clay (Potts & Burland 2000)

9

Figure 11 shows the results obtained in two analyses with different shear stiffness G; viz. $G/s_u = 10$ and $G/s_u = 1000$; in both cases the strength of the soil s_u was 80 kPa. The results show that failure occurs abruptly with little warning and that the weight of the tower at failure is dependent on the shear stiffness of the soil. The failure load with $G/s_u = 10$ is about half of that with $G/s_u = 1000$.

Figure 12 shows vectors of incremental displacement for the soft soil, in the last increment of the analysis. The displacements are located in a zone below the foundation and indicate a rotational type of failure. At first sight, this looks like a plastic type of collapse mechanism. However, examination of the zone in which the soil has gone plastic, also shown in Figure 12, indicates that it is very small and not consistent with a plastic failure mechanism. Consequently, this figure indicates a mechanism of failure consistent with leaning instability.

Vectors of incremental displacement just before collapse for the stiffer soil are shown in Figure 13. They indicate a traditional bearing capacity type mechanism, with the soil being pushed outwards on both sides. The plastic zone is very large and therefore the results clearly indicate a plastic bearing capacity type mechanism of failure.

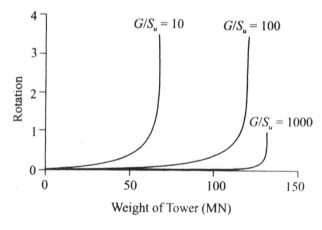

Figure 11. Load-settlement response of the tower on soft and stiff clay (Potts & Burland 2000)

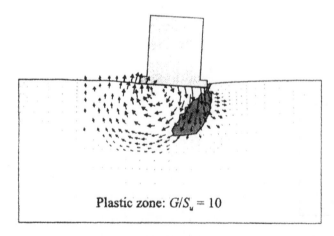

Figure 12. Vectors of incremental displacement at failure, and plastic zone for the soft soil (shaded) (Potts & Burland 2000)

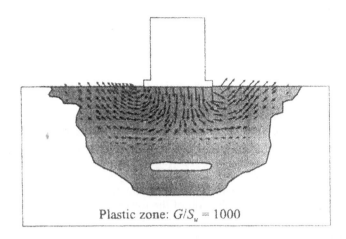

Plastic zone: $G/S_u = 1000$

Figure 13. Vectors of incremental displacement at failure and plastic zone for the stiff soil (Potts & Burland 2000)

3 SIMPLE ANALYTICAL MODELS

Modelling the tower as an inverted pendulum, the restraint exerted by the foundation may be evaluated assuming that the foundation is a circular plate (diameter D) resting on a linearly elastic medium. The latter may be either a Winkler type bed of springs of stiffness k or an elastic half space of constants E, ν. Calling W and $M_o = We$ the vertical load and the overturning moment (e = eccentricity of load), and ρ, α the settlement and rotation of the foundation, it may be shown that:

$$\begin{Bmatrix} \rho \\ \alpha \end{Bmatrix} = \begin{vmatrix} \dfrac{1}{k_\rho} & 0 \\ 0 & \dfrac{1}{k_\alpha} \end{vmatrix} \begin{Bmatrix} W \\ M_o \end{Bmatrix} \qquad (1)$$

The equation shows that there is no coupling between settlement and rotation. For a bed of Winkler type springs, the expression for the stiffness are:

$$k_\rho = \frac{\pi D^2}{4} k = Ak \; ; \quad k_\alpha = \frac{\pi D^4}{64} k = Jk \qquad (2)$$

with A and J being the area and the moment of inertia of the circular foundation respectively. For an elastic half space, the expressions are:

$$k_\rho = \frac{ED}{1-\upsilon^2} \; ; \quad k_\alpha = \frac{ED^3}{6(1-\upsilon^2)} \qquad (3)$$

In this simple linear elastic model, the stability of equilibrium is thus an intrinsic property of the ground – monument system. For instance, for the case of elastic half space:

$$FS = \frac{M_s}{M_o} = \frac{k_\alpha \alpha}{Wh \sin \alpha} \approx \frac{k_\alpha}{Wh} = \frac{E}{6(1-\upsilon^2)} \frac{D^3}{Wh} \qquad (4)$$

11

having assumed $\alpha \approx \sin\alpha$. For the case of Winkler springs:

$$FS = \frac{\pi D^4}{64} \frac{k}{Wh} \qquad (5)$$

In the case of the tower of Pisa, an evaluation of FS may be obtained based on the knowledge of the settlement of the tower, that is around 3 m. This gives k = 0.157 MN/m^3; E/(1-v^2) = 2.85 MN/m^2. Accordingly, FS = 0.36 for the Winkler springs and FS = 1.12 for the elastic half space. It may be concluded that the linear Winkler springs are less suitable than the elastic half space to model the tower. In any case, even the very simple linearly elastic models allow an important practical conclusion: the tower is very nearly in a state of neutral equilibrium.

For most of the 20th century, the inclination of the tower has been steadily increasing, but the changes of inclination have been extremely small compared with those that occurred during and immediately following the construction. The rate of inclination of the tower in 1990 was about six arc seconds per year. The cause of the continuing movement is believed to be a phenomenon of ratchetting, triggered by the fluctuations of the water table in Horizon A.

4 NON-LINEARITY

The relationship between M_s and α may be linearised over a short interval, but it is certainly non-linear and approaches a limiting value of M_s asymptotically. In a case like that of the tower of Pisa, that is on the verge of instability, consideration of non-linearity appears to be mandatory. To this end, the knowledge of the relationship $M_s = M_s(\alpha)$ is necessary.

In a merely phenomenological approach, the relation may be given *a priori* as a suitable expression, whose parameters may be determined by fitting the available data. In the case of the tower of Pisa an approach of this kind has been adopted by Schultze (1973) using a hyperbola, and by Lancellotta (1993) and Desideri & Viggiani (1994) using an exponential function.

Such a simple approach is valid only for a certain loading path, since the response in terms of moment – rotation relationship depends on the vertical load. The factor of safety has to be defined in incremental terms, as a ratio between the derivatives of M_s and M_o. It is not an intrinsic property of the ground monument system, but depends on the current value of the inclination and decreases with increasing inclination. Taking into account non-linearity, the stability of the tower of Pisa is still worse (Lancellotta 1993, Desideri & Viggiani 1994, Nova & Montrasio 1991).

Centrifuge modelling of the tower and its subsoil has been carried out at ISMES; the results obtained are reported and discussed by Pepe (1995). They gave further insight into the mechanisms of the instability and confirmed the elastoplastic character of the restraint exerted by the foundation and the subsoil on the motion of the tower.

In Figure 14, the properties of the foundation soils of the tower are compared with the properties obtained in the small scale model after consolidation under geostatic load in the centrifuge; the main features of the soil profile are satisfactorily reproduced in the model. Figure 15 reports the simulation of the construction of the tower, as obtained by one of the centrifuge tests. In a first stage of the construction, the model tower may only settle while the rotation is hindered. In a second stage, the model tower is left free to rotate. It may be seen that both the settlement and the rotation, scaled to the prototype, are in good overall agreement with those of the tower.

Finite element analyses of the behaviour of the tower and its subsoil have been carried out using a finite element geotechnical computer program developed at Imperial College and known as ICFEP (Potts & Gens 1984). The constitutive model is based on Critical State concepts and combines non-linear isotropy within the yield locus and work hardening plasticity. Fully

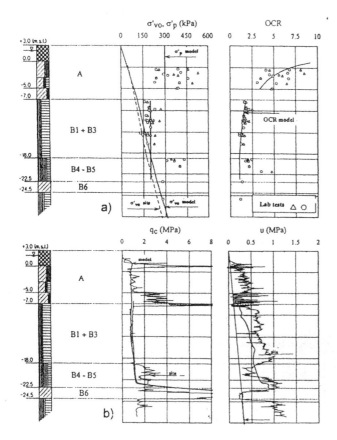

Figure 14. Simulation of the subsoil of the tower for centrifuge experiments

coupled consolidation is incorporated, so that time effects due to the drainage of pore water in the soil skeleton are included.

The prime object of the analysis was to improve the understanding of the mechanisms controlling the behaviour of the Tower (Burland & Potts 1994).

Accordingly, a plane strain approach was used for much of the work, and only later was three dimensional analysis used to explore certain detailed features.

The layers of the finite element mesh matched the soil sub-layering that had been established from soil exploration studies. Figure 16a shows the adopted mesh, while Figure 16b reports the detail of the mesh in the immediate vicinity of the foundation. In Horizon B the soil is assumed to be laterally homogeneous; however a tapered layer of slightly more compressible material was incorporated into the mesh for layer A_1 as shown by the shaded element in Figure 16b. This slightly more compressible region represents a more clayey material found beneath the south side of the foundation; in applied mechanics terms this slightly more compressible tapered layer may be considered as an "imperfection". The overturning moment generated by the lateral movement of the centre of gravity of the tower was incorporated into the model as a function of the inclination of the foundation, as shown in Figure 16b.

The construction history of the tower was simulated by a series of load increments applied to the foundation at suitable time intervals. The excavation of the catino in 1838 was also simu-

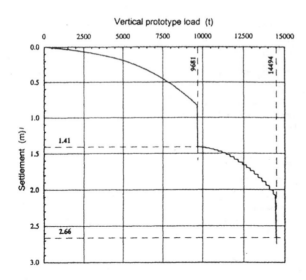

Figure 15. Result of a typical centrifuge simulation of the construction of the tower

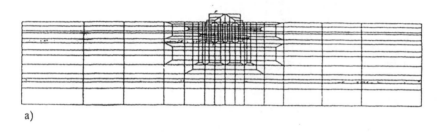

a)

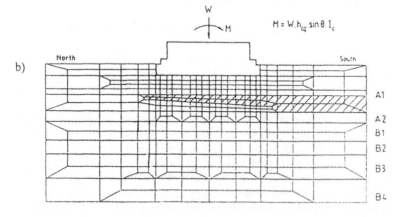

$$M = W.h_{cg}.\sin\theta.I_c$$

b)

Figure 16. Mesh adopted in the finite element model

lated in the analysis. Calibration of the model was carried out by adjusting the relationship between the overturning moment generated by the centre of gravity and the inclination of the foundation. A number of runs were carried out with successive adjustments being made until good agreement was obtained between the actual and the predicted present day value of the inclination.

Figure 17 shows a graph of the predicted changes in inclination of the tower against time, compared with the deduced historical values. From about 1272 onwards there is a remarkable agreement between the model and the historical inclination. Note the excavation of the catino in 1838 which results in a predicted rotation of about 0.75°. The final imposed inclination of the model tower is 5.44°, slightly less than the 1993 value of 5.5°. It was found that any further increase in the final inclination of the tower model resulted in instability: a clear indication that the tower is very close to falling over.

The analysis has demonstrated that the lean of the tower results from the phenomenon of settlement instability due to the high compressibility of the Pancone Clay. The principal effect of the layer of slightly increased compressibility beneath the south side of the foundation is to determine the direction of lean, rather than its magnitude. The main limitation is that the model does not deal with creep. Nevertheless the model provides important insights into the basic mechanisms of behaviour and has proved valuable later in assessing the effectiveness of various proposed stabilisation measures.

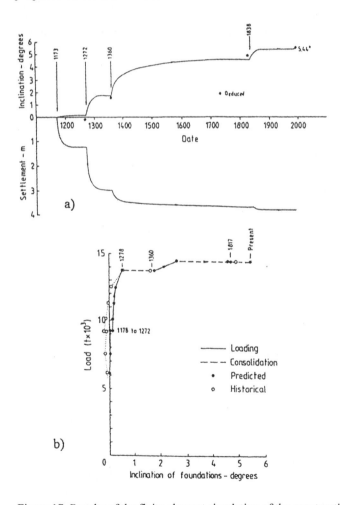

Figure 17. Results of the finite element simulation of the construction of the tower

5 STABILISING MEASURES

If an elasto-plastic model is assumed, than the relation between loads and displacement has to be written in incremental form:

$$\begin{Bmatrix} \partial\rho \\ \partial\alpha \end{Bmatrix} = \begin{vmatrix} \dfrac{1}{k_\rho} & \dfrac{1}{k_{\rho\alpha}} \\ \dfrac{1}{k_{\alpha\rho}} & \dfrac{1}{k_\alpha} \end{vmatrix} \begin{Bmatrix} \partial W \\ \partial M \end{Bmatrix} \tag{6}$$

The increment of displacement depends on the load increment, the current state of load and the load history. Hence the factor of safety depends on the current state of stress and stress history.

Centrifuge experiments by Cheney et al. (1991) (Fig. 18) show coupling between settlement and rotation, non-linearity and strain hardening plasticity. It may be seen that a decrease of inclination strongly increases the stiffness of the foundation – ground system. This generated the idea that a decrease of inclination increases M_s and FS, and can be used to stabilise the Tower. This has been the basic approach of the Committee to the stabilisation of the Tower; among other things, it is perfectly respectful of the integrity. In fact, it may be said that movements are in the DNA of the Tower!

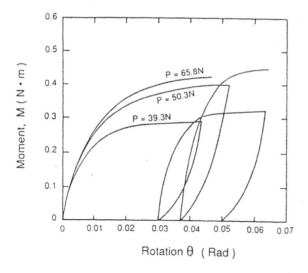

Figure 18. Centrifuge experiments by Cheney et al. (1991)

An important experimental confirmation of this approach came from a temporary safety measure against overturning of the tower. The Committee soon recognised the need for temporary and fully reversible interventions for the improvement of the safety against overturning by foundation failure, and for gaining the time to complete the investigations and analyses necessary to conceive and implement the final stabilisation measures. The observation that the northern side of the tower foundation had been steadily rising for most of the 20th century led to the suggestion of applying a north counterweight as a temporary measure. Accordingly, a design was developed consisting of a prestressed concrete ring cast around the base of the tower for supporting a number of lead ingots. A total of 6.9 MN of ingots were installed at an average distance from the axis of the foundation of around 6.3 m. This intervention was successfully implemented between May 1993 and February 1994. On 29th February 1994, one month after completion of loading, the northward change of inclination was 33"; by the end of July it had increased to 48". On 21st February 1994 the average settlement of the tower relative to the surrounding ground was about 2.5 mm. Figure 19 shows the observed changes in inclination of the tower during and after the application of the lead weights.

During the application of the counterweight, the rotational stiffness of the tower has been:

$$k_\alpha = \frac{43.3 \text{MNm}}{52''} = \frac{43.3}{2.5} 10^4 = 173,200 \text{ MNm} \tag{7}$$

The factor of safety has increased to:

$$FS = \frac{k_\alpha}{Wh} = \frac{173,000}{3200} = 54 \tag{8}$$

where h =22.6 m is the height of the centre of gravity of the tower.

It is noteworthy that, after a small decrease of the inclination, the tower has remained essentially motionless for over three years, apart from the seasonal cyclic movements and some accidents in September 1995.

The observed behaviour of the tower during the counterweight application was used to check and refine the finite element model further. Figure 20 shows a comparison of the Class A prediction and measurements of (a) the changes in inclination and (b) the average settlement of the tower during the application of the lead ingots. The computed settlement is in good agreement with the measured value; the predicted changes in inclination are about 80% of the measured values.

The refinement of the model involved a small reduction of the value of G/p'$_O$ in horizon A. After that, a better overall agreement between computed and observed values has been obtained (Fig. 21).

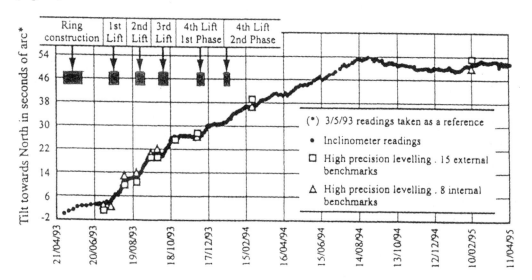

Figure 19. Behaviour of the tower during and after the application of lead weights

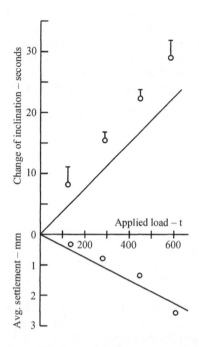

Figure 20. Comparison between prediction and measurements of changes in inclination and average settlement of the tower during the application of lead ingots

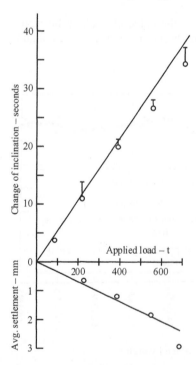

Figure 21. Comparison between prediction and measurements of changes in inclination and average settlement of the tower during the application of lead ingots after refinement of the model

6 UNDEREXCAVATION

Among different possible options to obtain a decrease of the inclination, the Committee has explored:

1. surface loading north of the tower,
2. vacuum pumping from the Pancone clay north of the tower,
3. electroosmosis,
4. removal of soil north of the foundation and below it (underexcavation).

All these solutions have been explored to different degrees by physical and numerical models, and electroosmosis and underexcavation were also investigated by full scale experiments. The underexcavation option was finally selected, and will be reported here.

The small scale model tests of leaning instability performed by Edmonds (1993) have been recalled in § 2. After those preliminary investigations, further model underexcavation tests were performed starting with a load of 165 N and a rotation of 5.5°. These conditions are believed to be representative of a tower on the verge of leaning instability.

The removal of soil was carried out by inserting a stainless steel tube with an outer diameter of 6 mm, and inside it, an inner suction tube connected to a vacuum pump. The inner tube, with an outer diameter of 2.1 mm, removes the sand from inside the larger tube, that is thus advanced into the soil by self boring, without significant disturbance of the surrounding soil. The whole probe is advanced to the desired location and then retracted, leaving a cavity that closes instantaneously.

The underexcavation tubes are held in position by external guides and penetrate the sand at an angle to the horizontal of 18° 26' (3/1). Five radial tubes have been adopted, covering a sector of 90° centred on the north side of the model tower.

A total of 14 model underexcavation tests have been performed with different combinations of probe positions and penetration sequences. The most important indications obtained are as follows:

-   underexcavation can be used to reduce the tilt of the model in a controllable manner. A reduction of tilt up to 1° has been obtained,
-   the movement of the tower can be steered using different probes inserted around the tower,
-   the results are reproducible, at least qualitatively,
-   a *critical point* exists some 10 mm north of the central axis of the tower, in the ground beneath it, beyond which ground removal aggravates the tilt, but behind which underexcavation produces a decrease in tilt,
-   repeated use of one probe in isolation rapidly ceases to affect the Tower's tilt significantly.

These findings have been found to apply qualitatively to the case of the tower of Pisa.

Thilakasiri (1993) modelled the subsoil of the tower as a set of elastic-perfectly plastic Winkler springs, and determined the spring constants by fitting the observed behaviour of the tower during construction. The analysis confirms that the inclination of the tower started during the second stage of construction, because of leaning instability; and accurately reproduces the present situation of the tower in terms of settlement and inclination. Thilakasiri simulated underexcavation by removing single strips of reaction stress at the soil-foundation interface. He found that underexcavation has a positive effect, provided it is confined north of the position of the load resultant. This is obvious from elementary statics; in this sense it can be stated that no critical point has been predicted. The effectiveness of the operation depends on the position from which the stress is removed; the optimum position has been determined at about half radius from the northern edge of the foundation.

Desideri & Viggiani (1994) modelled the overall behaviour of the subsoil and the tower foundation by an elasto-plastic strain hardening restraint, and simulated the underexcavation intervention by a reduction in overturning moment with constant vertical force. Desideri et al. (1997) modelled the subsoil as a bed of Winkler type elastic-strain hardening plastic springs; they found that the overall behaviour of the tower and the subsoil can be described by a set of yield loci eventually merging into a failure locus, and by an associative flow rule. All these spring models do not predict the occurrence of a critical line.

The centrifuge was also used to assess the effectiveness of underexcavation as a means to stabilise the tower. The process of soil extraction was modelled by inserting flexible tubes with wires inside into the ground beneath the model tower, prior to the commencement of the experiment. Once the model tower had come to equilibrium at an appropriate inclination under increased gravity, the wires were pulled out of the flexible tubes by an appropriate amount, while the model was in flight, causing the tubes to close simulating the closure of the cavity produced by a drill probe.

Figure 22 reports the results of a typical experiment. The test results confirmed the existence of a critical line and showed that soil extraction north of this line always gave a positive response.

The finite element model has been used to simulate the extraction of soil from beneath the north side of the foundation. It should be emphasised that the finite element mesh had not been developed with a view to modelling underexcavation; the individual elements are rather large for representing regions of extraction. Thus the purpose of the modelling was to throw light on the mechanisms of behaviour rather than attempt a somewhat illusory "precise" analysis. The soil extraction has been simulated by reducing the volume of any chosen element of ground incrementally, so as to achieve a predetermined reduction in volume of that element.

The first objective of the numerical analysis was to check whether the concept of a critical line, whose existence was revealed by the small scale tests by Edmonds (1993), was valid. Figure 23 shows the finite element mesh in the vicinity of the tower. Elements numbered 1, 2, 3, 4 and 5 are shown, extending southwards from beneath the north edge of the foundation. Five analyses were carried out in which each of the elements was individually excavated to give full cavity closure and the response of the tower computed. For excavation of elements 1, 2 and 3,

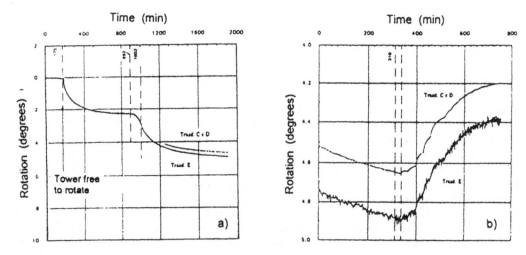

Figure 22. Simulation of the underexcavation in a centrifuge experiment. a) Construction of the tower, b) underexcavation

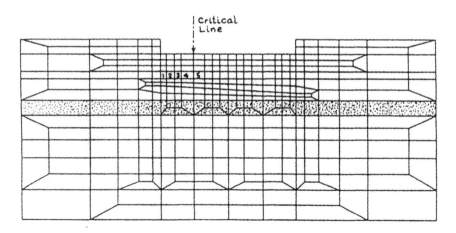

Figure 23. Finite element search for a critical line

the inclination of the tower reduces, so that the response is positive. For element 4, the response is approximately neutral, with an initial slight reduction in inclination which, with further excavation, was reversed. For element 5 the inclination of the tower increased as a result of excavation.

The above analyses confirm the concept of a critical line separating a positive response from a negative one. The results from the plane strain computer model show that the location of the critical line is towards the south end of element 4, which is at a distance of 4.8 m underneath the foundation of the tower, i.e. about one half the radius of the foundation.

It was noted that, as the location of excavation moved further and further south beneath the foundation, the settlement of the south side steadily increases as a proportion of the settlement of the north side. Excavation of elements 1 and 2 give a proportion of less than one quarter.

Having demonstrated that localised soil extraction gives rise to a positive response, the next stage was to model a complete underexcavation intervention aimed at safely reducing the inclination of the tower by a significant amount.

A preliminary study was carried out of extraction using a shallow inclined drill hole, extracting soil from just beneath the foundation. Although the response of the tower in terms of decrease of inclination was favourable, the stress changes beneath the foundation were large, consequently a deeper inclined extraction hole was investigated.

Figure 24 shows the finite element mesh in the vicinity of the foundation on the north side. The elements numbered 6 to 12 were used for carrying out the intervention and are intended to model an inclined drill hole. It should be noted that element 12 lies south of the critical line established by localised soil extraction as described above. The procedure for simulating the underexcavation intervention was as follows:

- the stiffness of element 6 is reduced to zero,
- equal and opposite vertical nodal forces are applied progressively to the upper and lower faces of the element until its volume reduces by about 5%. The stiffness of the element is then restored,
- the same procedure is then applied successively to the elements 7, 8, 9, 10 and 11 thereby modelling the progressive insertion of the drill probe. For each step the inclination of the tower reduces,
- when element 12 is excavated the inclination of the tower increases, confirming that excavation south of the critical line gives a negative response. The analysis is therefore restarted after excavating element 11,

21

- the retraction of the drill probe is then modelled by excavating elements 10, 9, 8, 7 and 6 successively. For each step the response of the tower is positive,
- the whole process of insertion and retraction of the drill probe is then repeated. Once again excavation of element 12 gives a negative response.

The computed displacements of the tower are plotted in Figure 24. The sequence of excavation of the elements is given on the horizontal axis; the upper diagram shows the change of inclination of the tower due to underexcavation; the lower diagram shows the settlement of the north and south sides of the foundation.

As underexcavation progresses from elements 6 through 11, the rate of change of northward inclination increases, as do the settlements. As the drill is retracted, the rate decreases. At the end of the first cycle of insertion and retraction of the drill, the inclination of the tower is decreased by 0.1°. The settlement of the south side is rather more than one half of the north side. A similar response is obtained for the second cycle but the change of inclination is somewhat larger. After the third insertion of the drill, the resultant northward rotation was 0.36°. The cor-

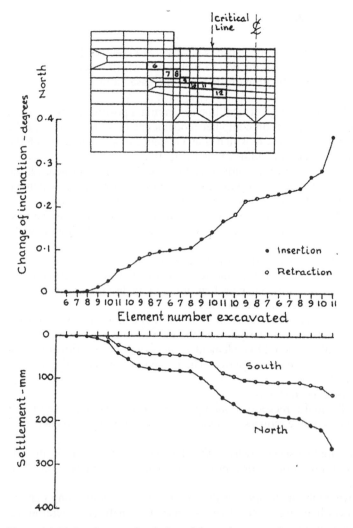

Figure 24. Finite element simulation of the underexcavation

responding settlements of the north and south sides of the foundation were 260 mm and 140 mm respectively.

As for the contact stress distribution, the process results in a slight reduction of stress beneath the south side. Beneath the north side, some fluctuations in contact stress take place, as was expected, but the stress changes are small. In general, the stress distribution after retraction of the drill are smoother than after insertion.

The results of the physical and numerical modelling work on underexcavation were sufficiently encouraging to undertake a large scale development trial of the field equipment. The objectives of the trial were:

- to develop a suitable method of forming a cavity without disturbing the surrounding ground during drilling,
- to study the time involved in the cavity closure,
- to measure the changes in contact stress and pore water pressures beneath the trial footing,
- to evaluate the effectiveness of the method in changing the inclination of the trial footing,
- to explore methods of "steering" the trial footing by adjusting the drilling sequence,
- to study the time effects to be expected between excavation events and after the operation was completed.

For this purpose, a 7 m diameter eccentrically loaded circular reinforced concrete footing was constructed in the Piazza north of the Baptistry, as shown in Figure 25. Both the footing and the underlying soil were instrumented to monitor settlement, rotation, contact pressure and pore pressure.

Figure 25. Large scale experiment of underexcavation

23

After a waiting period of a few months, allowing the settlement rate to come to a steady value, the ground extraction commenced by means of inclined borings, as shown schematically in Figure 26. Drilling was carried out using a hollow stemmed continuous flight auger inside a contra-rotating casing.

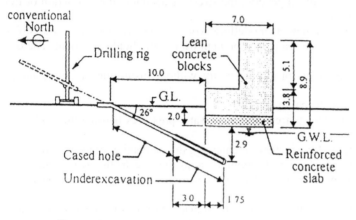

Drawing not to scale - all dimensions in meters

Figure 26. Vertical section of the large scale experiment of underexcavation

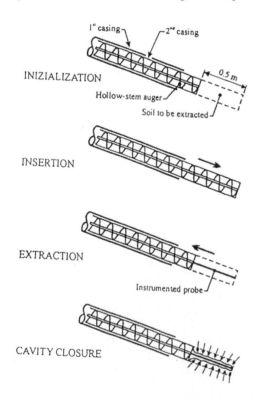

Figure 27. Scheme of the technique adopted to remove the soil

24

The trial was very successful. When the drill was withdrawn to form the cavity, an instrumented probe located in the hollow stem was left in place to monitor its closure (Fig. 27). A cavity formed in the Horizon A material has been found to close smoothly and rapidly. Figure 28 reports the measurements of the contact stress at the soil-foundation interface along the north - south axis, before underexcavation (19.09.95) and after a substantial rotation of the footing (01.12.95). The stress changes beneath the foundation were found to be very small. The trial footing was successfully rotated by about 0.25° and directional control was maintained even though the ground conditions were somewhat non-uniform. Rotational response to soil extraction was rapid, taking a few hours. At the completion of the underexcavation, on February 1996, the plinth came to rest and since then it has exhibited negligible further movements. Very importantly, an effective system of communication, decision taking and implementation was developed.

It is of importance to note that, early in the trial, overenthusiastic drilling resulted in soil extraction from excess penetration beneath the footing, causing a counter rotation (Fig. 29). Therefore the trial also confirmed the concept of a critical line.

The results of all the investigations carried out on the underexcavation were positive, but the Commission was well aware that they might be not completely representative of the possible response of a tower affected by leaning instability. Therefore it was decided to implement preliminary ground extraction beneath the tower itself, with the objective of observing its response to a limited and localised intervention. This preliminary intervention consisted of 12 holes

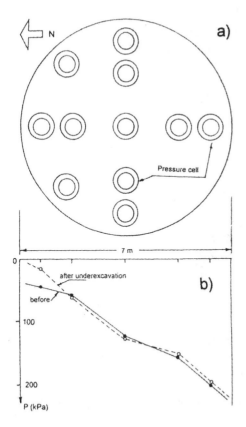

Figure 28. Variations of contact stress during the underexcavation experiment

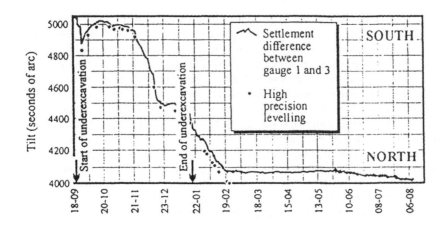

Figure 29. Results of large scale experiment of underexcavation

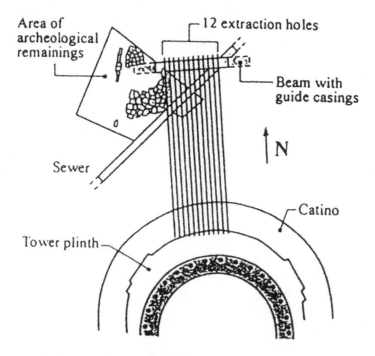

Figure 30. Preliminary underexcavation of the tower: plan

(Figs 30, 31) to extract soil from Horizon A to the north of the tower foundations, penetrating beneath the foundation not more than 1 m. The goal was to decrease the inclination of the tower by a significant amount, in order to check the feasibility of underexcavation as a means of permanently stabilising the tower, and to adjust the extraction and measurement techniques.

To protect the tower from any unexpected adverse movement during this or any other interventions aimed at the final stabilisation of the monument, a safeguard structure was considered mandatory. The structure finally chosen consists of two sub-horizontal steel stays connected to the tower at the level of the third floor and to two anchoring steel frames located behind the building of the Opera Primaziale, to the north of the tower. The scheme showing the safeguard structure is reported in Figure 32; it was installed and connected to the tower in December 1998.

Each stay is capable of applying a maximum force of 1500 kN, with a safety factor equal to 2. The force may be applied by dead weights or by hydraulic jacks; the value of the applied load is monitored continuously. At present, the load applied to each stay is equal to about 72 kN, just enough to keep it in position.

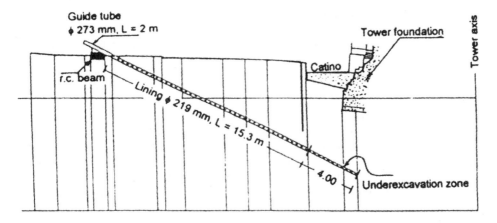

Figure 31. Preliminary underexcavation of the tower: section

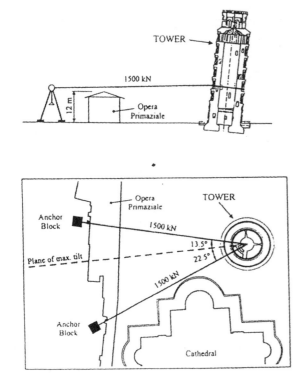

Figure 32. Scheme of the safeguard structure by steel cable stays

27

The underexcavation experiment has been carried out between February and June, 1999. The results obtained are reported in Figure 33. During the underexcavation period, the tower rotated northwards at an increasing rate, as the extraction holes were drilled gradually ahead near the north boundary of the foundation and below it. At the beginning of June 1999, when the operation ceased, the northwards rotation of the tower was 90"; by mid September it had increased to 130". At that time, three of the 97 lead ingots (weighing about 10 t each) acting on the north side of the tower were removed; since then the tower has exhibited negligible further movements. As a matter of fact, the preparatory operations for the final underexcavation (removal of the 12 guide casings of the preliminary underexcavation, installation of the 41 guide casings for the final underexcavation) have produced a slight further northward rotation, bringing the overall decrease of inclination in March, 2000 to 135".

The total volume of soil removed during the preliminary underexcavation has been 7 m^3, 86% of which was from the north of the tower and 14 from below the foundation. The rotation in the east - west plane has been much smaller, reaching a final value of about 10" westwards, as intended. Due to preliminary underexcavation, the north side of the tower foundation underwent an overall settlement equal to 1.3 cm; in the meantime, the south side first heaved by 2 mm, and then gradually settled by the same amount, showing that the axis of rotation is located between the two points.

To put these preliminary stabilising results in perspective, the evolution of the tilt of the tower base since 1993 is reported in Figure 34. The effect of the underexcavation experiment

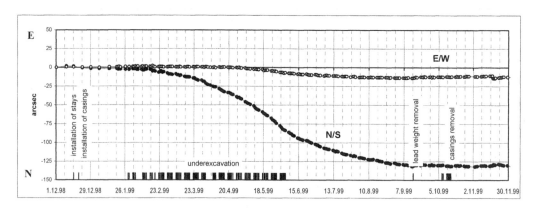

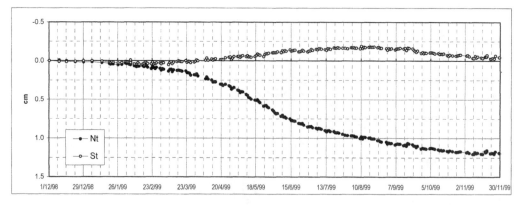

Figure 33. Results of the preliminary underexcavation

has been far greater than that of the counterweight and outweighs the seasonal cyclic movements.

A longer time perspective is gained by the diagram in Figure 35, reporting the inclination since 1935, as measured by a pendulum inclinometer installed at that time. It may be seen that the effect of the preliminary underexcavation has been to bring the tower "back to future" by over 30 years. Incidentally, it has been brought back to 1968, that has been considered not that bad a year to reach!

After the very encouraging results of the experiment with the preliminary underexcavation experiment, the Committee went on to the final underexcavation, that has been carried out between February 2000 and February 2001. This time, 41 holes have been drilled; the layout in plan and in a vertical section is reported in Figures 36 and 37. Some lateral holes had been prepared to extract soil just below the bottom of the catino, to allow it to follow the movements of the tower without cracking. As a matter of fact, they have not been employed at all. During the final underexcavation, all the lead ingots and the concrete beam on which they rested have been removed. In June 2001, the steel cable stays have been also dismantled, without having needed to be operated once.

The results obtained from the final underexcavation are reported in Figures 38 and 39. The target of decreasing the tilt of the tower by half a degree has been achieved. During the final stage, 38 m^3 of soil has been removed, 31% of which came from north of the tower and 69% from below the foundation. The maximum penetration below the foundation has been 2.5 m.

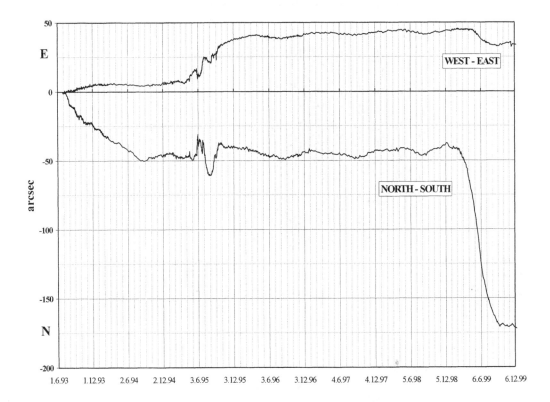

Figure 34. Results of the preliminary underexcavation (1993 – 1999)

29

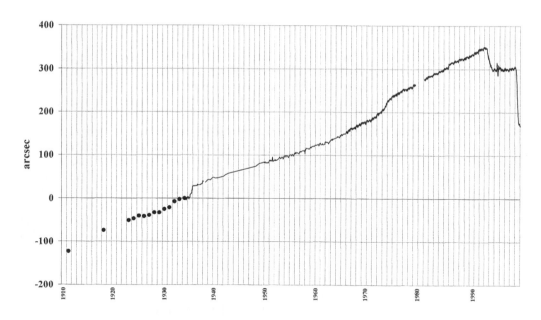

Figure 35. Results of the preliminary underexcavation (1910 – 1999)

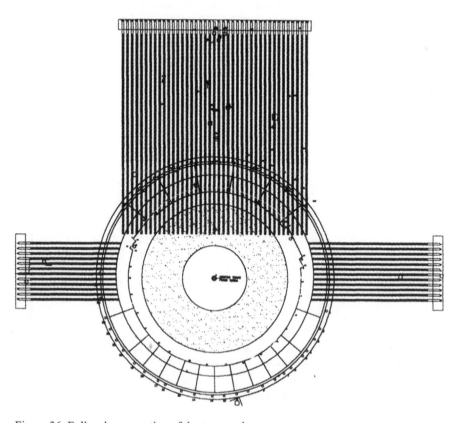

Figure 36. Full underexcavation of the tower: plan

30

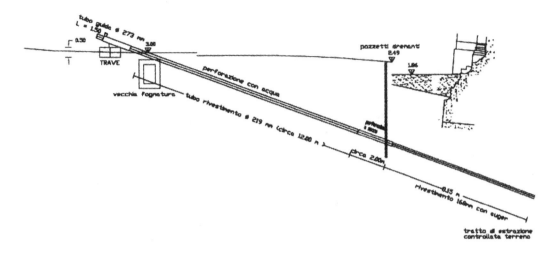

Figure 37. Full underexcavation of the tower: section

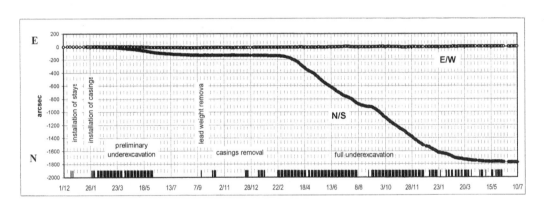

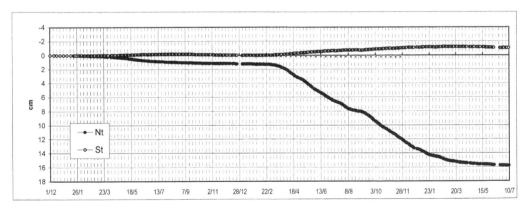

Figure 38. Results of the full underexcavation: rotations and settlements (1999 – 2001)

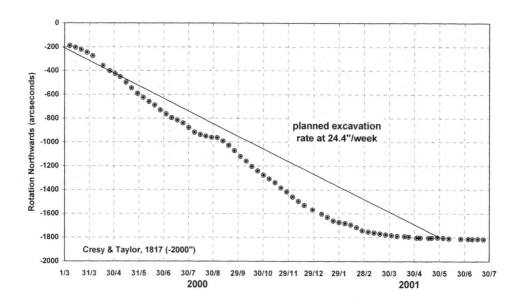

Figure 39. Results of the full underexcavation: rotations (2000 – 2001)

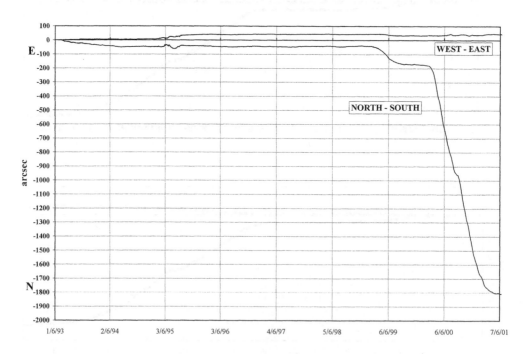

Figure 40. Results of the full underexcavation: rotations (1993 – 2001)

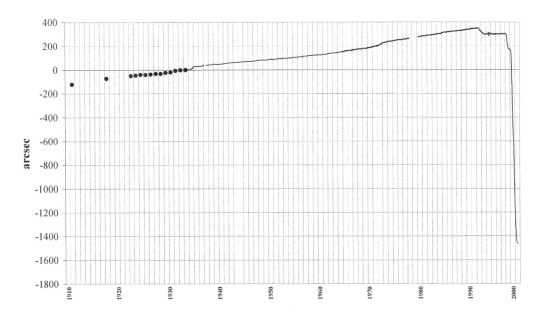

Figure 41. Results of the full underexcavation: rotations (1910 – 2001)

To put these results in perspective, the evolution of the tilt of the tower base since 1993 is reported in Figure 40. A longer time perspective is gained by the diagram in Figure 41, reporting the inclination since 1911, when the modern style of measurement began. It may be seen that the overall effect of the underexcavation has been to bring the tower "back to future" just before the excavation of the catino. A well conceived and executed excavation programme has compensated for the effects of the old excavation of the catino!

ACKNOWLEDGEMENTS

This paper reports the results of teamwork carried out by the whole Committee for the Safeguarding of the Tower of Pisa. Within the Committee, the Soil Mechanics aspects have been dealt with by Prof. J.B. Burland, Prof. M. Jamiolkowski (who also chairs the Committee) and the author.

The invaluable contribution of dr. N. Squeglia in supervising the site operations and the monitoring during underexcavation, and in preparing the present paper, is gratefully acknowledged.

REFERENCES

Burland, J.B., Jamiolkowski, M. & Viggiani, C. 1999. The restoration of the leaning tower: present situation and perspectives. Geotechnical aspects. *Workshop on the Tower of Pisa, Pisa* 1: 29-110.
Burland, J.B. & Potts, D.M. 1994. Development and application of a numerical model for the leaning tower of Pisa. *IS Pre-failure deformation characteristics of geomaterials. Hokkaido, Japan* 2: 715-738.
Cheney, J.A., Abghari, A. & Kutter, B.L. 1991. Leaning instability of tall structures. *Journ. Geot. Eng., Proc. ASCE,* 117(2): 297-318.
Desideri, A. & Viggiani, C. 1994. Some remarks on the stability of towers. *Symp. on Development in Geotechnical Engineering, Bangkok.*

Desideri, A., Russo, G. & Viggiani, C. 1997. Stability of towers on compressible ground. *Rivista Italiana di Geotecnica.*

Gorbunov-Possadov, M.I. & Serebrajani, R. V. 1961. Design of structures upon elastic foundations. *Proc. 5th ICSMFE, Paris* 1: 643–648.

Edmonds, H. 1993. The use of underexcavation as a means of stabilising the leaning tower of Pisa: scale model tests. *MSc Thesis, Imperial College, London.*

Habib, P. & Puyo, A. 1970. Stabilité des fondations des constructions de grande hauteur. *Annales IBTP*, 117–124.

Hambly, E. C. 1985. Soil buckling and leaning instability of tall structures. *Struct. Eng.* 63(3): 78-85.

Lancellotta, R. 1993. Stability of a rigid column with non linear restraint. *Géotechnique* 43(2): 331-332.

Nova, R. & Montrasio, R. 1991. Settlement of shallow foundations on sand. *Géotechnique* 41(2): 243-256.

Pepe, M. 1995. La Torre Pendente di Pisa. Analisi teorico-sperimentale della stabilità dell'equilibrio. *Tesi di Dottorato in Ingegneria Geotecnica, Politecnico di Torino.*

Potts, D.M. & Burland, J.B. 2000. Development and application of a numerical model for simulating the stabilisation of the Leaning Tower of Pisa. In Smith, Carter (eds), *Proc. Developments in Theoretical Geomechanics, the John Booker Memorial Symposium*: 737-758. Rotterdam: Balkema.

Potts, D.M. & Gens, A. 1984. The effect of the plastic potential in boundary value problems involving plane strain deformation. *Int. Journ. Analytical and Numerical Methods in Geomechanics* 8: 259-286.

Schultze, E. 1973. Der Schiefe Turm von Pisa. *Mitteilungen Technische Hochschule Aachen.*

Thilakasiri, H. S. 1993. A theoretical study of underexcavation for stabilising the tower of Pisa. *MSc Thesis, Imperial College of Science, Technology and Medicine, London.*

Constitutive and Centrifuge Modelling: Two Extremes, Springman (ed.)
© *2002 Swets & Zeitlinger, Lisse, ISBN 90 5809 361 1*

Constitutive cladistics: the progeny of Critical State Soil Mechanics

D. Muir Wood
Department of Civil Engineering, University of Bristol, United Kingdom

ABSTRACT: Discussion of the historical context of the development of the Cam clay model is used as a vehicle for description of the characteristics required for membership of the family of critical state models and of the genus of Cam clay models. It is shown that quite different routes to generation of constitutive models can produce essentially equivalent results. The lasting significance and extensive penetration of the framework of Critical State Soil Mechanics is emphasised.

1 INTRODUCTION

This paper is concerned with the classification of constitutive models for soils and the relationships between different soil models. Naming of soil models, and reference to soil models, especially so called 'critical state' models and Cam clay has not always been very precise and some clarification may be helpful. A biological analogy is used to develop a taxonomy of soil models. This is anything but pure and links across between genera are evident – the boundaries between species are not always clear. Such a study demonstrates rather rapidly that there may often not be a uniqueness of the constitutive modelling solution to a particular set of material data.

Certain assumptions are being made at the start of this discussion:

Principle of effective stress. The mechanical behaviour of soils is controlled by effective stresses and unless otherwise stated all stress quantities are effective stress quantities. This now seems an entirely unobjectionable assumption but historically it appears that undrained response has been seen as something special and distinct from drained response.

Particle-continuum duality. Soils are multiphase materials composed of particles and voids. The particles interact with mechanical and electrostatic forces. The voids are filled with one or more pore fluid. There is a particle-continuum duality that has to be accepted for soils. Clearly the mechanical response that is observed in a laboratory test or a physical model is influenced by the detail of the behaviour at the particle level. For example, dilatancy characteristics are controlled by the relative movements of adjacent particles. We expect such relative movement to be rather easy at low stress levels, as particles are able to ride up over each other, whereas at higher stress levels we can expect that crushing at the particle contacts will tend to become more important thus reducing the scope for volumetric increase. Nevertheless, geotechnical engineering applications are working at a scale that is large by comparison with the size of individual particles and it will be assumed that it is possible to define representative volume elements which are sufficiently large that particle effects can be smeared out and a continuum description will be valid. It then becomes acceptable to treat the behaviour in terms of continuum concepts such as stresses and strains. What 'sufficiently large' might be is a question for which an answer will not be attempted here (Muir Wood 2001). Suffice it to say that there are lengths which are much larger than the size of individual particles (but which may be dependent on the size of individual particles) which may come into play: such as thickness of interface regions, thickness of shear bands, mobilisation lengths for shear bands, patterning of localisations, pattern of cells of force chains in particulate assemblies.

Elastic-plastic models. Selection of theoretical framework for development of constitutive models for soils is very much a matter of personal preference. There are several possible choices. For the purposes of the present paper we will restrict ourselves to more or less conventional elastic-plastic models for which, in the end, the standard ingredients are assumptions about elastic response, yield criterion, flow rule and hardening rules. Though such a framework may not be regarded as completely thermodynamically rigorous it does have the advantage of being developed around some clear physical concepts which, experience has shown, it is feasible to explain to typical undergraduates and practising engineers. If the goal of research in constitutive modelling is to emerge with models that are to be used, then the acceptance by the potential users is important.

Our concern is with the description and classification of soil models. All calculation of performance of geotechnical systems relies on models of soil behaviour. All testing of soil elements or soil models is performed against a background of an expected pattern of response: an existing model. All models are figments of the imagination, but they may, through having been tried and tested, have been found to be of use in certain more or less limited classes of situation. Truth lies in observation whether of laboratory experiments or of field trials or of existing systems. However, interpretation of these observations, which again requires a model as a template, may be defective if the model is inappropriate. It is tempting to assume that things that we choose not to observe (because our model does not lead us to expect them) do not exist: absence of evidence is not the same as evidence of absence. Scientific understanding progresses by a repeated process of: hypothesis, test, revise – or prediction, observation, reflection (Fig. 1). Ideally the tests or observations should be 'orthogonal' to the model so that its hypotheses are truly tested – testing that sets out merely to confirm hypotheses is less helpful. Hypotheses or conjectures cannot be proved true, they can only be falsified.

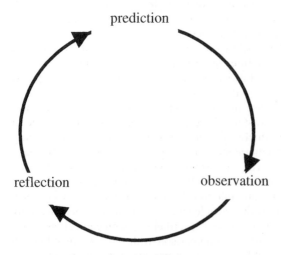

Figure 1. Reflective practice loop

2 HISTORICAL DEVELOPMENT: GROPING TOWARDS THE LIGHT

We will assert that the period from 1945 to 1970 was crucially significant for the development of soil mechanics in general and, in the present context, of constitutive models for soils in particular. This period runs from the end of the Second World War until the untimely death of Ken Roscoe. By the end of this period the ideas of critical state soil mechanics had established a niche in geotechnical engineering and the Cam clay model had emerged as the precursor of many soil models and probably, for all its shortcomings, the first to pull together many of the observations of soil response that had been perplexing researchers and practitioners for many years before. So far as the development of soil mechanics in the United Kingdom is concerned, it is a key period during which active research was being conducted in a rather small number of universities and research centres – the effect of the major expansion of higher education in the 1960s had not yet had much effect.

Of course, this period does not sit in isolation. Apart from the principle of effective stress – which may be regarded as a demonstrably useful hypothesis, a conjecture which has not been significantly refuted so far as application in geotechnical engineering is concerned – there are two other precursors which are important.

Rendulić (1937) was one of the first to perform triaxial tests with pore pressure control and measurement. His pioneering tests on Vienna clay (Wiener Tegel) showed that the effective stress paths followed in undrained compression and extension tests on normally consolidated samples matched closely with contours of specific volume or water content deduced from a series of drained compression and extension tests. His findings were confirmed and extended by the work of Henkel (1960a, b) demonstrating that the pattern could be extended to overconsolidated samples as well (Fig. 2).

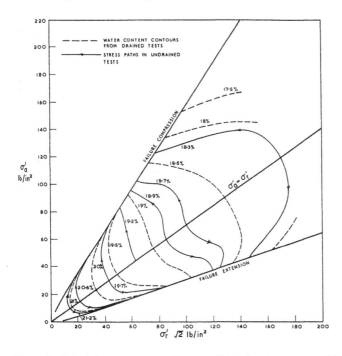

Figure 2. Triaxial tests on normally consolidated and overconsolidated Weald clay (maximum consolidation pressure 827 kPa (120 lb/in^2): effective stress paths in undrained tests (solid lines); water content contours from drained tests (dashed lines) (Henkel 1960b)

To us it seems natural to see drained and undrained response as part of a single picture of soil behaviour understood in terms of effective stresses. A soil sample has no way of knowing what is about to happen to it – whether the drainage line is going to be open or closed – and its constitutive behaviour is unaffected by decisions that are made by the operator of the testing apparatus. The triaxial apparatus has two degrees of stress or deformation freedom: radial stress σ_r and axial stress σ_a, or mean stress p and distortional stress q; axial strain ε_a and radial strain ε_r, or volumetric strain ε_p and distortional strain ε_q. (Effective stresses are denoted, as usual by $'$.)

$$p = \left(\frac{\sigma_a + 2\sigma_r}{3} \right); \quad q = \sigma_a - \sigma_r; \quad \delta\varepsilon_p = \delta\varepsilon_a + 2\delta\varepsilon_r; \quad \delta\varepsilon_q = \tfrac{2}{3}\left(\delta\varepsilon_a - \delta\varepsilon_r \right) \qquad (1)$$

We perform 'conventional' triaxial tests in which we keep the cell pressure constant and increase (or decrease) the axial strain because these are easy to perform. But of course, at any stage we can choose to control the stresses as we will – and in particular could choose to vary radial and axial stress (while keeping the drainage open) in such a way that the volume of the sample had no desire to change. This then becomes a drained constant volume process which, as we can deduce from an extension of Rendulić's work, will produce the same effective stress response as a more standard undrained test (in which pore pressure adjusts the mean effective stress to keep the volume constant).

Although the conventional method of performing triaxial tests uses a combination of stress control (constant cell pressure) and strain control (increasing axial strain), the undrained test is just one special case of a strain controlled test with constant ratio of imposed increments of volumetric strain and distortional strain:

$$\frac{\delta\varepsilon_p}{\delta\varepsilon_q} = \zeta \qquad (2)$$

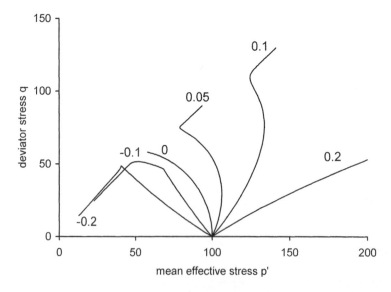

Figure 3. Constant strain increment paths imposed on normally consolidated (modified) Cam clay: values of ζ shown

If $\zeta = 0$ then this produces an undrained increment. If $\zeta = 3/2$ then this produces one-dimensional compression. For positive values of ζ, and with $\delta\varepsilon_q > 0$, this produces paths which tend to anisotropic compression paths of which the one-dimensional compression is a special case. Experimental evidence (and the Cam clay model) tells us that whatever the starting point, such paths will tend asymptotically to a constant stress ratio, but with continuing increase in stresses (Fig. 3). Evidence for the response to be expected with $\zeta < 0$ and $\delta\varepsilon_q > 0$ is less clear. In any event, if $\zeta \neq 0$ then the imposed strain path implies an indefinitely continuing change in volume and the case $\zeta = 0$ produces a very special path which can be continued indefinitely as an asymptotic state without running into volumetric limits. Incrementally, however, this path loses its special character – there are infinitely many strain controlled increments that can be imposed.

Casagrande (1936) proposed the concept of a critical void ratio as a volumetric packing towards which samples (of granular soils) would tend with continued shearing. 'Every cohesionless soil has a certain critical [void ratio], in which state it can undergo any amount of deformation or actual flow without volume change.' This seems to be a unifying concept which provides some order to the apparently contrasting behaviours of dense and loose granular materials. Although Casagrande seems to be thinking of critical void ratio as a single value of void ratio for a given soil, it is clear that the critical void ratio must in fact be dependent on the stress level at which the test is conducted. Taylor (1948) shows evidence for this pressure dependence.

The important step taken by Roscoe et al. (1958) was then to propose, on the basis of extensive simple shear testing at Cambridge and on the basis of reanalysis of triaxial tests performed at Imperial College (principally), that the idea of a regular set of critical states to which all samples tended in monotonic shearing was common to all soils. The idea of the *state* of the soil, its volumetric packing and its effective stress condition is important. At a critical state the state remains constant even though distortion (which is not part of the state of the soil) continues:

$$\frac{\partial q}{\partial \varepsilon_q} = \frac{\partial p'}{\partial \varepsilon_q} = \frac{\partial v}{\partial \varepsilon_q} = \frac{\partial \varepsilon_p}{\partial \varepsilon_q} = 0 \tag{3}$$

State is composed of limited quantities – distortional strain is unlimited. A critical state is indeed an asymptotic ultimate state for the soil. One of the key features of the presentation by Roscoe et al. was the demonstration that ultimate critical states reached in undrained (compression) tests and in drained (compression) tests on samples with different degrees of overconsolidation were essentially identical – the need to pull together the separate response of drained and undrained samples was appreciated.

Roscoe et al. showed that the stresses at critical states defined an essentially frictional relationship – so we have the idea of looking at soil response in parallel in terms of stresses and in terms of volumetric packing in an effective stress plane and a compression plane (Fig. 4). This already starts to provide further unification and permits semi-quantitative estimates of soil response to be made – broad estimates of character of volume change and expected ultimate strengths etc. (Muir Wood 1990). However, the ability to unify the distortional response in detail is still lacking.

Researchers in the 1950s reporting experimental observations and trying to bring together results of drained and undrained tests were clear that some allowance had to be made for the volume changes that were occurring in the drained tests and for the work that was being done by the stresses in producing these volume changes. The notion of *boundary energy correction* was introduced as a way of correcting the measured stress – usually the distortional stress q. The discussion of boundary energy corrections becomes progressively more Byzantine through the 1950s. For example, Penman (1953) suggested that the corrected distortional stress q_{cp} should be:

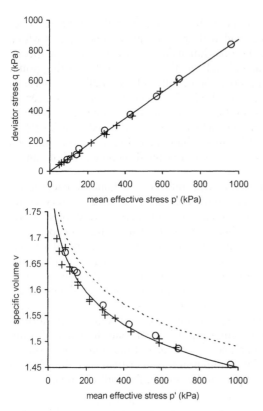

Figure 4. Analysis of data of end points of triaxial tests: (o normally consolidated samples; + overconsolidated samples; solid line: critical state line deduced from undrained tests; dashed line: normal compression line) (after Roscoe et al. 1958)

$$q_{cp} = q + p' \frac{\delta \varepsilon_p}{\delta \varepsilon_a} \tag{4}$$

Bishop (1954) objected to this on the grounds that the work input to the sample had been incorrectly calculated and suggested that the corrected distortional stress q_{cb} should be:

$$q_{cb} = q + \sigma'_r \frac{\delta \varepsilon_p}{\delta \varepsilon_a} \tag{5}$$

Poorooshasb & Roscoe (1961) indicate that the energy input to a deforming sample can only be calculated correctly if the strain increment and stress quantities form correctly work conjugate pairs. With this undoubtedly correct statement about work input they produce an alternative corrected distortional stress q_{cr}:

$$q_{cr} = q + p' \frac{\delta \varepsilon_p}{\delta \varepsilon_q} \tag{6}$$

The next stage is to propose some functional expression for the variation of this corrected stress and Roscoe et al. (1963) make the essentially arbitrary assumption – for which experimental data provide some support – that the corrected distortional stress from (6) should at all times bear a frictional relationship with the mean stress p' (Fig. 5):

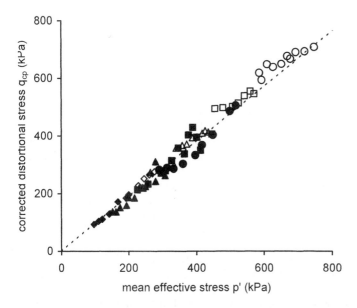

Figure 5. Distortional stress corrected for volume changes using (6) (including elastic volumetric energy); solid symbols: undrained tests; open symbols: drained tests; radial stresses: ◆ 207kPa; ▲ 310kPa; ■ 414kPa; ● 552kPa (after Roscoe et al. 1963)

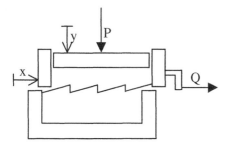

Figure 6. Direct shear box

$$q_{cr} = q + p'\frac{\delta\varepsilon_p}{\delta\varepsilon_q} = Mp' \tag{7}$$

(In fact Roscoe et al. make a correction for elastic volumetric strain so that the strain ratio in this expression is regarded as a ratio of *plastic* strain increments. This will be discussed below.) There is of course a logic to this assumption: it resonates with the analysis carried out by Taylor (1948) of the volume changes occurring in direct shear box tests on sand. There the definition of work input is clear cut from the start.

When sands are sheared, for example in a shear box, they change in volume. The work done in a shear box sample which is supporting vertical load P and horizontal load Q and is undergoing horizontal (shearing) displacement δx and vertical (volumetric) displacement δy (Fig. 6) is:

$$\delta W = P\delta y + Q\delta x \tag{8}$$

Taylor (1948) assumed that this work was entirely dissipated in friction at all stages of a shear test so that:

41

$$P\delta y + Q\delta x = \mu P\delta x \qquad (9)$$

or

$$\frac{\delta y}{\delta x} = \mu - \frac{Q}{P} \qquad (10)$$

Taylor's shear box data for sands having two different initial densities are presented in Figure 7 together with the replotting of the data in the form of this friction:dilatancy relationship. The linkage of deformation mechanism to stress state (Q/P is an indication of mobilised friction) provides at least a first approximation to the observed response – especially when one considers the range of x displacement over which it seems to apply. When the mobilised friction Q/P is less than μ the sand is contracting; when the mobilised friction is greater than μ the sand is expanding. The ratio -δy/δx gives an indication of the tendency to volume increase for the sand and is known as dilatancy. An expression such as (10) which links dilatancy with mobilised friction is called a stress-dilatancy relationship or flow rule and suggests that mobilised friction (Q/P) minus dilatancy (-δy/δx) should be a constant.

The data from Taylor's shear box tests on dense (initial specific volume 1.562) and loose (initial specific volume 1.652) Ottawa sand indicate μ ≈ 0.49 corresponding to a critical state angle of friction of about 26°.

A direct analogy can be drawn between the behaviour observed in the shear box and the behaviour to be expected in the triaxial apparatus. Treating the correspondence within the framework of elastic-plastic modelling, the displacement increments become the strain increments and the shear and normal loads become the deviator and mean effective stresses:

$$\delta x \rightarrow \delta\varepsilon_q \qquad (11a)$$

$$\delta y \rightarrow \delta\varepsilon_p \qquad (11b)$$

$$Q \rightarrow q \qquad (11c)$$

$$P \rightarrow p' \qquad (11d)$$

and the stress-dilatancy expression or flow rule becomes:

$$\frac{\delta\varepsilon_p}{\delta\varepsilon_q} = M - \frac{q}{p'} = M - \eta \qquad (12)$$

where M is the critical state stress ratio at which constant volume shearing can occur and η is the stress ratio q/p'. This is of course just a rewriting of (7). As we will see, this forms the basis of the original Cam clay model (Roscoe & Schofield 1963).

We can convert the other proposed boundary energy corrections ((4), (5)) into flow rules by making the same assumption as Roscoe et al. (1963) that the corrected distortional stress is always frictionally related to the current mean stress, and then rearranging the expressions as required. The Penman flow rule is:

$$\frac{\delta\varepsilon_p}{\delta\varepsilon_q} = \frac{3(M-\eta)}{(3-M+\eta)} \qquad (13)$$

and the Bishop flow rule is:

$$\frac{\delta\varepsilon_p}{\delta\varepsilon_q} = \frac{3(M-\eta)}{3-M} \qquad (14)$$

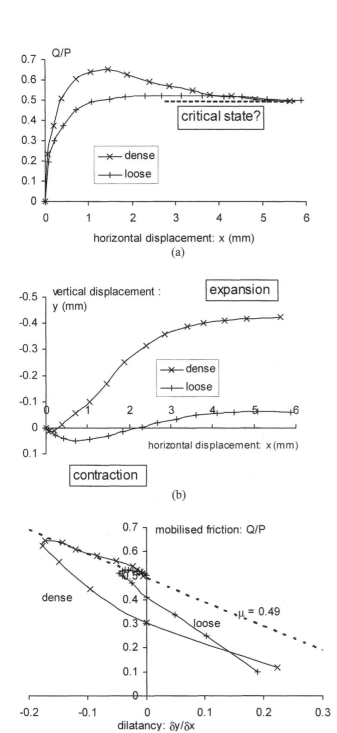

Figure 7. Direct shear tests on Ottawa sand (dense v_o 1.562; loose v_o 1.652): (a) mobilised friction; (b) vertical displacement; (c) stress:dilatancy response (data from Taylor 1948)

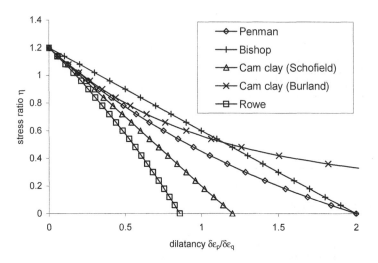

Figure 8. Comparison of boundary energy corrections interpreted as flow rules

These different flow rules are plotted in Figure 8 together with the flow rule for the modified Cam clay model (Roscoe & Burland 1968) and, for good measure, the stress:dilatancy relationship of Rowe (1962). All these flow rules show critical states: when $\eta = M$ the rate of volume change falls to zero. The differences emerge at lower values of stress ratio – in particular modified Cam clay asserts that the rate of distortion should be zero (the rate of dilation infinite) at zero stress ratio. The discussions in the literature of the relative merits of these different ways of interpreting test data are pretty heated at times. Figure 8 shows that the differences are really, in the end, in the detail.

3 CAM CLAY AS PRIMOGENITOR?

The basic ingredients of an elastic-plastic model are:

elastic properties – describing recoverable response for any change in stress

yield function – defining the boundary of elastically attainable states of stress

flow rule – defining the mechanism of plastic deformation for any increment of stress causing yield and irrecoverable, plastic strain

hardening rule – linking the magnitude of plastic strains with the change in size of the yield surface

(It may also be necessary to define a *failure criterion*.)

The end point of the debate about boundary energy corrections is essentially a flow rule – we still need to define the other ingredients. For a flow rule to be useful in an elastic-plastic model it has to define a ratio of *plastic* strain increments. In writing a statement about the distribution of energy this requires the consideration of the possibility of some energy being stored elastically. Roscoe et al. (1963) assumed that there was elastic volumetric deformation:

$$\delta\varepsilon_p^e = \frac{\kappa\delta p'}{vp'} \qquad (15)$$

but that elastic distortional energy was essentially negligible. From a numerical point of view it is helpful to have non-infinite elastic stiffnesses and at the very least we might introduce an elastic distortional deformation:

$$\delta\varepsilon_q^e = \frac{\delta q}{3G} \tag{16}$$

noting that anything other than a constant value of shear modulus G can easily run into thermo-dynamic problems.

Failure of soils usually conjures up images of Mohr-Coulomb failure criteria. Early attempts to devise elastic-plastic models for soils (Drucker et al. 1957) got themselves into a fankle by assuming on the one hand that the Mohr-Coulomb failure criterion must also be a yield criterion and on the other hand that Drucker's (1954) stability principle, which requires plastic strain increments to be directed along the outward normal to the yield surface (Fig. 9), was an essential ingredient of any soil model. Drucker's discussion defines *a* class of stable materials but defining a class of stable materials does not constrain real materials to be stable. In fact, as we shall see, even invoking Drucker's normality rule does not guarantee material stability.

A qualitative solution was provided by Calladine (1963) who concluded that 'normally consolidated clay may be regarded as an "isotropic hardening" strain hardening plastic material for which the hardening depends on the plastic changes in volume'. The key was to invoke normality but to link it to yield loci (in the $p':q$ compression plane) of some essentially arbitrary shape (Fig. 10) chosen to ensure that plastic distortion was zero for isotropic state of stress (the plastic potentials/yield loci cut the p' axis orthogonally) and that plastic compression was zero for critical state stress ratios (the plastic potentials/yield loci have a horizontal tangent for $\eta = q/p' = M$). Yielding under changes in p', and hence expansion of the yield loci, was linked with an idealised form of the normal compression and unloading of clays – noting that the character of this response was essentially similar to the yielding of metals.

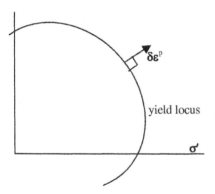

Figure 9. Yield locus in stress space and normality of work conjugate plastic strain increment vector

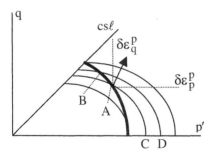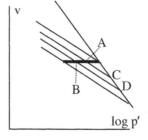

Figure 10. Plastic strain increments normal to yield loci linked with elastic lines in compression plane: consequent effective stress path for undrained test (after Calladine 1963)

From there it is just a small step to combine the work equation/flow rule (12) with the assumptions of normality and of volumetric hardening to produce the original Cam clay model (Roscoe & Schofield 1963) which can reasonably claim to be the first complete elastic-plastic constitutive model for soils.

Mathematically we can illustrate the process of model building as follows. We work in terms of a vector of stresses:

$$\sigma = \begin{pmatrix} p' \\ q \end{pmatrix} \tag{17}$$

(but this vector could include all six components of the effective stress tensor) and vector of work conjugate strain increments:

$$\delta\varepsilon = \begin{pmatrix} \delta\varepsilon_p \\ \delta\varepsilon_q \end{pmatrix} \tag{18}$$

(this could also include all six components). We assume an additive decomposition of strain increments into elastic and plastic parts:

$$\delta\varepsilon = \delta\varepsilon^e + \delta\varepsilon^p \tag{19}$$

The *elastic response* is written using an elastic stiffness matrix **D** – elastic strains occur for all changes in stress so the stress change is directly related to the elastic component of strain increment:

$$\delta\sigma = \mathbf{D}\delta\varepsilon^e \tag{20}$$

The *yield function* introduces one or more hardening variables χ: for yielding to occur, and for irrecoverable plastic strains to be generated, the stress state must satisfy the yield criterion:

$$f(\sigma, \chi) = 0 \tag{21}$$

The *flow rule* defines a mechanism of plastic straining – in other words the ratio of separate components of plastic strain when yielding is occurring:

$$\delta\varepsilon^p = \mu \frac{\partial g}{\partial \sigma} \tag{22}$$

where μ is a 'scalar multiplier' indicating that the flow rule specifies relative but not absolute magnitudes of plastic strain increments. The function g is a plastic potential function which is only required in the form of its derivatives.

The *hardening rule* links the hardening variable(s) χ with plastic strain:

$$\chi = \chi(\varepsilon^p) \tag{23}$$

Combining these equations together with the 'consistency' assumption that if plastic strains are occurring:

$$f = \delta f = 0 \tag{24}$$

we find

$$\delta\sigma = \left[\mathbf{D} - \frac{\mathbf{D}\dfrac{\partial g}{\partial \sigma}\dfrac{\partial f}{\partial \sigma}^{\mathrm{T}}\mathbf{D}}{\dfrac{\partial f}{\partial \sigma}^{\mathrm{T}}\mathbf{D}\dfrac{\partial g}{\partial \sigma} - \dfrac{\partial f}{\partial \chi}^{\mathrm{T}}\dfrac{\partial \chi}{\partial \varepsilon^{p}}\dfrac{\partial g}{\partial \sigma}} \right] \delta\varepsilon = \mathbf{D}^{ep}\delta\varepsilon \tag{25}$$

generating the elastic plastic stiffness matrix $\mathbf{D}^{ep}$ linking strain increments and stress increments. This naturally presents itself as an elastic predictor – the first term in (25) – and a plastic corrector – the second major term in (25) – which is necessary if plastic strains are occurring and, evidently, the stiffness is correspondingly reduced. The advantage of using a non-infinite value of elastic stiffness is perhaps clear.

Definition of an elastic-plastic model then becomes a matter of specifying $\mathbf{D}$, f, g, χ. For original Cam clay:

$$\mathbf{D} = \begin{pmatrix} \dfrac{vp'}{\kappa} & 0 \\ 0 & 3G \end{pmatrix}; \quad f = g = q - Mp'\ell n\left(\dfrac{p'_o}{p'}\right); \quad \chi = p'_o\,; \quad \dfrac{\partial p'_o}{\partial \varepsilon_p^p} = \dfrac{vp'_o}{(\lambda - \kappa)} \tag{26}$$

and the elastic-plastic stiffness matrix is:

$$\mathbf{D}^{ep} = \frac{\begin{pmatrix} 1 + \dfrac{Mp'v(M-\eta)}{3G(\lambda-\kappa)} & -(M-\eta) \\ -(M-\eta) & (M-\eta)^2 + \dfrac{M\kappa(M-\eta)}{(\lambda-\kappa)} \end{pmatrix}}{\dfrac{(M-\eta)^2}{3G} + \dfrac{\kappa}{vp'} + \dfrac{M\kappa(M-\eta)}{3G(\lambda-\kappa)}} \tag{27}$$

One of the elegant features of Cam clay is that it is not necessary (though it is possible) to specify a failure criterion: failure conditions are an emergent property of the model. The model also discovers ultimate critical states because of the combined forms of the flow rule and hardening law: hardening requires plastic volumetric strain which ceases when $\eta = M$. The elastic-plastic response of (27) can be inspected for this value of stress ratio. The elastic-plastic relationship becomes:

$$\begin{pmatrix} \delta p' \\ \delta q \end{pmatrix} = \begin{pmatrix} \dfrac{vp'}{\kappa} & 0 \\ 0 & 0 \end{pmatrix} \begin{pmatrix} \delta\varepsilon_p^p \\ \delta\varepsilon_q^p \end{pmatrix} \tag{28}$$

showing, as expected, that the only plastic stress changes that are admissible, once the critical state stress ratio has been attained plastically, are constant q increments along the horizontal tangent to the yield locus.

The elastic-plastic stiffness matrix (27) appears cumbersome but is necessary. The compliance matrix, calculating strain increments from stress increments is simpler but runs into the problem that not all stress increments are accessible. There are regions of stress space that are clearly forbidden territory. As a trivial example, (28) demonstrates that, as we know, if we are at a critical state generating plastic strains with $\eta = M$, then it is not possible to apply stress increments which imply $\delta q > 0$. On the other hand, no matter what the state of stress, *all* strain increments are possible.

Figure 3 and equation (28) show that to be on the current yield locus at the critical state stress ratio is not necessarily a disaster – this is not necessarily an inescapable attracting black hole, nor does it necessarily require infinite plastic strain to reach this condition. With appropriate control – most conveniently, strain rate control – we can pass through this condition from the 'wet' side to the 'dry' side. The effective stress path (either controlled – but this may be tricky – or emerging) will have a horizontal tangent, as revealed by (28). The critical state stress ratio remains a singularity – we have to pass through it, at that instant, elastically (along the yield locus) – but to have the correct effective stresses and volumetric packing is a necessary but not a sufficient condition to have reached an ultimate asymptotic state of perfect plasticity.

An elegant way of revealing some of the restrictions on stress accessibility and also illustrating key elements of the response of the model is to construct stress response envelopes (Gudehus 1979). (These, by the way, are also an efficient way of organising programmes of testing and gathering experimental data for validation of constitutive models and constitutive frameworks.) Two are shown for Cam clay in Figure 11. At any particular state of stress a rosette of strain increments of equal magnitude but different direction is imposed. The resulting stress increments, calculated using the elastic or elastic-plastic stiffness matrix as appropriate, are plot-

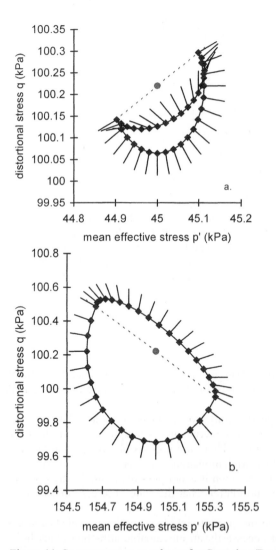

Figure 11. Stress response envelopes for Cam clay (Burland) (a) η > M; (b) η < M

48

ted and an envelope of all these stress increments drawn. With the stress variables and work conjugate strain increment variables chosen for the triaxial p′:q plane, for a stress state inside the current yield locus, producing purely elastic response for all strain increments, the stress response envelope is an ellipse. For a stress state on the yield locus the stress response envelope consists of two portions: an elliptical arc for strain increments which unload inside the yield locus and a much flatter elliptical arc for elastic-plastic increments – obviously the stiffness for such increments is much lower so that, for a given strain increment, the stress cannot get so far from the starting point.

Figure 11b shows the stress response envelope for a state of stress on the 'wet' side of the yield locus, with $\eta < M$: an orderly response is obtained. Figure 11a shows the envelope for a state of stress on the 'dry' side of the yield locus with $\eta > M$. Now, no matter the direction of the strain increment, no matter whether the response is elastic or elastic-plastic, the stress increment retreats inside the current yield locus: we either have elastic unloading or plastic softening.

The advantage of using the stiffness formulation is now clear. Working from stress increments to strain increments we are confronted both with the problem of inaccessibility – outward directed stress increments are not admissible – and uncertainty – the same stress response can imply two quite different strain responses. Practically this is an indication of potential for bifurcation of response, likelihood of deformation instabilities and development of localisation within a sample – and indeed we expect 'dry' samples to tend to fail by generation of failure planes. The mathematical treatment of localisation is complex and the numerical analysis using models which show tendencies to localise requires care. We may note simply that, even though Drucker's stability criterion was invoked in the generation of the Cam clay model, it does not guarantee stable response under all states of stress.

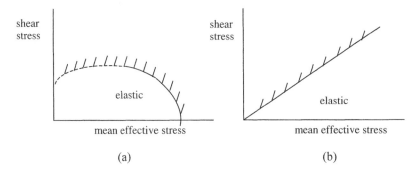

Figure 12. (a) Yield surface closed on mean stress axis: weak definition of Cam clay? (b) Contrasted with Mohr-Coulomb model having open yield surface

This is *the* original Cam clay model but the Cam clay name has been used rather less precisely. What then is Cam clay? Can we establish a definition of *a* Cam clay model? Can we propose a taxonomy of soil models which permits some flexibility of requirements for membership of the Cam clay family? Given that in commercial geotechnical numerical analysis programs the standard models that are available might include Cam clay and Mohr-Coulomb, one might suggest that the weakest requirement for membership of the Cam clay family is that the yield surface is closed across the mean stress axis (Fig. 12). A somewhat stronger requirement would be to insist on associated flow together with volumetric hardening:

$$f = g; \quad \frac{\partial \chi}{\partial \varepsilon_q^p} = 0 \tag{29}$$

49

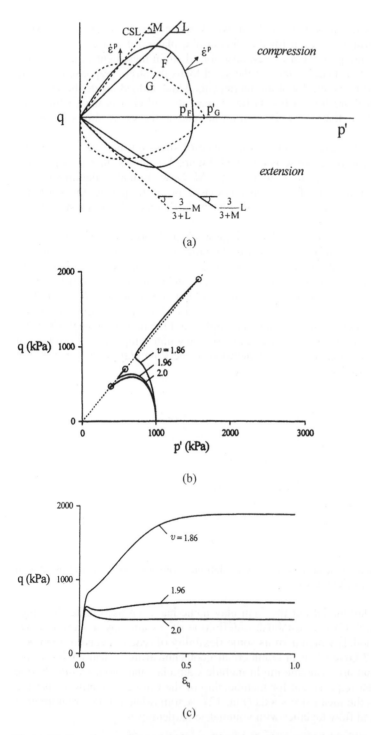

(a)

(b)

(c)

Figure 13. Superior sand: (a) yield locus (F) and plastic potential (G); (b) undrained effective stress paths; (c) undrained stress:strain response (after Boukpeti & Drescher 2000)

The strictest requirement would insist on these things *and* the particular form of flow rule/yield locus assumed in (26).

Some models which fall within the weakest definition of Cam clay are listed in Table 1. Cam clay (Burland) gm is the (genetically modified) version of Cam clay described by Roscoe & Burland (1968) and called simply Cam clay by Muir Wood (1990) because this is the form of Cam clay model that has seen widest implementation and application in numerical analysis. This form of Cam clay model has proved to be quite suitable to subspeciation: one might mention the **subspecies** 'bubble' model (Al-Tabbaa & Muir Wood 1989) with subsequent **sub-subspecies** 3SKH model (Stallebrass & Taylor 1997) and fabric softening model (Rouainia & Muir Wood 2000, Gajo & Muir Wood 2001).

The Nova & Wood (1979) model relaxes many of the pure Cam clay assumptions but is recognisably of the same genus. Nonassociated flow is necessary in order to reproduce the undrained effective stress paths that are typical of dense and loose sands. The combination of distortional and volumetric hardening provides additional modelling capability but has the result that a unique critical state line no longer emerges unless other stratagems are adopted. Nova & Wood use the failure criterion as a switch to turn off the distortional hardening. This is a pragmatic but mathematically inelegant solution which introduces discontinuities into the results of simulations. Boukpeti & Drescher (2000), in the recent versions of Superior sand, take the neater approach of making the distortional contribution to hardening gradually disappear as the state of the soil approaches the critical state line. A typical yield locus and plastic potential curve for Superior sand are shown in Figure 13 together with simulations of undrained triaxial compression tests on sand samples of various densities.

Table 1. Taxonomy of Cam clay models

Class	elastic-plastic models			
Family	critical state models			
Genus	Cam clay			
Species	Cam clay (Schofield)	Cam clay (Burland) gm	Nova-Wood sand model	Superior sand
associated flow?	✓	✓	✗	✗
volumetric hardening?	✓	✓	✓	✓
distortional hardening?	✗	✗	✓	✓
failure criterion?	✗	✗	✓	✗
Subspecies		'bubble'		
Sub-subspecies		3SKH fabric softening		

4 NON-UNIQUENESS OF CONSTITUTIVE APPROACHES

These last two examples of Cam clay models are attempting to reproduce the behaviour of sand and it may seem curious to be using a framework for this purpose which was originally devised for modelling clay. Turned round another way, Cam clay seems to lend itself to formulations where the hardening is predominantly volumetric which seems appropriate to clays but perhaps less appropriate to sands where, except at high stress levels where the degree of particle crushing becomes significant (though there will always be some crushing at the highly stressed particle contacts), the irrecoverable deformations that are seen are linked primarily with rearrangements of particles and hence primarily with *distortional* deformations.

The language of friction and Mohr-Coulomb models sounds relevant to sands and it turns out that a rather simple model for sand can be generated by hybridising Mohr-Coulomb and Cam clay concepts. The resulting model (Severn-Trent sand: Gajo & Muir Wood 1999) is clearly based within the framework of critical state soil mechanics and generates ultimate critical states but is equally clearly not a Cam clay model.

Table 2. Severn-Trent sand classification

Class	elastic-plastic models
Family	critical state models
Genus (hybrid)	Mohr-Coulomb × Cam clay
Species	Severn-Trent sand
associated flow?	×
volumetric hardening?	×
distortional hardening?	✓
failure criterion?	×

We start with a simple hardening Mohr-Coulomb form in which the yield function is a straight line (Fig. 14):

$$f = q - \eta_y p' = 0 \qquad (30)$$

and current yield stress ratio η_y is assumed to harden hyperbolically with distortional strain, eventually reaching a peak value η_p:

$$\frac{\eta_y}{\eta_p} = \frac{\varepsilon_q^p}{b + \varepsilon_q^p} \qquad (31)$$

Assumption of associated flow with such a model will give excessive dilatancy – and continuing volumetric expansion as the soil heads for failure with $\eta = \eta_p$ (Fig. 14a). Let us, however, borrow the plastic potential from the original Cam clay model:

$$\frac{\delta\varepsilon_p^p}{\delta\varepsilon_q^p} = M - \eta_y \qquad (32)(12 \text{ bis})$$

This, we know, will show critical states and a decreasing rate of dilation as the stress ratio increases (Fig. 14b).

The hyperbolic hardening rule (31) is simple but clearly inadequate on its own – it shows monotonic increase of stress ratio with increasing strain whereas typical shear tests on sand – especially dense sand – show a peak stress ratio and subsequent softening. Mathematical description of nonmonotonic functions is to be avoided if possible. We can do this by noting that the peak strength of sand is not a soil constant but will depend on stress level and density. A simple way of introducing this dependence is through the state parameter ψ (Fig. 15a) which expresses the volume distance of the current state from the critical state at the same stress level. Here is a second clear critical state reference within the Severn-Trent sand model. We might, for example, assume that the peak stress ratio η_p in (31) varies linearly with ψ (Fig. 15b):

$$\eta_p = M - k\psi \qquad (33)$$

Now, as the stress ratio heads towards the current peak (31), volume changes are occurring, as a result of the flow rule (32), which lead to changes in the peak strength (33). The tight interrelationship between these mechanisms produces peak stress ratios and ultimate critical states

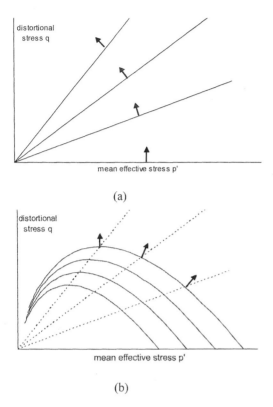

(a)

(b)

Figure 14. (a) Normality to Mohr-Coulomb yield loci giving excessive dilatancy; (b) combination of Mohr-Coulomb yield loci and Cam clay plastic potentials (flow rule)

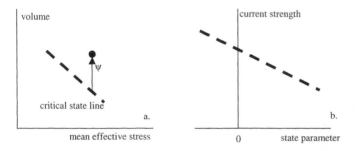

Figure 15. (a) Definition of state parameter ψ; (b) dependence of current strength on state parameter

with no modification to the simple monotonic hyperbolic hardening law (Muir Wood et al. 1994). Some typical simulations of undrained tests on Hostun sand are shown in Figure 16 (Gajo & Muir Wood 1999).

Superior sand and Severn-Trent sand have apparently reached the same end point by completely different routes. The models have evolved from different parentage towards a common ecological niche. One seems to be a Cam clay model the other a Mohr-Coulomb model. However, there are two strong elements of common ground. Both use flow rules which lead to critical states. Both use compression plane evidence – state parameter or equivalent – to control one or more of the driving functions. The hardening rules and yield functions are quite different.

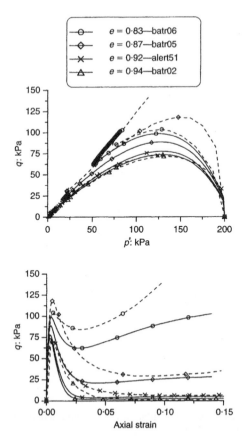

Figure 16. Severn-Trent sand simulations of undrained triaxial compression tests on Hostun sand with different initial void ratio: solid lines: experimental observation; dashed lines: simulations (from Gajo & Muir Wood 1999)

5 CLOSURE

Critical state soil mechanics can be regarded as 'a deeply running theme that volume changes in soils are at least as important as changes in effective stresses in trying to build a general picture of soil behaviour' (Muir Wood 1990). The term 'critical state model' carries no precision so far as description of constitutive approach is concerned: the number of models which demonstrate ultimate asymptotic critical states is very large and we have shown here that this group of models includes models which have been developed with very different underlying philosophies. The term 'Cam clay' is slightly, but only slightly, more specific. Cam clay is a critical state model but not all critical state models are Cam clay. Clearly, 'critical state model' cannot imply Cam clay. Cam clay can be seen as a genus of models within the more general family of critical state models. The original Cam clay model is a species within this genus. Here Cam clay models are contrasted with Mohr-Coulomb models through their use of closed rather than open yield surfaces – but this is not a very deep distinction since it is quite possible to add a cap to the Mohr-Coulomb yield surface as an additional or linked mechanism of generation of plastic strains.

From the angst of the debate around boundary energy corrections, the Cam clay flow rule eventually emerged. This flow rule (and its somewhat close kin) is probably the most valuable element of the original Cam clay model. The insistence on application of Drucker's stability and hence normality criterion turns out not to be helpful: soils in general and sands in particular do not fall within Drucker's class of stable work hardening materials. But the construction of a more general class of models in which separate yield functions and plastic potentials have to be proposed presents no major theoretical challenge. There is a consequent implication for experimental studies. Testing needs not only to determine the (changing) mechanism of plastic deformation (which can be achieved on monotonic paths) but also to detect (objectively?) the onset of plastic deformation which requires subtle testing (Tatsuoka & Ishihara 1974) and/or multiple samples (Alawaji et al. 1990).

There is no uniqueness in constitutive solutions – the choice of model and the choice of underpinning theoretical framework are a matter of personal preference. Critical state soil mechanics does, however, seem to have established a permanent place – not only in the class of elastic-plastic models (e.g. Gudehus 1996, Bauer & Herle 2000). There has been a lot of cross-fertilisation. '[However,] it is becoming all too obvious that many constitutive models are too complicated to be applied. To this end, the evolution of constitutive equations can be grossly likened to Darwin's evolution of species. We should bear Darwin's principle in mind: survival of the fittest, not the best!' (Wu & Kolymbas 2000). Critical state soil mechanics seems to be a clear survivor.

REFERENCES

Alawaji, H., Alawi, M., Ko, H.-Y., Sture, S., Peters, J.F. & Muir Wood, D. 1990. Experimental observations of anisotropy in some stress-controlled tests on dry sand. In J.P. Boehler (ed.) *Yielding, damage, and failure of anisotropic solids* Mechanical Engineering Publications, London EGF5: 251-264.

Al-Tabbaa, A. & Muir Wood, D. 1989. An experimentally based "bubble" model for clay. In S. Pietruszczak & G.N. Pande (eds) *Numerical Models in Geomechanics NUMOG III* London: Elsevier Applied Science: 91-99.

Bauer, E. & Herle, I. 2000. Stationary states in hypoplasticity. In D. Kolymbas (ed.) *Constitutive modelling of granular materials:* 167-192. Berlin: Springer.

Bishop, A.W. 1954. Correspondence on 'Shear characteristics of a saturated silt' (A. Penman). *Géotechnique* 4(1): 43-45.

Boukpeti, N. & Drescher, A. 2000. Triaxial behaviour of refined Superior sand model. *Computers and Geotechnics* 26(1): 65-81.

Calladine, C.R. 1963. Correspondence 'The yielding of clay'. *Géotechnique* 13(3): 250-255.

Casagrande, A. 1936. Characteristics of cohesionless soils affecting the stability of slopes and earth fills. *J. Boston Soc. Civil Engineers* 23(1): 13-32.

Drucker, D.C. 1954. A definition of stable inelastic material. *J. Applied Mechanics, Trans. ASME* 26: 101-106.

Drucker, D.C., Gibson, R.E. & Henkel, D.J. 1957. Soil mechanics and work hardening theories of plasticity. *Trans. ASCE* 122: 338-346.

Gajo, A. & Muir Wood, D. 1999. A kinematic hardening constitutive model for sands: the multiaxial formulation. *International Journal for Numerical and Analytical Methods in Geomechanics* 23: 925-965.

Gajo, A. & Muir Wood, D. 2001. A new approach to anisotropic, bounding surface plasticity: general formulation and simulations of natural and reconstituted clay behaviour. *International Journal for Numerical and Analytical Methods in Geomechanics* 25(3): 207-241.

Gudehus, G. 1979. A comparison of some constitutive laws for soils under radially symmetric loading and unloading. In W. Wittke (ed.) *Proc. 3rd Int. Conf. on Numerical methods in geomechanics, Aachen*: 1309-1323.

Gudehus, G. 1996. A comprehensive constitutive equation for granular materials. *Soils & Foundations* 36(1): 1-12.

Henkel, D.J. 1960a. The relationships between the effective stresses and water content in saturated clays. *Géotechnique* 10(2): 41-54.

Henkel, D.J. 1960b. The shear strength of saturated remoulded clays. In *Proc. Research Conf. on Shear strength of cohesive soils* (Boulder) ASCE: 533-554.

Muir Wood, D. 1990. *Soil behaviour and critical state soil mechanics.* Cambridge: Cambridge University Press (462pp).

Muir Wood, D. 2001. Some observations of volumetric instabilities in soils. IUTAM Symposium on Material instabilities and the effect of microstructure, Austin, Texas.

Muir Wood, D., Belkheir, K. & Liu, D.F. 1994. Strain-softening and state parameter for sand modelling. *Géotechnique* 44(2): 335-339.

Nova, R. and Wood, D.M. 1979. A constitutive model for sand. *Int. J. for Numerical and Analytical Methods in Geomechanics* 3(3): 255-278.

Penman, A.D.M. 1953. Shear characteristics of a saturated silt. *Géotechnique* 3(8): 312-328.

Poorooshasb, H.B. & Roscoe, K.H. 1961. The correlation of the results of shear tests with varying degrees of dilatation. In *Proc. 5th Int Conf SMFE, Paris* 1: 297-304.

Rendulić, L. 1937. Ein Grundgesetz der Tonmechanik und sein experimenteller Beweis. *Bauingenieur* 18: 459-467.

Roscoe, K.H. & Burland, J.B. 1968. On the generalised stress-strain behaviour of 'wet' clay. In J. Heyman & F.A. Leckie (eds) *Engineering plasticity*: 535-609. Cambridge: Cambridge University Press.

Roscoe, K.H. & Schofield, A.N. 1963. Mechanical behaviour of an idealised 'wet' clay. In *Proc. 2nd Eur. Conf SMFE, Wiesbaden* 1: 47-54.

Roscoe, K.H., Schofield, A.N. & Thurairajah, A. 1963. Yielding of clays in states wetter than critical. *Géotechnique* 13(3): 211-240.

Roscoe, K.H., Schofield, A.N. & Wroth, C.P. 1958. On the yielding of soils. *Géotechnique* 8(1): 22-52.

Rouainia, M. & Muir Wood, D. 2000. A kinematic hardening constitutive model for natural clays with loss of structure. *Géotechnique* 50(2): 153-164.

Rowe, P.W. 1962. The stress-dilatancy relation for static equilibrium of an assembly of particles in contact. *Proc. Roy. Soc. London,* A269: 500-527.

Stallebrass, S.E. & Taylor, R.N. 1997. The development and evaluation of a constitutive model for the prediction of ground movements in overconsolidated clay. *Géotechnique* 47(2): 235-253.

Tatsuoka, F. & Ishihara, K. 1974. Yielding of sand in triaxial compression. *Soils & Foundations* 14(2): 63-76.

Taylor, D.W. 1948. *Fundamentals of soil mechanics.* New York: John Wiley.

Wu, W. & Kolymbas, D. 2000. Hypoplasticity then and now. In D. Kolymbas (ed.) *Constitutive modelling of granular materials*: 57-105. Berlin: Springer.

DISCUSSION

S.M. Springman & T. Weber
Institute for Geotechnical Engineering, Swiss Federal Institute of Technology, Zurich, Switzerland

In opening this first discussion, Sarah Springman commented that many people have used versions of Cam clay to develop their constitutive ideas on soils far removed from the initial focus (e.g. on the wet side of the Critical State Line). She noted that examples for some of the many evolutionary daughters and sons included Cam clay based models for structured and cemented soils and that there were also quite a number from rival species, such as damage and localisation models, which have evolved subsequently.

Ivo Herle expressed interest in the reference to Dirac, quoting Richard Feinman's anti-hierarchical stance in saying that 'the most efficient way in developing material models is by guessing and not by hierarchical development'. David Muir Wood countered *quot homines tot sententiae* (there are as many opinions as there are people). He said that guesswork may be helpful, but experimentally inspired guesswork would be rapidly more productive than pure guesswork and that well-tuned experiments, which were designed to pick flaws in existing models, would lead to rather faster inspiration. This would be the most productive way of making progress in the modelling, and would link into the loop of observation and reflection and prediction again.

Fook Hou Lee asked about the significance of the taxonomy of critical state models and Cam clay models. He noted that the Cam clay model includes essentially the ideas of critical state, of associated flow and of dilatancy, and modellers have relaxed those ideas in somewhat different ways. He was not convinced by the usefulness and significance of defining the species, genus and hierarchies of models. Muir Wood replied that this had been an instinctive idea, born out of concern about the sloppiness in the terminology of use of the terms 'Cam clay' and 'critical state models'. He felt that there was a difference between using a term, what people understood about what that term actually meant and what that term actually implied. Lee argued that there was some room for differences in the definition about what was a strong formulation and what was a weak formulation of a Cam clay model with which Muir Wood agreed: 'let a thousand flowers flourish'.

The Chair called for offerings from the localisation fans, to which Cino Viggiani responded by posing a question about the dramatic softening apparent on a deviator stress versus axial strain curve. He observed that the simulations showed amazing coincidence between the model and with the experimental observation and asked about the influence of strain localisation on this stress-strain curve? Muir Wood agreed that there would be inhomogeneity in the soil in a sample undergoing a triaxial test, but the aim was to try to show that a model of this class would be able to produce certain characteristic types of response. The model could be combined with a more detailed localisation analysis – the point is that it would be able to say something physically useful about what would happen within the region of strain localisation. Viggiani commented provocatively that the model could not be a good model, because it apparently produced a homogeneous response, which was coincident with an inhomogeneous reality. Muir Wood replied that obviously the tuning of parameters was incorrect for this very reason and that an improved analysis could take these aspects into account. However, he was concerned that there was always a danger of constitutive modellers taking published results at their face value, without understanding the details of the ways in which they had been obtained and the potential (possibly unreported) inhomogeneities lurking behind them!

Andrzej Niemunis asked about whether anyone had had further experience with the Superior sand models after a disappointing attempt had been made to program them, because the model had showed some instability, resulting from the large degree of non-associativity between the loading and flow directions. He concluded that this had created numerical problems. Muir Wood indicated that Natalie Boukpeti was programming the Superior sand model using FLAC. He also said that he, Alessandro Gajo and Andrew Drescher wanted to see to what extent a range of sand behaviour could be reproduced with limited experimental input using these two contrasting model approaches.

Springman closed the discussion by commenting on the positive signs in the search for truth. Although it was still quite a long way off, she thought we were nearer than we had been a few years ago!

Constitutive and Centrifuge Modelling: Two Extremes, Springman (ed.)
© *2002 Swets & Zeitlinger, Lisse, ISBN 90 5809 361 1*

Micro-geomechanics

M.D. Bolton & Y.P. Cheng
Cambridge University Engineering Department, United Kingdom

ABSTRACT: The paper discusses the neglect of micro-mechanics in soil mechanics, and seeks to establish a role that will benefit both the research worker and the practitioner. In support of the mathematical construct of "plasticity", micro-mechanics introduces observations of grain crushing and re-arrangement. Not only does this help to explain the dimensionally inconsistent concept of the normal compression line, it goes some way to unifying our understanding of sands and clays. Indeed, bridging the grain-continuum duality is the key to raising the confidence of practitioners both in the meaningfulness of certain constitutive modelling parameters and in the scaling rules applied to the behaviour of small scale physical models.

1 INTRODUCTION

A gulf has arisen between research and practice in geotechnical engineering. Practitioners need to take decisions about real geotechnical phenomena. They have at their disposal a large body of empirical correlations and rules of thumb with which to filter observations they make on site and in the laboratory. They also have the principles and routines of soil mechanics set out by Terzaghi and his followers, which they learned at University. Research workers, on the other hand, have developed the art of creating elaborate geotechnical simulations, both numerically using finite element packages and physically in centrifuge models. Given the required effort to fix parameters in the mathematical models, or to create representative soil profiles in centrifuge models, devotees of each technology may feel it is possible to predict what will happen in the field. The contrast in research ideology between the mathematical and the physical model is the subject of this Workshop. But we should not forget the credibility gap between the pair of them and engineering practice. Practitioners are often deeply sceptical about the capacity of any simulation technique to recreate the essential features of the behaviour of real soil profiles. The reasons for this are instructive.

The fundamental mechanical behaviour of soils has largely been investigated through the testing of homogeneous reconstituted samples of uniform grain size and mineralogy. Soils with a wide dispersion of particle sizes are empirically known to behave quite differently from uniform soils, but repeatable samples with dispersed particle sizes are very difficult to create in the laboratory. Dispersed granulometries have generally been avoided. This is as true of the databases of compressibility and strength of sands in triaxial tests used to create mathematical models, as it is of centrifuge models used to represent real soil strata. In almost every case, sands are "clean" and clays are "pure". And yet we know that the phenomena of internal erosion, piping and subsidence which afflict gap-graded soils, for example, are of great practical significance in the design of earthworks. We should reflect that they are also precisely the same class of phenomenon as the segregation of grain sizes which deflects us from using dispersed granulometries in our research. This is obviously perverse. It is also characteristic of a wider carelessness of soil fabric amongst research workers that surprises and disappoints the practitioner.

Although every practitioner has been taught the importance of grain size ratios for uniformity coefficient and filter criteria, they find that constitutive models ignore particulate classifications and employ unfamiliar abstract parameters instead, sometimes by the dozen. Real soil profiles are generally defined using a distribution of grading curves, water contents, plastic and liquid limits, and Standard Penetration Tests or Cone Penetration Tests. Having defined some soil layers using these "empirical" index tests, there follows an obscure process leading to the selection of certain modelling parameters representing the mechanical behaviour of each of stratum. Often, these modelling parameters conform to certain features on the non-linear responses of triaxial tests conducted on a relatively sparse distribution of "undisturbed" tube samples taken from within the various strata. Enlightened modellers may seek to cross-correlate modelling parameters with index test values in order to maximise the use of this plentiful empirical information. This laudable objective is presently hampered by ignorance of the physical meaning both of our standard index tests, and of mathematical modelling parameters: these numbers come from universes of knowledge that barely overlap for most engineers.

In other branches of material science, the link between material characterisation and engineering behaviour was made through microscopy. Solids are easy to section and view, in electron microscopes for example, so that idiosyncrasies in macroscopic behaviour could be related to micro-structural features. Soils are much more difficult to "fix" under the microscope so their micro-structural responses to changes in stress, or vice versa, are largely unknown. Soil mechanics, and especially constitutive modelling, has therefore developed in the absence of an understanding of the micro-mechanisms which control macro-behaviour - whether in index tests or in research-quality triaxial tests. International Society of Soil Mechanics and Geotechnical Engineering Technical Committee (ITC35) aims to foster this understanding.

2 UNDERSTANDING COMPRESSIBILITY

Compressibility is conventionally represented on plots of voids ratio versus the logarithm of effective stress using functions that are not even dimensionally consistent (McDowell and Bolton 1998). These familiar bi-linear plots have an "elastic" gradient C_s (using logarithms to base 10) or κ (using natural logarithms) and corresponding "plastic" gradients C_c or λ. An example for dry Dog's Bay carbonate sand is shown in Figure 1.

The relationship between this data and the mathematical models taught to engineers is revealing. There are two distinct phases of *virgin* compression, with 1 MPa as the transition stress. The first phase is far from linear on this logarithmic plot, but much more nearly linear when stress is plotted on a linear scale, see Figure 2. However, it is also almost completely irrecoverable, or "plastic" in nature. The second phase is linear on a log plot, and equally irrecoverable.

The unload-reload loops (not unique lines) are comparatively recoverable, but not exactly so, as the expanded version in Figure 3 demonstrates. Each unload path is more nearly linear against log stress, while the reload path is more linear versus stress. The re-invigoration of plastic irrecoverable strain is seen to occur before the soil reaches its previous maximum stress, and this is seen to create on-going compaction when the cycle is repeated.

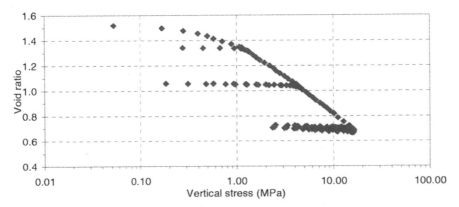

Figure 1. One-dimensional compression curves for Dog's Bay sand

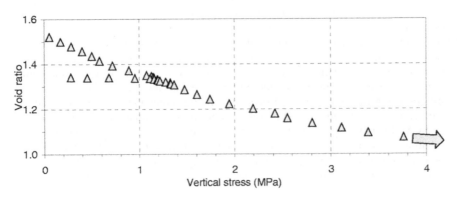

Figure 2. Virgin compression: initially linear, and irrecoverable

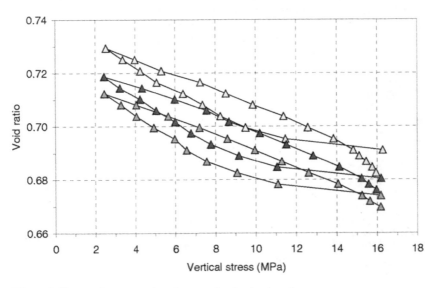

Figure 3. Hysteretic compaction due to unload-reload cycles

Some of the discrepancies between the data and the conventional teaching of geotechnical engineers is understandable in that specialists, at least, would be aware of them. Some discrepancies would seem strange even to engineers who had taken a specialist graduate course, however; this would include the two-phase virgin compression and the question of whether to use gradients based on natural or logarithmic scales. In these circumstances the engineer might look back to start of this section and blame the trouble on the fact that the data is of carbonate sand. That is how we have developed the geologist's habit of assuming that every ground type (carbonate soils, residual soils, quick clays etc.) requires its own specialists and conferences. The agenda of this paper is that we can discover micro-mechanical interpretations of data which should have wider currency than a particular region, stratum, mineralogy or particle size.

The one-dimensional tests reported above were conducted in a mini-oedometer with a glass platen through which photographs were taken with a digital camera - see Figure 4. These images, with Figures 1 to 3, show different regimes of micro-mechanical behaviour to set against the different regimes of macroscopic data. Figure 5 shows the progression of microstructures. At zero stress there is an ambiguity about the definition of the top surface of the sand sample, which is not yet in contact with the glass; there is a significant compression required on the first application of load as projecting particles rotate to permit the glass to bear on a greater number. As stress increases in phase 1 of virgin compression there is an occasional breakage of a few grains, accompanied by rotations and rearrangements of many other grains. Voids are reduced in size mainly by particle rearrangements. Beyond 1 MPa, however, it becomes clear that compression is occurring by the fracture of grains, including those that are already broken. Larger voids close as split grains rearrange; as this continues, the smaller voids themselves are filled by the readjustment of even smaller fragments. This self-similar process is qualitatively the same as that first described by McDowell et al. (1996), where it was pointed out that the mechanism possesses natural hardening due to the increased brittle strength of finer fragments. The data of Figures 1 to 3 can be made much more intelligible through this application of clastic mechanics - the recognition that soils comprise brittle grains that can crush, rearrange and deform elastically.

Images such as Figure 5, when taken together with particle size analyses showing statistical evidence of the consequences of grain crushing, can raise confidence in the data of Figure 1 so as to support the ultimate selection of compressibility parameters. Instead of relying on some mathematical model that was fitted blindly to some particular range of stress applied to some different soil, the selection of plots and parameters can be referenced against the observed re-

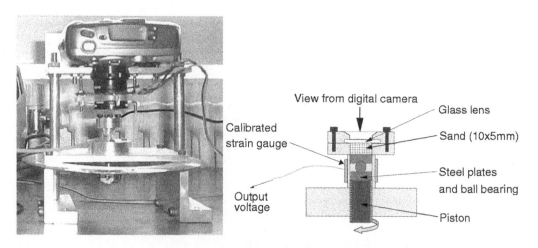

Figure 4. Mini-oedometer for capturing images of microstructure during 1D compression

sponse of the microstructure. The compressibility of sand is determined by the degree to which the grains can rearrange with and without fracture. On an unload-reload cycle, as observed in Figure 5, there is very little observable change in structure and correspondingly very little change in voids ratio. Nevertheless, the small cyclic compression seen in Figure 3 may have very significant engineering effects. If a similar change in structure were to occur at constant volume, there would have to be a strong reduction in effective stress. One possible cause of these small changes on cyclic loading is demonstrated in Figure 6 which focuses on the response of a single carbonate grain. It is seen to fragment up to a stress of 5.6 MPa, and the fragments then remain fixed in number on unloading and reloading, while one fragment in particular finds it can rotate and readjust its position. Perhaps these slight rearrangements of fine fragments explain the cyclic degradation of sands, which give rise to liquefaction.

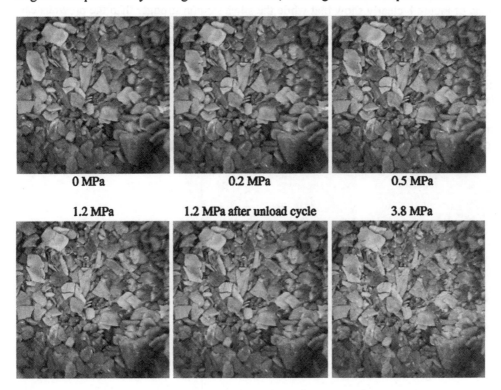

Figure 5. Platy carbonate sand seen through a platen loaded at advancing stress levels

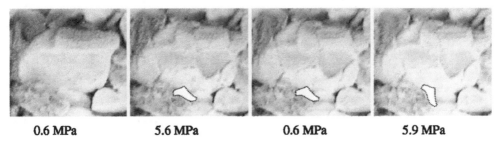

Figure 6. The migration of a small fragment during an over-consolidation cycle

It immediately becomes obvious that platy, shelly sands such as Dog's Bay sand will probably behave quite differently from sands comprising rotund silica grains. This point is proved in Figure 7, which shows the result of compression in a uniform sub-rounded silica sand. After some initial rearrangement as the grains successively carry strong contact forces, and then rotate and release them, the aggregate stiffens up. A very interesting crushing mechanism is then observed. One grain crushes to small fragments against the glass platen; then its neighbours are overloaded and similarly crush. Eventually, the majority of grains at that level are crushed into a "snow" of fines which can be seen to fall downwards into the cavernous voids which remain between the original grains below the elevation at which crushing first occurred. While the fines are percolating downwards, the soil seems rather compressible. It then tightens up in possible preparation for another "snowfall".

The data of Figure 8 clearly show that while the silica sand is stronger than the previous carbonate sand, its "yielding" is more sudden, and its eventual "normal compression line" is much steeper and occasionally unstable. There are a few occasions when the stress actually falls; this is possible since the apparatus is displacement controlled. At points of instability, a heavily loaded grain crushes into a myriad of fragments and the sample unloads since neighbouring grains fail to take up the lost contact force. The easy migration of tiny fragments of silica sand into relatively large parent voids contrasts with the relative immobility of the split grains of carbonate sand; hence the relative large plastic compressibility of the former. The observation of microstructure has not only explained the scatter evident in Figure 8 but has provided a rich stream of ideas relevant to soil crushability and compressibility in general.

0 MPa   15 MPa

Figure 7. Severe fragmentation during compression of a silica sand

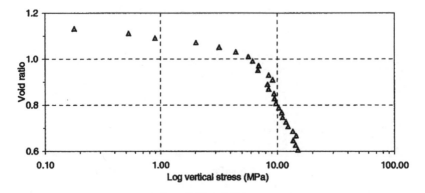

Figure 8. Compression curve for the silica sand

3 UNDERSTANDING PENETRATION RESISTANCE

Friction and dilatancy are understood in terms of grains sliding, rotating and over-riding at points of contact. Engineers broadly recognise that dilation in dense, interlocked soils can produce double the shear strength measured in stress-controlled element tests on loose soil, which does not dilate. The volumetric expansion causes more work to be done against confining pressure at the boundaries. It is also understood that the penetration resistance (CPT, SPT) of dense sands can be greater by an order of magnitude, due to the additional stresses induced around the penetrometer by suppressed dilation. Engineers should also follow the author (Bolton 1986) in understanding that dilation caused by interlocking is very large at small stresses, and reduces towards zero as stresses increase leading to the progressive crushing of interlocking asperities and grains.

Such an understanding in the context of penetrometer data, for example, is of immense value in correlating CPT or SPT data with more fundamental parameters. Klotz & Coop (2001) show that "state" incorporating both density and stress, is required for good correlations of centrifuge model pile test results, and Jamiolkowski et al. (1988) show, from calibration chamber tests, that increased compressibility gives reduced penetration resistance. We are unable to distinguish independently, from SPT or CPT data alone, the effects of reduced sand density and those of increased compressibility. However, one might have thought that we could up-scale from CPT to pile penetration resistance simply using area ratios, since the same mechanisms would be at play. This is not the case, however.

Two factors of uncertain origin emerge from test data. Firstly, a reduction factor of up to 2 or 3 must be applied to q_c from a CPT to obtain the tip resistance q_b of driven piles. Chow (1997) attributes this to the ratio in diameter between the two, explained by the enhanced dilation in shear bands around the smaller penetrometer. White (2002) reassesses the available data and shows that no simple scaling law emerges when grain size is taken into account. Secondly, Heerema (1980) coined the term "friction fatigue" to describe the reduction in skin friction as piles are driven. A variety of field and centrifuge model test data chart the loss of skin friction at a given point in space as a pile is driven past it. Randolph et al. (1994) explained this as a loss of radial effective stress due to local volume reductions accompanying the rearrangement of grains on the surface of a rough pile.

White's micro-mechanical observations of the penetration process provide a rather clearer explanation, and go some way to resolving the contrast between observations made at different scales, and in different soils. Figure 9 indicates the various zones of sand influenced by penetration seen through the window of a calibration chamber. As the model pile advances, the stress beneath it increases beyond that required to crush grains. A "nose cone" of crushed and compacted sand is driven ahead of the blunt pile, creating a quasi-penetrometer. Sand crushes as it is displaced laterally by the nose-cone, and this ends up coating the skin of the pile with fine fragments, once the tip has gone past. These fines can easily migrate into the voids of the neighbouring parent material in the flanks: see Figure 10. When this occurs, the fines effectively disappear, permitting the sand at the flanks to relax towards the pile. One might imagine the asperities on the pile shaft shaking the fines as they travel past, encouraging them to diffuse away. The further the pile is driven, the more relaxation there is, up to the point when the fines have been absorbed entirely. Not only does this micro-mechanical observation support the tenor of Randolph's argument, it also suggests that the granulometry of the parent soils and the manner of its grain fragmentation (i.e. Figure 7 compared with Figure 5) will dominate the rate at which fines can migrate, and at which lateral stresses can relax. Broadly graded sands, comprising grains that split rather than crumble, might be expected to display slower frictional degradation than the rest. This hypothesis needs validation, of course.

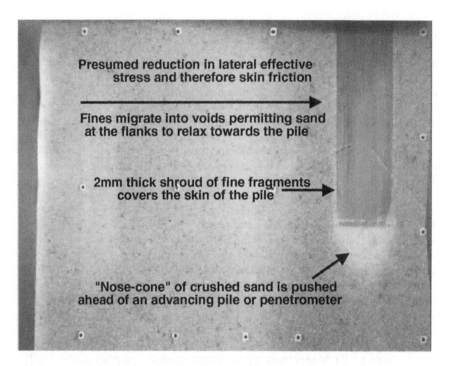

Figure 9. Photograph of sand penetration in a calibration chamber: White (2002)

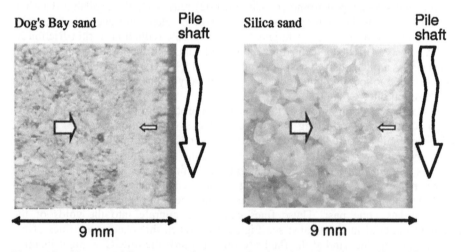

Figure 10. Close-up of fines rearranging to permit relaxation of sand near pile: White (2002)

Micro-mechanics provides an interesting new commentary on research priorities. It suggests, for example, that it will be impossible *in principle* to obtain a good prediction of pile penetration resistance in sand from a finite element simulation, even when finite deformations have been allowed for. Instead, it points to the availability of physically meaningful correlations between soil grain characteristics, CPT data and pile performance, raising confidence both in elementary mechanical models and in the meaning of empirical data-bases.

4 UNDERSTANDING PLASTICITY

There is no better example of confused characterisation than in the understanding of soil plasticity. For geotechnical engineers, the term relates to the range of water contents at which a clayey soil can be moulded, specifically between Atterberg's liquid and plastic limits. The drop-cone test provides an arbitrary "liquid limit" as the water content of a paste that displays a shear strength of about 1 kPa "measured" via the depth of penetration of an 80 gram steel cone. The meaning of the "plastic limit" is more obscure. The test requires that a clay thread must crumble in tension as it is rolled and remoulded. One may imagine that the mechanism must depend on the creation of indirect tension, as in the Brazilian test for concrete, but clay cracking is not really a "solids" phenomenon. The clay particles are utterly cracked and separated to begin with. The separation of two pieces of a rolling thread is solely due to the separation of the once continuous fluid phase. Separation of fluid implies either cavitation or air entry, and the air entry suction for clays is about 100 to 1000 kPa. If the clay were not completely saturated, then gassing would be likely to occur at 100 kPa suction in any event. If the minor effective stress in saturated clay at air entry is 100 to 1000 kPa, then the major effective stress would be 230 to 2300 kPa, if the angle of internal friction was about 23°, typical of clays. This provides a range of 65 to 650 kPa shear strength for clay at air entry and, by implication, at the plastic limit. It is interesting that the shear strength of clay at the plastic limit is quoted in the same range, and increases with decreasing particle size: (Wood 1990).

University research workers in the constitutive relations of soils have a rather different perspective: recoverable strains are called "elastic" whereas irrecoverable strains are called "plastic". Many constitutive modellers base their idealised relationships on the concepts of plasticity theory, created to describe the behaviour of ductile metals and then adapted by Schofield & Wroth (1968) to describe some of the key features of saturated soils - sand or clay, drained or undrained, normally consolidated or over-consolidated - in models which were typified by Cam Clay. The fundamental plasticity concepts that were used to derive Cam Clay were:

- the decomposition of total measurable strains into elastic and plastic components,
- the existence of a closed yield surface in effective stress space which delineates purely elastic from elastic-plastic behaviour,
- a flow rule which fixes the proportions of plastic strain increments induced at yield (in Cam Clay, strain increments to be normal to the yield surface, when consistent definitions of stress and strain are used),
- a hardening rule governing the expansion or contraction of the yield surface (in Cam Clay, simply the current voids ratio).

The Cam Clay concept of soil plasticity was connected with the engineer's plasticity index only through the adoption of an empirical plastic compressibility. As their title suggests, Schofield & Wroth's book, Critical State Soil Mechanics, invoked a critical state line as a unique path of voids compressing according to the logarithm of mean effective stress, as the soil is continually sheared. They could as easily have introduced a line of normal compression. Cam Clay's first prediction is that the existence of one must imply the existence of the other, parallel to it on a log stress plot.

Let us not forget, however, that plasticity to the driller performing ground investigations, or for that matter to the sculptor, simply refers to the "cohesive" mouldability of fine-grained soils. All that is necessary is that the soil can trap water in suction up to about 100 kPa, and that the corresponding effective stresses are not lost on remoulding. Sands can neither trap nor retain such suctions. Their pore sizes are too large for surface tension to produce more than a few kPa of pore suction. Quartz silts may be able to hold suctions but, like sands, their mean effective stress can be annihilated by gentle shaking, or strongly reinforced by remoulding. When confined at constant water content, the "undrained" strength of sand or silt is variable. The driller places a lump of fine-grained soil, "silt" or "clay", on the back of his hand and shakes it: if it liquefies into a puddle with a bright top surface, it is "silt", if not it is probably "clay". He then steadily squeezes the lump: if the once liquefied soil now cracks, the well-trained driller declares that it is simultaneously at its liquid and plastic limits and confirms the diagnosis "non-plastic fine soil".

The stereotypical driller of the last paragraph has, in a sense, set the agenda for soil constitutive relations for the last 25 years. Constitutive modellers have been looking for a more universal soil model, that will replicate many features of Cam Clay, but which will also offer ambiguous values of the undrained strength of saturated sandy soils, leading to phenomena such as liquefaction. Models have certainly been produced, but based simply on curves with more than a dozen parameters fitted over the complex non-linear responses of certain triaxial stress-path tests. Decision-makers generally express doubts about the meaning and value of the many curve-fitting parameters, and designers generally ignore such complex models.

It is proposed here that micro-mechanics offers the way out of the impasse that constitutive modelling has got itself into. The first step is to attempt to place a micro-mechanism against each facet of soil behaviour, and then to relate these mechanisms to constitutive equations. This adventure has hardly begun. Consider the Environmental Scanning Electron Microscope (ESEM) pictures of kaolin mud, shown at three different scales in Figure 11. One must surely been drawn by the similarity with carbonate sand seen in Figures 5 and 6. In each case there is a jumble of apparently broken and breakable pieces of different sizes. And of course, the shape of the compression curve of Speswhite kaolin (Al Tabbaa 1987) shown in Figure 12 bears a striking resemblance to that shown in Figure 1 for Dog's Bay sand. The best fit λ value is almost identical, κ is about twice as large, but the ratchetting compaction after an over-consolidation cycle is also similar. Perhaps there is no qualitative difference between the mechanical behaviour of a platy carbonate sand and a platy kaolin clay. Quantitatively, of course, the clay is about 10 times weaker at the same voids ratio as the sand, and many times less permeable due to its different void size.

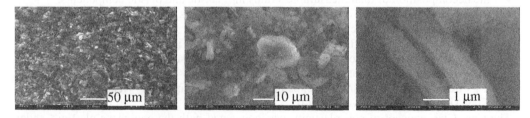

Figure 11. ESEM pictures of kaolin mud

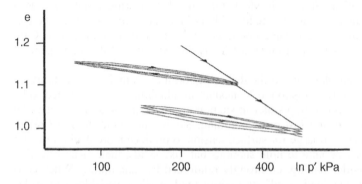

Figure 12. Compression of kaolin (Al Tabbaa 1987)

68

Where does this leave the engineer's concept of plasticity as defined by Atterberg's limits, or by Critical State Soil Mechanics? Consider the following thought-experiment. If a fluid were found with 10^4 times the surface tension of water, and used as a pore fluid for Dog's Bay sand in liquid and plastic limit tests qualitatively similar to Atterberg's but scaled up in accordance with the greater grain size and strength, would the sand appear to be "plastic"? Instead of having an air entry value of about 10 kPa, the sand saturated with this hyper-water would have an air entry value of 100 MPa. So left out on the bench to evaporate, a "wet" sand would shrink, crush and rearrange its grains after the fashion of Figures 5 and 6 and compress after the fashion of Figure 1, whilst remaining saturated. If one continually remoulded it by rolling it like dough for a French loaf, would it not display a "plastic limit" with a "cohesion" of about 100 MPa? And if, on the other hand, one performed a drop test with a cone of 80 kg mass, might one not define an arbitrary "liquid limit" for a shear strength of 1MPa? Figure 1 clearly shows that the voids ratios at the plastic and liquid limits defined in this way would be about 1.35 and 0.30. The corresponding hyper-water contents would be 50% and 11%, and the plasticity index would be 39%. This would match very reasonably with the observed plastic compression index $C_c = 0.53$, or $\lambda = 0.23$, which are values appropriate not only to Dog's Bay sand but also to kaolin clay. Perhaps the key question is whether the sand saturated with hyper-water would show stable "cohesion" like a clay. Casagrande's original test protocol for the liquid limit called for a groove to be cut in the soil, and then shaken vertically until it closed. Perhaps it would simply be necessary to for the hyper-fluid also to have 10^4 times the viscosity of water, to slow down particle rearrangements, if the sand were to appear "plastic" in Atterberg's sense.

The general concepts of plastic yielding - used in modern constitutive models - are completely unclear with regard to the relative influence of grain crushing and grain rearrangement. Constitutive modellers have seen nothing wrong in defining a yield stress that they can not normalise in terms of any inherent material characteristic of the grains. They rely on the concept of "stress history" without asking what manner of "book" a soil uses to record its autobiography. They expect to be taken seriously by decision-makers even though they can only fit curves using mathematical expressions, in the absence of physical understanding. There are some advances now being made in computer simulation of granular materials by Discrete Element Models. DEM simulations of crushable grains (Robertson & Bolton 2001) may well prove to be crucial in resolving outstanding questions about the true nature of soil plasticity.

5 CONCLUSIONS

A particulate approach to soil behaviour is advocated by ITC35 with the aim of drawing together two contrasting streams of thought - the empirical approach to soil classification, and the continuum approach to constitutive modelling. It will be essential to obtain a new data stream of micro-structural information to complement existing data such as CPT resistance, or the stress-strain data of triaxial stress paths. It will be equally essential to develop micro-mechanical models corresponding to the main classes of macro-behaviour (compressibility, hypo-elasticity, stiffness degradation, strength, dilation, creep, phase transformation, liquefaction etc.) so that the influences of particle morphology, fabric and grain strength can be understood. This should also help physical modellers convince engineers that they have selected appropriate soil types to represent the full-scale field profile, and that the influence of larger relative particle sizes can be accounted for properly.

Some promising developments have been shown. A view through the platen of a one-dimensional compression apparatus has demonstrated that the initial virgin loading of a carbonate sand induces strong rearrangement of the grains, with occasional grain fractures that indicate local stress concentrations. As the stress increases, more grains break. As the virgin compression becomes quasi linear versus the logarithm of stress, one sees the fracture even of previously broken fragments, with a corresponding reduction of volume due to the increased efficiency of packing. On the other hand, unloading and reloading induces a quasi-elastic response, though closer attention reveals a hysteresis loop which does not actually close after each cycle. Further work is required to elucidate whether cyclic compaction (tantamount to loss of effective stress in a cyclic undrained test) is due to irreversible grain displacements, or to further grain breakage. Equally, there is the question of whether a view through a "window", which is analogous to a very large particle, is representative of internal micro-mechanisms. Nevertheless, the growing evidence that plastic hardening is due to grain breakage must be of great future value, and is certainly provocative for those who wish to include the progressive crushing of clay agglomerates in their constitutive understanding of clays.

Evidence has also been presented of the different roles of grain crushing and re-arrangement in the penetration of sands by piles and penetrometers. It is well-known, of course, that cavity expansion analysis provides a reasonable justification for the penetration resistance of sands; the precise kinematics accordingly may have seemed less significant. However, it has now been demonstrated that a model pile becomes coated in fine, crushed sand as it is driven. The apparent diffusion of these fines laterally, into the voids of the parent soil, provides an explanation for the tendency of this flanking soil to fall towards the shaft as this drives past. This lateral soil extension can in turn explain the very large reduction in lateral stress at a given point in space as the pile tip advances a few further diameters. If it were not for this very strong relaxation, the skin friction of driven piles would be enormous. Engineers use empirical factors to predict this reduction, which can almost eliminate skin friction in carbonate sands. However, the precise factors which should influence this reduction factor have not been understood. Observation of the micro-mechanics now suggests that "fretting" is a more accurate description than "friction fatigue" which has been used heretofore, and offers a rational basis for selecting influence factors in a database.

There has been a corresponding success in the modelling of a soil fabric using Discrete Elements, in the shape of agglomerates formed of bonded micro-spheres. The crushability of these grains has been shown to transform the behaviour of the soil element they comprise. Realistic virgin compression and over-consolidated soil behaviour have been shown to depend on grain breakage. This occurs at external stress levels orders of magnitude smaller than would have been expected considering the strength of the individual grains, due to the strong deviations in contact forces from place to place. The precise role of grain crushing (and therefore grain shape and mineralogy) in the gamut of soil constitutive behaviour is now much more open to scientific investigation. There is even the strong indication that concepts such as soil plasticity will be clarified in a way that will unify our understanding of soils and strengthen the ability of practising engineers to make rational judgements about soil properties.

REFERENCES

Al-Tabbaa, A. 1987. *Permeability and stress-strain response of Speswhite kaolin*. PhD dissertation, University of Cambridge.

Bolton, M.D. 1986. The strength and dilatancy of sands. *Géotechnique* 36(1): 65-78.

Chow F.C. 1997. Investigations into the behaviour of displacement piles for offshore foundations. PhD dissertation, University of London (Imperial College).

Heerema, E.P. 1980. Predicting pile driveability: Heather as an illustration of the friction fatigue theory. *Ground Engineering* 13(Apr.): 15-37.

Jamiolkowski, M., Ghionna, V.N., Lancellota, R. & Pasqualini, E. 1988. New correlations of penetration tests for design practice. *1st Int. Symposium on Penetration Testing, Orlando,* 1:263-296.

Klotz, E.U. & Coop, M.R. 2001. An investigation of the effect of soil state on the capacity of driven piles in sands. *Géotechnique* 51(9): 733-751.

McDowell, G.R., Bolton, M.D. & Robertson D. 1996. The fractal crushing of granular materials. *International Journal of the Mechanics and Physics of Solids* 44(12): 2079-2102.

McDowell, G.R. & Bolton, M.D. 1998. On the micro-mechanics of crushable aggregates. *Géotechnique* 48(5): 667-679.

Randolph, M.F., Dolwin, J. & Beck, R. 1994. Design of driven piles in sand. Géotechnique 44(3): 427-448.

Robertson, D. & Bolton, M.D. 2001. DEM simulations of crushable grains and soils. *4th Int. Conference on Micromechanics of Granular Media, Sendai*: 623-626.

Schofield, A.N. & Wroth, C.P. 1968. *Critical State Soil Mechanics*. New York: McGraw-Hill.

White, D.J. 2002. *An investigation into the behaviour of pressed-in piles*. PhD dissertation, Cambridge University.

Wood, D.M. 1990. *Soil behaviour and critical state soil mechanics*. Cambridge: Cambridge University Press.

DISCUSSION

S.M. Springman & T. Weber
Institute for Geotechnical Engineering, Swiss Federal Institute of Technology, Zurich, Switzerland

Sarah Springman commented that David Muir Wood had proposed continuum modelling as the preferred mode for practice, whereas Malcolm Bolton had advanced the discrete element modelling (DEM) cause and Cino Viggiani had worried about localisation effects. Viggiani questioned the DEM numerical results in that a monogranular material had been investigated, with particles of the same diameter (uniform grading) and spherical shapes. He criticised these severe limitations, which would hide some phenomena, and in particular, shear banding. He wondered how strong the limitations would be in terms of shape and grading. Bolton replied that, to his knowledge, this was the first attempt to simulate real crushable soil. He noted that it would be necessary to model very dispersed sizes to get anything looking like soil behaviour, and this research had considered dispersed sizes in the sense that there were grains within the larger aggregates. However he accepted that it would be a significant improvement if the grains themselves had an initial fractal distribution of sizes, and if each of those grains contained micrograins that could permit that grain, however large, to break. He felt that an even bigger restriction was the roughness of the grains modelled as agglomerates of bonded spheres, but held that the representation of grain crushability was paramount.

Ivo Herle asked about the similarity between clay and sand and disagreed that the only difference was the physico-chemical bond and that these were weak for clays. He pointed out that these bonds were extremely strong, and they depended on the distance between particles, which was much smaller for clay particles. He felt that making these bonds weaker could not explain the similarity in behaviour between sand and clay because the physical background was totally different! Bolton replied that some weak breakable bonds should be present to reproduce the characteristics of a crushable material and that clay particles could be modelled as agglomerates with with strong bonds almost everywhere and weak bonds in some places. He noted that this would be similar to modelling in rock mechanics with fissures, joints and bedding planes, with micro-cracks developing from various directions and points in the agglomerates. Bolton thought that current modelling of clay as breakable agglomerates had captured most aspects of monotonic or even cyclic behaviour. Herle commented that the behaviour of clay was not represented correctly by basing it on some concepts of modelling sand. He questioned that the supposition that the breakage of the clay particle was brittle since these particles were held together by the capillary, physical and chemical forces. Bolton wondered if the same thing could be said for rocks, which were certainly brittle, and added that researchers would be convinced when somebody could observe an experiment to load an agglomerate of clay particles, and to watch it break!

Herle also took up the point about focusing on the parameters rather than the models. He said that models must develop prior to parameters, so parameters were always linked to a model and that this was also connected with using the DEM for constitutive modelling. He noted that this was not a continuum, but that continuum mechanics were required for modelling the grains and asked whether linear elasticity really was suitable for this. Bolton agreed without exception and added that engineers and constitutive modellers were using parameters without knowing what they meant and he mentioned the example of the "elastic" and "plastic" gradients κ and λ of a one dimensional compression test on virgin sand. Bolton pointed out that a mathematical modeller could fit the data quite well with κ and λ and an apparent pre-compression, even though observations showed that the initial compression was more nearly linear, almost entirely "plastic" due to grain breakage and rearrangement, and although no pre-compression had been applied.

Springman commented that there was a big gulf between the academics and the practitioners and that the latter would be using continuum analysis for some time to come, principally because various forms of constitutive modelling have been implemented in most of the geotechnical finite element codes. She noted that in talking about models and parameters, there was still no skeleton upon which future implementations (of micro-mechanical models) in numerical models could be hung and wondered where the future lay. Bolton felt that there were exciting prospects for linking micro-mechanics to continuum models and parameters. Muir Wood tended to agree with Herle that the challenge was to try to model at the particle interaction level, by including the real nature of the physico-chemical interaction between these particles. He thought that this would become more feasible with the increasing availability of massive parallel computing resources but was sceptical about the assumptions made presently by the particle modellers concerning the contact between two particles, and whether the particle contacts were spherical or non-spherical.

Andrzej Niemunis thought that the influence of crushing was exaggerated by carrying out an experiment in which such a large fluctuation of stress was applied on a relatively small sample of sand with a wavy surface confined by a planar glass plate. He said that thinking in megaPascals was inappropriate and thought that the onset of crushing was much later than was shown. Bolton replied that it was open to discussion how stress was distributed through a soil body, but most people seemed to accept the early photo-elastic simulations of De Josselin De Jong and Verruijt, and Allersma and others more recently, as well as DEM simulations. He asserted that stress was highly dispersed within the granular medium, especially in any small test sample. He said that it was not necessary to throw away the constitutive models or to despair of continuum analysis, because this had developed by leaps and bounds, but the difficulty now was to persuade a senior decision maker of the appropriateness of the constitutive model. He felt that the question of confidence in engineering parameters was paramount, and that micro-structural correlations would assist in the characterisation of geomaterials just as they had with other branches of materials science.

Jean Sulem remarked on the suitability of adopting continuum or discontinuum approaches for geomechanics in that definitions of stress and strain were necessary for continuum mechanics, as average values over some representative volume. He commented that classical constitutive models were well suited to most homogeneous states of stress and strain and asked if it was justified when these quantities varied significantly in space. He referred to the examples mentioned as well as interface interaction, strain localisation, fracture, stress singularities, where the stress or strain field varied rapidly in the neighbourhood of constitutive points. He thought that other means of modelling could be selected, within the framework of continuum mechanics, by taking large strain and stress gradients into account, before reverting to a discontinuum approach. He mentioned gradient models, such as gradient plasticity, especially for modelling problems with locally high heterogeneity.

Bolton agreed and referred to work presented at a recent conference on 'Powders and Grains' in Sendai, following which Ioannis Vardoulakis (2001) commented that even the disappearance of small broken grains into the voids between large grains could be accounted for within continuum mechanics. Bolton emphasised that solid mechanics models would be made even more abstruse if solid diffusion were to be included and was afraid that complex constitutive models were no help to industry, becoming a private matter between consenting research workers! He mooted that the purpose of having civil engineering departments and geotechnical research groups was to try to relate to real construction problems experienced by real engineers. He had seen soil grains in David White's pile driving experiment diffuse into the pores between other particles, and thought that generation of some rules closer to fluid mechanics and mixing theory were required in which the grain size distribution of the parent soil would be very important. A practising engineer could decide, based on these rules and the data, whether they would allow for more or slightly less "friction fatigue" on piles and would design accordingly. He concluded by saying that prediction was subordinate in engineering practice to decision.

Viggiani felt that notwithstanding the very interesting and provocative nature of these ideas, the speaker was mixing the language from two different concepts in saying that 'stress was very dispersed in a granular medium'. He said that this contradicted the notion that stress was a continuum mechanics concept and this approach didn't help to understand the phenomenon. Bolton replied that the sand grains themselves could still be regarded as a piece of continuum.

Claudio Tamagnini remained unconvinced by Bolton's ideas, in that the discrete and the continuum approach were parallel. He opined that people would develop their continuum field approaches to improve their theories and make better predictions, because some ingredients of the theories (e.g. Muir Wood 2002), could be developed not just from the scratch, but from observations made at the microscopic level. He commented further that a practising engineer needed a workable tool and that micro-mechanics could not fulfil this role. He doubted that a foundation problem could be solved using micro-mechanics, which would be analogous to a structural engineer designing a beam using quantum field theories. Bolton denied that this was the thrust of what had been said, and reminded everyone about the challenge of up-scaling. He cited the soil particles, which might be regarded as continuum matter, whereas a finer model would be the crystallite to which continuum mechanics could be applied too. He mentioned the principles of rock mechanics in relation to the breakage of a sand grain, which was simply a rock mechanics problem with unusual boundary conditions. He concluded that micro-mechanics offered the opportunity to get the scaling right between two different views of the continuum, with the scale of continuum mechanics describing the behaviour of a sand grain, and the scale of continuum mechanics describing a rockfill dam.

Chris Martin commented that the research presented was only relevant to fully drained conditions and that the problems of including pore fluids or even gas into the models would appear to be immense, in particular in three dimensions. He felt that this could be a problem with the DEM so there seemed to be a major advantage in using the continuum approach, since pore fluid, drainage and effective stress were all conveniently modelled. However, Muir Wood thought that it would not be an insurmountable problem to incorporate pore fluid in the discrete element analysis and with pressures too, as a boundary to the whole problem. Bolton added that there were two groups attempting to do this, including Colin Thornton in Aston University in the UK and another group in the United States. He concluded that it may not be necessary, but it would certainly be interesting to try.

REFERENCES

Muir Wood, D. 2002. Constitutive cladistics: the progeny of Critical State Soil Mechanics. In S.M. Springman (ed.) *Workshop on Constitutive and Centrifuge Modelling, Monte Verità.* Lisse: Swets & Zeitlinger.
Vardoulakis, I. 2001. Thermo-poro-mechanical analysis of rapid fault deformation. In Y. Kishino (ed.) *Powders and Grains 2001, IV International Conference on Micromechanics of Granular Media, Sendai, Japan*: 273-280. Lisse: Balkema.

Constitutive and Centrifuge Modelling: Two Extremes, Springman (ed.)
© *2002 Swets & Zeitlinger, Lisse, ISBN 90 5809 361 1*

Centrifuge technology

J. Laue
Institute for Geotechnical Engineering, Swiss Federal Institute of Technology, Zurich, Switzerland

ABSTRACT: Geotechnical centrifuge modelling is used for a wide range of investigations, which cover almost the whole field of geotechnical engineering problems. Independently of the question to be answered, the basis of the centrifuge modelling technique is the assembly of a test set up in a suitable environment. This paper deals mainly with sample preparation and the choice of material to be used, site investigation and characterisation of the soil properties, loading options and transducers, and other ways of obtaining test results as well as scaling laws and differences in the design of geotechnical centrifuges. It finishes with a section on ongoing advances in the development of realistic modelling procedures.

1 INTRODUCTION

Centrifuge modelling facilitates the investigation into a wide range of geotechnical engineering problems. Many of these problems cannot be modelled effectively using other forms of physical modelling. Some problems cannot be modelled at all because of size considerations and the unknown likeliness of occurrence (e.g. landslides or most natural hazards) or because time effects (consolidation processes) are involved. This contribution outlines some of the possibilities in terms of use of and requirements for centrifuge machines, scaling laws, model preparation, load input and measurement techniques suitable for use or potentially useful in geotechnical centrifuges. The last section deals with the importance of adapting appropriate modelling procedures together with recent developments in this area.

Physical modelling is required in cases where:

a) no experience of the expected mechanisms is available,
b) an idea of the system behaviour exists and some components are known, but the whole system has not been studied,
c) some studies or measurements have been carried out in the field, but where the influencing parameters have not been investigated in detail,
d) studies are available but data are required under defined boundary conditions for the whole system,
e) they can be used as a tool, providing an engineer with physical data to verify calculation methods,
f) they can be applied as a proof to check existing calculation mechanisms by reproducing data for an equivalent physical modelling test.

Table 1. Keywords from the CLEOPATRE data base (Thorel 2001)

Devices and methods used	Observation and model characterisation	Validation of centrifugal modelling	Material used	Structure tested	Earth structure tested	Stimulus applied	Contaminant transport	Other phenomena	Other topics
Centrifuge	Observation	Similitude	Clay	Footing	Reinforced earth	Static Loading	LNAPL	Flow	Bibliography
Data acquisition	Transducer	Experimental verification	Sand	Pile	Nailing	Cyclic Loading	DNAPL	Consolidation	1g model
Device inflight	Marker	Comparison with prototype	Rock	Group of Piles	Other Reinforcement	Dynamic Loading	Tracer	Heat Transfer	Geology
Actuator or Shaking Table	Inflight test	Comparison with numerical method	Equivalent material	Retaining wall	Geotextile or Geosynthetic	Seismic	Adsorption	Creep	Out of civil engineering
Sand preparation	Video	Scale effect	Reinforced soil	Sheet pile or Diaphragm Wall	Embankment	Explosion	Diffusion-advection	Cratering	Silo
Clay preparation	Camera	Modelling of model	Unsaturated soils	Culvert	Dam	Knock	"Fingering"	Faulting	Structure
Robot	Radiography	Grain size Effect	Saturated soil	Tunnel, Cavity	Slope	Vertical Loading	Solute transport	Electro-kinetics or Electro-osmosis	Fluid
Laminar container	Image Processing	Dimensional Analysis	Frozen soil	Anchor	Trench	Horizontal Loading	Fluid Entrapment	Liquefaction	Teaching
	Cone test & Piezocone		Ice	Suction anchor pile	Liner (soil)	Inclined Loading	Remediation	Soil Improvement	Hydraulic gradient
	Vane test		Silica flour	Gravity Caisson	Sand Compaction Pile (SCP) or Sand Drain	(Load) Eccentricity		Damping	Base Friction Table
	Presiometer		Glass Beads						Calibration Chamber
	T-Bar		Crushed Glass						
			Intermediate Soil						

The planning and conducting of physical model testing requires a wide range of decisions to be taken at various stages. This is especially valid for modelling in centrifuges where, aside from technical problems and challenges to be solved, safety requirements must also be taken into account.

Table 1 shows an overview of the keywords used in the CLEOPATRE database located at the Laboratoire des Ponts et Chaussées (LCPC) in Nantes, which is described in more detail by Thorel et al. (2000). This database consisted (as of 1999) of more than 1500 papers in seven languages about centrifuge modelling. These include the proceedings of the four conferences of TC2 (Technical Committee on Physical Modelling of the International Society of Soil Mechanics and Geotechnical Engineering) with respect to centrifuge modelling and testing, which make up 32% of all contributions (Corte 1988, Ko & McLean 1991, Leung et al. 1994, Kimura et al. 1998). Also included and recommended for reading are keynote contributions from the Rankine lecture given by Schofield (1980) and the book edited by Taylor (1995).

Many contributions deal with machines, measurement and data acquisition systems. Geotechnical investigations have focused mainly on footings and pile loading tests (144 and 141 respectively). Almost 126 references deal with similitude of models, and around 144 with comparisons between numerical and physical modelling of prototype problems. The database also shows that most tests are conducted in pure, mostly uniformly graded sand (290 references) or in clay (205 references). The extent of these references show that this overview of centrifuge technology cannot hope to be complete. This overview will highlight some points of interest and will focus on the necessity of idealisation with respect to the available modelling possibilities.

2 GEOTECHNICAL CENTRIFUGES

Geotechnical centrifuges benefit from the additional centripetal forces acting on a model while the centrifuge is rotating, to increase the self-weight of the soil and thus create a stress distribution in the soil sample, which is comparable with reality. This is shown in the force diagram given in Figure 1 for the two different types of centrifuges – the beam with a swinging basket where the acceleration field acts on the model axis and the drum centrifuges with a fixed channel. It should be noted that, in general, $\ddot{r} = 0$.

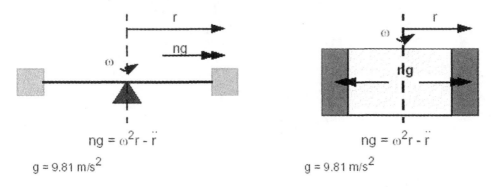

Centrifuge modelling: BEAM

$$ng = \omega^2 r - \ddot{r}$$

$$g = 9.81 \ m/s^2$$

Centrifuge modelling: DRUM

$$ng = \omega^2 r - \ddot{r}$$

$$g = 9.81 \ m/s^2$$

Figure 1. Sketch of the mechanics relating to the two types of centrifuges (Springman 2001)

Figure 2. The Bochum Geotechnical Centrifuge Z1

Figure 3. ETHZ Drum Centrifuge, view of the safety shield

Most of the centrifuge centres worldwide use beam centrifuges. One of the beam centrifuges, the Bochum geotechnical beam centrifuge Z1 (Jessberger & Güttler 1988), became operational in 1987 and is shown in Figure 2. In a beam centrifuge, a model is usually placed on a swinging basket on one side of the arm and balanced by a counterweight on the other side. The model swings up while the centrifuge rotates: gravitational forces act vertically, and centripetal forces horizontally, on the swinging basket. This causes an artificial acceleration field to act on the sample, while the basket rotates into the direction of the sum of the acting accelerations. This acceleration field provides the increase of gravity on the sample. The effect of this increase of gravity will be shown in the next chapter.

There are fewer geotechnical drum centrifuges in operation worldwide (Table 2). A full list of all geotechnical centrifuges can be found at http://dutcgeo.ct.tudelft.nl/allersma/tc2/cusers.htm. For better comparison, the capacity (g-tonnes) of each centrifuge is derived by multiplying the maximum possible tonnage (the amount of soil) in the drum with the maximum g-level.

Figure 3 shows a view of the drum centrifuge of the ETH Zurich (ETHZ), while Figure 4 details the main features of the machine. The soil sample is placed in the channel (drum) over the whole circumference of the machine. The advantage of a drum centrifuge over a beam is the possibility of accessing the soil model by the tool plate (or tool table). It is possible to start, stop and move the tool plate independently of the drum itself. Thus tools mounted e.g. actuators sitting on this tool plate, can be changed during a test without disturbing the sample in the drum. This improves flexibility of modelling in most respects and limits the number of stress cycles or excursions that the model must be exposed to between 1 and ng.

Table 2. List of drum centrifuges as of 2000: Drum centrifuge capacity (Springman et al. 2001)

Date of commissioning	Drum dimensions m Depth x Perimeter x Width	Payload t	Max g level g	Capacity g-tonnes	Location
1971	0.025 x (0.25 x π) x 0.12 ~ 0.003 m³	0.006°	1000*	6.1°	UMIST, UK (Schofield 1976)
1979	0.13 x (1.2 x π) x 0.23 ~ 0.11 m³	0.2	650*	145	Davis, USA (Fragaszy & Cheney 1981)
1986	0.1 x (0.8 x π) x 0.3 ~ 0.075 m³	0.195°	150*	43°	Utsonomiya, Japan (Kusakabe et al. 1988)
1988	0.15 x (2 x π) x 1 ~ 0.95 m³	~1.7	400	~ 675	Cambridge, UK (Dean et al. 1990)
1995	0.115 x (0.74 x π) x 0.185 ~ 0.05 m³	0.13°	416	~ 50°	Hiroshima, Japan form. Cambridge minidrum Mk1 (Kusakabe & Gurung 1997)
1995	0.12 x (0.74 x π) x 0.18 ~ 0.05 m³	0.13°	400	~ 50°	Cambridge minidrum MkII (Barker 1998)
1996	0.17 x (1 x π) x 0.25 ~ 0.13 m³	0.2	450	90	COOPE, Brazil[+] (Gurung et al. 1998)
1997	0.15 x (1.2 x π) x 0.3 ~ 0.23 m³	0.6°	484	290°	UWA, Perth[!] (Stewart et al. 1998)
1998	0.3 x (2.2 x π) x 0.8 ~ 1.65 m³	3.7	440	~ 1600	Toyo, Japan (Miyake & Yanagihara 1999)
1999	0.3 x (2.2 x π) x 0.7 ~ 1.45 m³	2.0	440	880	ETHZ, (Springman et al. 2001)

(* horizontal axis, ° estimated for better comparison with a fill of density of 2600 kg/m³ as no detailed data was available, [+] a similar machine is working at the MIT, [!] An identical machine is in operation by Kiso Jiban Consultants in Tokyo and another with the same size drum at the TIT Tokyo)

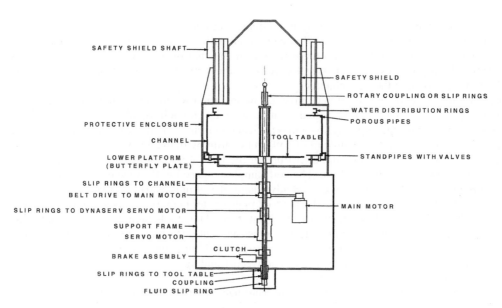

Figure 4. Sketch of the ETHZ Drum Centrifuge (Springman et al. 2001)

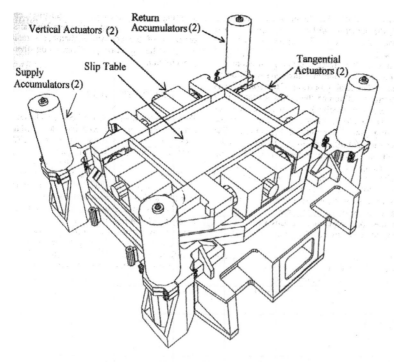

Figure 5. Shaking table of the new HKUST beam centrifuge (Shen et al. 1998)

Other new centrifuge centres follow the path of building new machines for focusing upon specific topics, like the Hong Kong University of Science and Technology (Shen et al. 1998) and others who choose new machines especially designed for carrying out earthquake tests.

80

These machines need to be more robust to withstand the shaking of one of the swinging baskets during an earthquake. They also need to be designed for higher load capacities, since multiaxis shaking tables (e.g. Fig. 5) have to be mounted on top of the swinging basket. Mitchell (1998) describes other centrifuges designed for environmental topics. One of these centrifuges is a two-beam centrifuge (arranged in a cross), allowing for four models to be tested in one centrifuge run with less technical equipment in terms of actuation and instrumentation (Fig. 6).

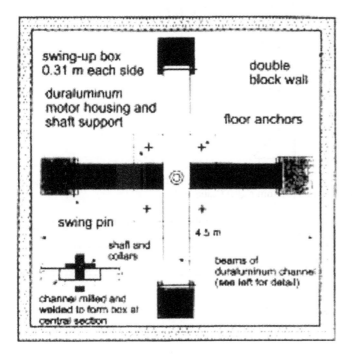

Figure 6. Environmental centrifuge (Mitchell 1998)

3 SCALING LAWS AND INFLUENCE OF INCREASED GRAVITY

The principle of centrifugal modelling in geotechnics is that the stresses in the subsoil can be replicated in a small-scale model (scaled by a value of 1/n), with equivalent values and distribution compared to a prototype or real situation by increasing the level of gravity by a factor of n times g. This interplay is shown in Figure 7 whereby the stresses in the subsoil will be the same after scaling the model by 1/n and increasing the gravity to ng. The scaling laws can then be derived using dimensionless numbers. The scaling relationships are shown in Table 3 and Table 4, the latter showing the scaling of time.

A main advantage of using a geotechnical centrifuge is the scaling of time for problems that include consolidation (diffusion) processes. However, for some problems, the scaling of time is crucial since different processes scale by different amounts, as shown in Table 4. In case of e.g. earthquakes or other dynamic influences, the quicker scaling of diffusion processes conflicts with the time scaling of the application of load (load frequency or velocity). Thus measures must be taken to slow down the flow in the diffusion process – for example to increase the viscosity of the fluid by, for instance, adding glycerine or other fluids, or to reduce the permeability of the soil. Both of these options have been examined, e.g. by Laue (1997), who found good

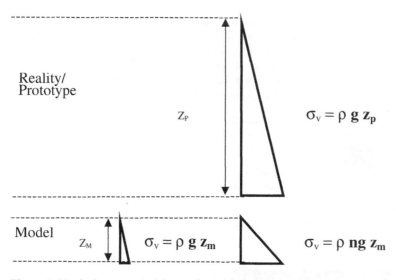

Figure 7. Vertical stresses (σ_v) in a soil model (index m) comparing reality (index p) (top) with 1 g small scale model (bottom left) and centrifuge model (bottom right) (Laue 1996)

Table 3. Scaling law relationships (based on Schofield 1980) following $\sigma_v = \rho_m(ng)_m(z/n)_m = \rho_p g_p z_p$, where subscripts m & p refer to model and prototype

Parameter	Unit	Scale (model/prototype)
Acceleration	m/s^2	n
Linear dimension	m	1/n
Stress	kPa	1
Strain	-	1
Density	kg/m^3	1
Mass or Volume	kg or m^3	1/n^3
Unit weight	N/m^3	n
Force	N	1/n^2
Bending Moment	Nm	1/n^3
Bending moment /unit width	Nm/m	1/n^2
Flexural Stiffness/ unit width	Nm^2/m	1/n^3

Table 4. Scaling law relationship for time effects (based on Schofield 1980)

Parameter	Unit	Scale (model/prototype)
Time: diffusion	s	n^2
Time: dynamic	s	n
Frequency	1/s	n

consistency of both techniques in a series of modelling of models tests. Ellis et al. (1998) describe the necessities of additional laboratory testing in the respect of minimising the influence of these measures on the behaviour of the soil.

A correction to the applied stresses in the soil has to be made for the influence of the radius of the centrifuge with depth of the model as shown in Figure 8. The development of g depends on the distance of the given depth in the sample from the central axis. The smaller the centrifuge, and thus the distance between the central axis and the sample, the larger the difference between the prototype stress distribution (assumed linear) and the stress distribution in a centrifuge model. Using a corrected depth may solve the problem (e.g. Renzi et al. 1994), but care has to be taken as to which radius is used to derive the effective increase of g for a certain problem. For research on shallow foundations, the applied g-level at the surface should be used. For most tests, the g level at 2/3rds of the model depth may be adopted as this allows for the lowest over-all error (Schofield 1980).

Other points have to be taken into account for the test design, e.g. the curved shape of the re-sulting g in a model leading to horizontal force components on structures or to curved water level surfaces. This is less of a problem in a drum centrifuge where the surface of the soil is normally built in a curved shape. The curved surface will cause other problems for e.g. larger foundations if the surface curvature is not considered in the design of the structure.

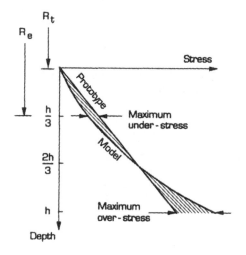

Figure 8. Comparison of stress variation with depth in a centrifuge model and the prototype (Taylor 1995)

4 MODEL-MAKING AND MATERIAL

One of the main questions is whether a real soil (probably required for practical applied re-search) or a laboratory soil should be used. Usually, a laboratory soil, for which an extensive da-tabase of properties exists, will be selected. In any case, it has to be decided before the actual test whether the model is to be built outside the centrifuge in the laboratory or in the centrifuge itself. The decision depends on the kind of centrifuge in which the test will be conducted. In beam centrifuges, the decision will most likely be to prepare the soil stratum in advance, whereas in a drum centrifuge, the sample will be most likely to be built inflight. The preparation of models and the choice of the material to be used depend on whether the soils are fine or coarse grained. Fine and coarse grained soils are therefore treated separately in the following sections.

4.1 Coarse grained (granular) material

In the case of granular material, the scaling laws have to be taken into account. Considering that the geometrical scaling is related to a factor of n, a grain with a particle size of 2 mm will be equivalent to a stone with 20 cm diameter at 100g. Thus in most cases, granular soils are tested using fine-grained sand with maximum particle size of less than 0.3 mm. The influence of the scaling of sand on the test result has to be proven by modelling of model tests.

Building granular samples in the laboratory is mostly done by pluviation. The sand is pluviated using a large hopper (sand container) and a nozzle outlet, which is moved by hand over the sample. A specific density (or relative density) can be reached by controlling the distance between the nozzle and the soil sample as well as the volume flow. In some institutes, pluviation is done automatically using a programmable moving raining apparatus where, as an additional control, the velocity in one of the plane directions may be varied. The reproducibility of a sample with medium-high density will be higher when using a motor controlled pluviation device; the areas of influence of the side of the model containers also become larger. The influence of the boundary is caused by particles deflecting off the container wall, loosening the sample.

Preparing the sample in the laboratory by means of a sand hopper allows the production of medium to high densities, with good reproducibility. The creation of low-density samples is also possible, but the quality of low-density samples suffers under the transport conditions from the laboratory to the centrifuge basket. The movement of the container may densify the sample.

Conducting more than one test in the same sample, which often requires stopping and starting of the centrifuge, also leads to changes in density. Starting and stopping the centrifuge 3 to 5 times results in cyclic loading causing volumetric compression of the soil and minimises additional compression due to the self-weight of the sample throughout the main test cycles.

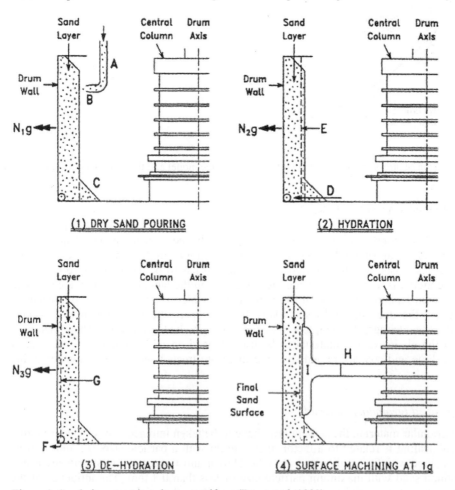

Figure 9. Sand placement in a drum centrifuge (Dean et al. 1990)

When using a beam centrifuge, the building of samples inflight requires a hopper, which is placed on top of the model container. This solution is suitable for providing additional loads such as small embankments, but it will not allow a creation of a full model unless the container is small. Using a hopper, the influences on the path of the particles inflight also have to be taken into account. When a particle is dropped from the base of the hopper, it is subjected to both radial and circumferential accelerations (e.g. the Coriolis effect), which must be allowed for in predicting the particle trajectory and location at rest.

Samples can be built inflight in a drum centrifuge more easily. One possibility (Dean et al. 1990) is shown in Figure 9. Dry sand is poured into an external hopper to fall down through a tube (A) before being accelerated by compressed air through an outside nozzle (B) (all parts are only exposed to the 1g field), which allows the particles to be deposited in the drum in the rotating field. The density achieved is a function of the flow rate of the sand, air pressure and acceleration of the centrifuge. The sand is saturated (E) after infill by supplying water to the base drain (D) and then dehydrated by opening the drain to cause suction in the sample (G). This suction allows the centrifuge to be stopped without a slip failure developing through the sample. Once the drum is stopped, the surface can be prepared depending on the requirements of the test by using a scraper (I) mounted on the central column (H).

Another possibility for filling the channel of the drum is to use a spinning disk. The spinning disk is shown in Figure 10 while filling the drum with sand. Sand is poured with a controlled flow rate through a stationary tube onto the spinning disk, whereupon the particles accelerate radially outwards towards the drum gradually building up a sand deposit. The particle trajectories are mainly influenced by windage, and thus by grain size, in addition to the starting velocity outside the spinning disk (Graemiger 2001, Laue et al. 2002). Very low densities (pluviating the sand on a water surface) up to medium high densities can be reached using this method. The further treatment of the sample follows the same procedures of hydrating, dehydrating and surface treatment as described before. Figure 11 shows the surface treatment in the ETHZ Centrifuge.

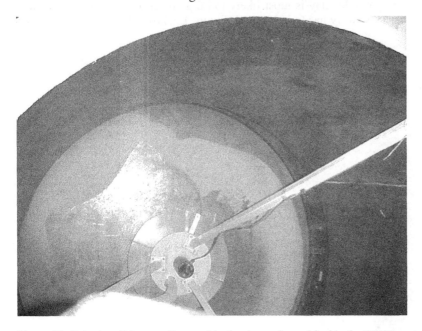

Figure 10. Spinning disk spreading sand in the drum-channel inside the ETHZ centrifuge

Figure 11. Scraping the sand surface in the ETHZ centrifuge

4.2 *Fine grained (clayey) material*

Fine grained soil samples are prepared in the laboratory and consolidated in large oedometer units outside the centrifuge or inflight in the container or channel of the drum. Even if the testing of real soil samples is possible, clay is most likely to be prepared by mixing a slurry to a predefined water content. This slurry has to be deaired. Figure 12 shows the ETHZ mixing unit

Figure 12. ETHZ clay preparation unit for mixing up to 200 litres of clay slurry

designed for preparing up to 200 litres of clay slurry in a vacuum environment of ~-1 bar. In most tests, a laboratory soil, such as industrially produced kaolin powder, is used.

For sample preparation outside the centrifuge, the slurry is poured into a strongbox and placed in the large oedometer. Stresses are applied in increasing steps (Steps 1-4 in Fig. 13) to consolidate the sample, a process that can take between 1 and 4 weeks. After finishing the consolidation, the sample has to be unloaded and transferred to the centrifuge. Suction develops in the sample (Steps 5 and 6 in Fig. 13). After acceleration of the centrifuge, the sample consolidates again at the imposed stress level. In this case, at least the surface of the sample will be over-consolidated clay: an example of the stresses and the related over-consolidation ratio (OCR) is given in Figure 14. A stress of 100 kPa is applied to the clay sample during the consolidation period in the oedometer (left figure, almost vertical line at 100 kPa) and the applied stress from self-weight during consolidation in the centrifuge increases linearly from zero at the surface (left figure). The resulting over-consolidation ratio is shown in the middle plot and the resulting distribution of undrained shear strength on the right hand side.

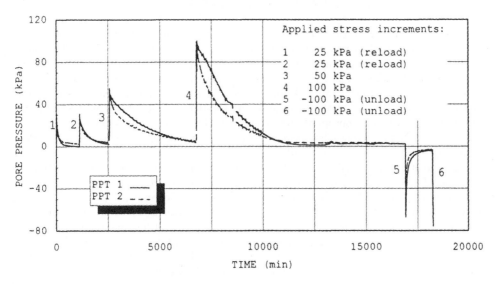

Figure 13. Pore pressures in a clay sample under 1 g consolidation (König et al. 1994)

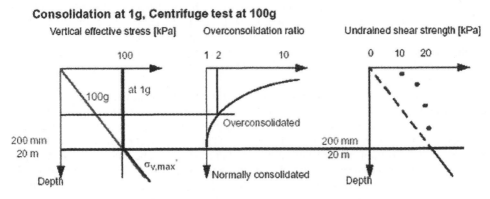

Figure 14. Over-consolidation ratio (OCR) and vertical effective stress in a clay sample during preloading and consolidation in the centrifuge (Springman 2001)

87

The preparation of normally consolidated samples requires the clay to be consolidated from a slurry inflight. The slurry is placed above a drainage layer (most likely sand or a ribbed plate) and accelerated in the centrifuge. Due to the drainage layer, consolidation takes place in two directions. After consolidation, an additional surface load may be applied to the sample for a period of time to create a stiffer crust (through partial consolidation) or to allow for a small OCR at the top of the sample (e.g. Almeida 1984). Care must be taken in modelling the surface of the sample in drum centrifuges. The full saturation of the clay will be lost during the standing phase of the drum, while the surface water may run out due to gravity or evaporation and the surface dries. Given that, tests on the degree of saturation must be conducted in the future.

The material properties of a clay sample may be changed using different clays and adding rock flour to change the stress ratios at failure, the permeability, or the stiffness (Rossato et al. 1992, Springman 1993).

5 CLASSIFICATION AND INVESTIGATION OF SAMPLES

The classification and investigation of the soil samples can be conducted before, during and after the test. This section will focus on the sample investigations carried out inflight. The classification of the soil before and after a centrifuge test will be most likely to follow the same procedures as for full scale field test situations. In situ investigation tools have been scaled down to conduct comparable tests inflight, although for some of the tools, complete scaling is not possible. A CPT device with a diameter of 3.5 cm in reality should be scaled down by the factor of n for full similitude. This is not appropriate because a new device has to be built for each test at different values of n, and becomes impossible for practical modelling reasons, because transducers of the required sizes may be not available.

Four inflight investigations methods will be introduced; vane tests, T-bar tests, cone penetration tests and the derivation of small strain stiffness by means of accelerometers.

Vane tests can be conducted in the centrifuge with slight modifications to the procedures used in the field. These modifications are e.g. described in Springman (2001). Following measurements of torque required to turn the vane, the undrained shear strength s_u can be obtained. Care has to be taken over the size of the vane. A vane with a height of 1 cm in the centrifuge when tested at 100g will give a value of the undrained shear strength for a layer of 1 m averaged over the two horizontal planes (circular) and the vertical plane (cylindrical).

T-Bar tests are developed especially to derive the strength of soft, normally consolidated materials. The derivation of the undrained stiffness follows a classical analysis of a clearly defined failure mechanism around a rough round bar with smooth ends. Details about this technique can be found in Stewart & Randolph (1991). Figure 15 shows the equivalent ETHZ T-Bar with a diameter of d = 5 mm and a length of l = 20 mm. The connection rod to the T-Bar has a smaller diameter at the junction with the bar to reduce the disturbance of shaft on the measurement of resistance (d = 4 to 6.4 mm, l = 60 mm). Thus T-Bar testing is only possible for very soft samples at lower g-levels.

The tip of the Cone Penetration Test device (CPT) of the ETHZ is shown in Figure 16. CPTs allow the measurement of the forces acting on the 60° cone as well as on the shaft behind the tip. A pore pressure transducer is located directly behind the tip to measure the pore pressures developing during the penetration process. The use of CPTs in geotechnical centrifuges is described for instance by Bolton et al. (1999) and is related more to the behaviour of a driven pile due to the scaling of the geometrical units. A centrifuge test with the 11.3 mm diameter ETHZ cone at a g-level of 50 represents a prototype cone of 56.5 cm. Nevertheless the results of these CPT tests can be used for evaluating the soil in terms of relative density (e.g. Graemiger 2001) or strength or location of boundaries between any differences in material (e.g. finer and coarser

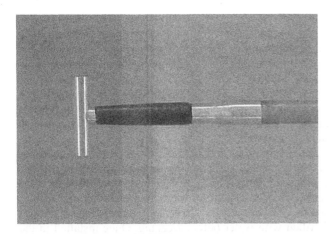

Figure 15. T-Bar of ETHZ (d = 5 mm, l = 20 mm, connection to T-Bar d = 4 to 6.4 mm, l = 60 mm)

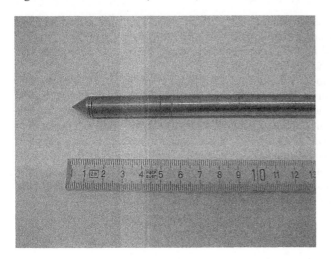

Figure 16. Tip of the cone of ETHZ built by ISMES spa (d = 11.3 mm)

grained materials, e.g. Bucher 1999). CPT tests also guarantee good comparison between different samples for a wide range of soils. The efficiency of CPT's has been proven in a European research programme between 5 different geotechnical centrifuge facilities (Renzi et al. 1994, Bolton et al. 1998).

A method not often used for inflight sample investigation, but with a lot of potential, is the measurement of the wave velocities after an impact-loading event. This method has been investigated by Siemer (1996) and is published in English e.g. by Laue et al. (1996) or Siemer (2000). Small accelerometers (Bruel & Kjaer, 4371 and 4393) have been placed along vertical and horizontal lines beneath each other in the soil model. Measurements have been taken after an impact, which has generated a pressure wave through the model. The small strain shear modulus derived shows good comparison with well-known relationships (e.g. by Iwasaki & Tatsuoka 1977), which supports the application of this method in determining high quality inflight material properties.

6 LOADING OF SAMPLES

Very often loads must be applied to the soil sample in the centrifuge inflight. The loading is usually applied with the help of actuators, but there are several other possibilities. These may introduce a surcharge load consisting of soil or other materials, as well as the filling of an enclosed rubber membrane with water or air pressure. The latter has to be used with restrictions on the expansion of the containment balloon. Also, "environmental load scenarios" such as heating and cooling of parts of the sample, or the simulation of thawing of frozen material, or the provision of water (e.g. rain), or polluted water, are possible.

Loading has to be conducted under defined conditions and should always be applied based on the specific requirements of the test. Multifunctional actuators for defined loading conditions can supply loads using air pressures, hydraulic oil pressures, or stepper motors. Simpler techniques can be used for specific conditions. Allersma (1998b) used a spring and an ice block to simulate the actuator for a slow pullout test. This actuator is shown in Figure 17. The spring is prestressed (compressed) and the gap is filled with water before freezing. Once installed in the centrifuge, the ice melts slowly and the spring applies tension to an anchor built into the soil sample.

Pneumatic devices can usually deliver pressures up to 10 bar (~1 MPa). Such actuators have to be quite big to apply sufficient loading at higher g-levels, stiff samples or large/deep foundations. The load control can be affected by the compressibility of air, leakage of the system and substantial changes in the stiffness of the loaded system.

Hydraulic devices have a higher supply pressure (up to 300 bar, ~30 MPa). The high fluid pressure requires special rotary couplings. Load ranges and maximum frequencies of hydraulic actuators are higher than those of pneumatic units. For example the hydraulic system used at Ruhr - University Bochum (Fig. 18) can deliver frequencies of 100 Hz for small displacements using a pressure of 230 bar and servo-valve control. To provide this frequency at high oil pressure, storage tanks are installed next to the actuator to prevent a rapid increase of flow through the rotary coupling. Hydraulic actuators are also often used to apply the earthquake shaking tests. The disadvantage of hydraulic actuators in cyclic tests is the sensitivity of the hydraulic system to changes of the stiffness in the test set-up.

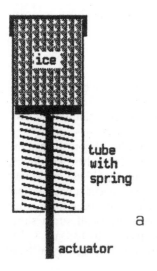

Figure 17. "Ice block" actuator for tension loads (Allersma 1998b)

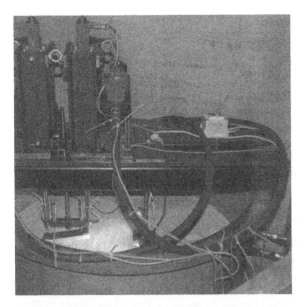

Figure 18. Hydraulic actuators located on a beam for a test on an eccentrically loaded shallow foundation (Bay-Gress et al. 1999)

Figure 19. Actuator tower of the ETHZ drum centrifuge (Springman et al. 2001)

With the development of computer technology, actuators consisting of stepping-motors or servomotors (mechatronics) are now being used more frequently. They consist of a motor applying torque to a lead screw, which loads the system. Communication with the actuators is achieved via PC interface modules (e.g. RS232). Figure 19 shows the ETHZ actuator towers,

Table 5. Specifications of the ETHZ actuators and tower

Specification	Value
Allowable continuous radial load rating	11 kN
Allowable continuous vertical load rating	2.5 kN
Maximum vertical velocity	100 mm/s
Range of vertical movement	450 mm
Maximum radial velocity	10 mm/s
Maximum radial movement	350 mm

which are placed on the toolplate in the drum centrifuge. Table 5 gives the specification of the actuators mounted in this tower on the toolplate.

The motor control will permit cyclic loading, but only at low frequencies. Also "creep effects" may occur for actuators with the displacement control conducted by counting the rotations, while the lead screw slips under constant load through the motor.

Freezing and heating can be supplied by a thermo-source applied directly into the system. In some centrifuges, heat or cold can be transferred directly from the stationary world through special rotary couplings. Prefrozen material may be insulated and then allowed to thaw inflight. Tests, in which sources of water, rain or contaminated solutions are needed require an onboard storage tank or another connection to the outer stationary world. Examples for tests regarding heat, cold, or pollutant may be found in Culligan-Hensley & Savvidou (1995) and Smith (1995).

7 TRANSDUCERS, MEASUREMENTS AND VISUALISATION

The aim of model tests in the centrifuge is to localise mechanisms and to get values or ratios for force, stress or displacement changes, as well as to observe the response of pore and water pressures, degrees of saturation or degrees of concentration. Many different types of transducers, measurement techniques and data processing methods are available and used in centrifuge modelling. They will be outlined briefly in the following sections. A subdivision is made into classical transducers, visualisation techniques and post-test investigations.

A crucial factor in obtaining data from centrifuge modelling is data acquisition and storage. Data acquisition had been a major concern in the past due to lack of computer capacity in respect of storage capabilities, speed of logging and potential of data post-processing, and due to the influence of noise during transfer of analogue signals through the sliprings. Nowadays data acquisition and quality is less of a problem, since analogue data can be digitised onboard and transferred to the control room for post-processing. For high frequency measurements, or in case of measuring and recording data at a rate in excess of the capacity of the data acquisition system, the data can be stored onboard and downloaded during or after the test, when no measurements are taking place.

Care has to be taken about the size and the weight of transducer as well as additional "on board" processing and amplification requirements. The transducer delivers signals to this "on board" custom-made amplification or junction box and they are transferred via the sliprings to the control room. Problems of weight arise mainly due to a restricted capacity and due to limited space in the centrifuge. Special care has to be taken for any transducer placed directly in the soil or at the interface between any structure and the soil sample. Transducers embedded in the soil will influence the behaviour of the soil stratum, as they represent intrusions that are scaled up in size and weight by the relevant scaling factor. Cables and extensions of e.g. LVDT's might also act as reinforcement to the soil. Transducers placed at the side or the bottom of a container or a structure may also influence the behaviour of the sample or the results, as they can cause arching because of different stiffness between the soil, the measurement diaphragm and the rigid base.

Methods of visualisation are related to the chosen model size and techniques available. Two-dimensional models allow insight into mechanisms via inspection through thick perspex windows, and with certain techniques, the evaluations of strain fields, interface and boundary effects can influence the results. In three-dimensional tests, the boundary effect does not influence the test result, as the test will be conducted far enough away from the sides, but visualisation is more difficult in terms of generating data representing these mechanisms.

7.1 *Measurement of forces and stress*

A major task in the modelling process is to get the load conditions right and to measure acting and reacting forces. Effort has also to be made in measuring (as far as possible) stresses acting in the soil as well as the stress distribution. The question of measuring stresses in soil has already be discussed in Burland (1967) where he pointed out that "stress is a philosophical concept" whereas "deformation is the physical reality". This raises the question whether stresses really can be measured - especially without influencing the stress distribution.

Forces can generally be measured by load cells connected between loading devices and actuators or between structures and a boundary. Standard force transducers are available and work under high g without problems (e.g. HBM U2A, available for a wide range of loads). The most commonly used measuring principle is by deflection of a stiff steel membrane instrumented by strain gauges. Other methods, such as the transfer of a load via an incompressible fluid pressure acting on a deflecting membrane are also possible. Only force transducers based on oscillating principles are unusable, as the additional g-level impedes the measurement.

It is always possible to build your own transducer, by applying strain gauges to a thin plate, to measure the deflection on a well-known static system. This is similar to the technique used in some standard transducers. Most additional force and moment distribution measurements are obtained by applying strain gauges to the interior or exterior of a model of a structure (e.g. retaining walls, piles, tunnel linings). Preference should be given to applying strain gauges to the interior of the structure, as they would otherwise affect the interface between the structure and the soil. In case of measurements with strain gauges, temperature effects should be minimised by using full Wheatstone bridge circuits and temperature compensating gauges. Alternatively, a half bridge can be used with a set of dummy strain gauges placed next to the sample on a similar piece of structure, where there is expected to be no load. Also strain gauges, their shielding, glue and heat dissipation have to be appropriate for the test condition as does the waterproofing.

As already mentioned, the measurement of forces and stresses inside the soil or at the boundary is highly affected by the differential stiffness of the soil, the structure and the transducer. The measurement of stresses in the soil or even at the interface between soil and structure are complex, as sizes of transducer are scaled up at the prototype scale by a factor of n. If they are stiffer than the soil, they are most likely to attract load. A continuous series of transducers might be placed at one boundary to obtain a stress distribution, but arching and stress rearrangements might also occur in this case. Some transducers were designed, manufactured and used by different institutes, e.g. the Stroud (1971) cell was developed at Cambridge University Engineering Department (CUED) as a contact stress transducer to measure the magnitude and the direction of shear and eccentric normal stresses (Fig. 20). The load is applied on a stiff surface causing movements and deformation in the bottom horizontal webs (for measuring shear) and in the vertical webs (for measuring normal loads). Another approach in measuring earth pressure inside the soil is given by Egan & Merrifield (1998). Garnier et al. (1999) also discussed the placement of embedded pressure cells.

A new technique, that has been applied recently to the field of geotechnical engineering (Paikowsky & Hajduk 1997), is the use of tactile pressure sensor mats. These mats were originally developed for surface pressure measurement for biomedical and other mechanical engineering

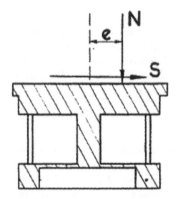

Figure 20. Cross section through the Stroud cell (Stroud 1971)

tasks. A photo of these pressure sensors is shown in Figure 21. First results of the use of the mats in the centrifuge under raised g conditions are given in Springman et al. (2002) and Laue et al. (2002).

The pressure mats consist of a matrix based tactile sensor protected by thin films with a thickness of 0.1 mm. Resistance is measured at the junction (sensels) of the orthogonal strips, which offer circa 2000 measurements over the area of the transducer. The data is transferred via a "handle", which needs to be secured for use at high g-levels, to an onboard microprocessor, which converts and digitalises the data. After transferring this digitalised data to the stationary world, it needs to be processed by a special program, so that stress measurement can be calculated over a specified area. Sensors of different shapes and sizes (in terms of plan area) are available.

Especially challenging is the calibration procedure, where different steps are necessary (see Laue et al. 2002 or Springman et al. 2002). After a warming up phase, the transducers have to be brought into equilibrium. This can be reached by supplying a constant pressure (e.g. with a prestressed air bag) or in a pressure cell. This allows the software of the equipment to set the pad to a constant starting equilibrium. Before conducting a test, calibration can take place using

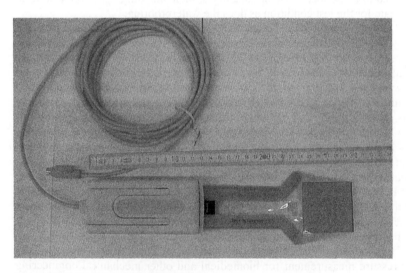

Figure 21. Pressure sensor with handle

weights with a flat base, which should be large enough geometrically to cover the loaded sensels and which may also be placed on an intermediate sand layer. The magnitude of the calibration should exceed the maximum expected load by a factor of 1.5.

Another technique employed to measure stresses in samples inflight is proposed by Allersma (1998a). This technique uses the photoelastic particles themselves as a sensor. These particles are exposed to polarised light, which is able to pass through the sample and primarily reflects the distribution of the principal stresses inside a two-dimensional sample. This can be photographed and the brightness can be evaluated using an elaborate data processing system and be used to quantify the principal stress values.

7.2 *Measurements of pore pressure, degree of saturation and concentration*

Fluid pressures, pore pressures, the degree of saturation and concentration can be determined, but this demands an increasing order of complexity. Pore pressure transducers provide an indication of the equivalent height of the free water table at that point and are used to determine the changes in pore water pressures due to consolidation or cyclic loading. In most centrifuge centres, pore pressure transducers of the type DRUCK PDCR 81, which are available for a wide working range of pressures, are used. Figure 22a shows a diagram of this transducer. The pressure is measured across a membrane as a differential to the outside pressure. The connecting cable is thicker than the usual electrical cable, but the size of the transducer is reasonably small (6 mm diameter, 12 mm length, see Fig. 22b). Other companies also provide small pore pressure transducers (Kyolite) but these transducers are not so widely used in geotechnical centrifuges.

In order to measure fluid rather than soil pressure, and to prevent damage to the transducer, the membrane of the PDCR 81 must be shielded from the soil. Porous filter stones have to be installed between the membrane and the transducer. These stones have to be selected according to the permeability of the soil and they have to be saturated totally to guarantee a sufficiently quick response to changes in pore pressures in the model. Problems specifically related to maintenance of saturation may occur during tests in silts and fine sand, which are being de-saturated because of the suction potential of the soil. In less permeable soils, this transducer might also be used to measure the suction (e.g. for clay, Muraleetharan & Granger 1999). The Network of European Centrifuges for Environmental Geotechnical Research (NECER) programme deals with the measurements of the degree of saturation and of the concentration of chemical con-

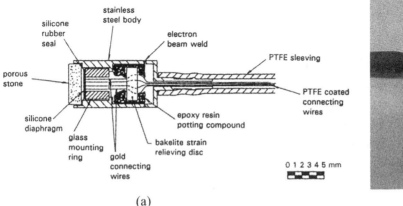

(a)   (b)

Figure 22. a) Sketch (Taylor 1995) and b) photo of PDCR81 transducer

tamination. Different approaches for these measurements may be found in the conference proceedings on physical modelling and testing in environmental geotechnics (Garnier et al. 2000).

Some other measurement techniques for measuring suction, water content or concentration of pollutants function effectively under the influence of g. Typically measurements may be made using the Time Domain Reflectometry (TDR) method consisting of a set of two electrodes fitted in the model (Crancon et al. 2000). Because of the length between 4 and 10 cm and a cable size diameter of at least 5 mm, sample disturbance may occur in a model test, as described before. Other methods use capacitive sensors (Dupas et al. 2000), electrodes or photometric sensors (Lynch et al. 2002, Fig. 23) based on photodiodes, when using coloured tracer fluids.

Resistivity probes can be used to monitor environmental changes by measuring chemical concentrations via resistivity variations of a pore fluid. An example of a resistivity transducer is given in Figure 24 (Culligan-Hensley & Savvidou 1995).

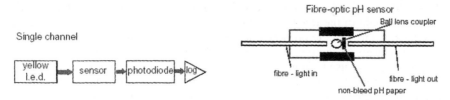

Figure 23. Schematic of photodiode system and sensor (Lynch & Bolton 1996)

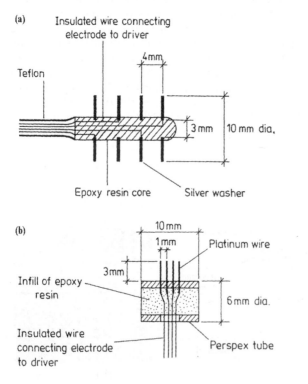

Figure 24. Miniature four electrode resistivity probe a) section through 4 washer probe b) section through four pin probe (Culligan-Hensley & Savvidou 1995)

7.3 Measurements of deformation and strains

Classical deformation measurements are made using LVDTs (Linear Variable Differential Transformers) and linear potentiometers. LVDTs can most easily be placed outside a model to make surface measurements on soil or structures, however extension rods sleeved in a straw may be connected to small bearing plates buried in the soil. The measurement range depends on the transducer. For small differential transducers it is possible to obtain readings in milli Volts, which deliver a high resolution of around 10 Volts subdivided by the range of the transducer e.g. 20 mm. The resolution will be reduced for transducers operating over a larger range. Care has to be taken for non-linearity effects at both ends of the measurement range next to the boundaries of about 2% of the measuring range. Linear potentiometers can be used for larger deformation.

Non-contact laser transducers are now used to investigate deformation at the surface (Fig. 25). Accuracy, linearity and quoted resolution are similar to the measurements using LVDTs, although the resolution is also dependent on the distance between the transducer and the measurement area. The shorter this distance, the better the probable resolution will be, which can reach an accuracy of 1μm for a transducer transmitting to a smooth steel surface. The resolution for granular soils is about a grain size. The measurement principle is based on triangulation measuring of the refraction of a laser beam. Laser transducers can be tracked over an area to survey the soil surface, to obtain the topography of the model, but needs more or less steady state deformation conditions since this measurement process takes time.

The screening of the soil surface may also be done with a system using three cameras. One such method is given by Taylor et al. (1998). The test setup is shown in Figure 26. Using two or three cameras it is possible to create vectors for marker points that can be processed to give an image of the surface deformations (Fig. 27).

Other measurements are adopted for 2-dimensional test set-ups. Strain fields and the movement of the soil mass caused by loading can be observed by the insertion of markers, whose positions may be analysed using the photos taken during the test. The easiest form of this analysis is projecting the slides of the photos on the wall and drawing the marker positions at different test steps. In the age of digital images, photos can be re-analysed using software by taking the different grey scales of the markers and soil into account. Thus an electronic evaluation is possible and the precision of the measurements is dependent upon the resolution of the cameras. A modern technique of digital image processing is proposed by White et al. (2001, Fig. 28). This method allows a precision of 1/25th of a pixel, so resolution is highly dependent on the size of the observation window. A maximum resolution of up to 10 μm can be reached. An advantage of the adoption of this image processing using Particle Image Velocimetry (PIV) is that markers, which may disturb sample behaviour, are not needed in the test.

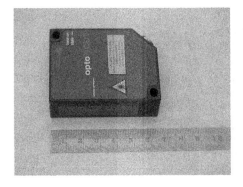

Figure 25. Size of a typical laser transducer

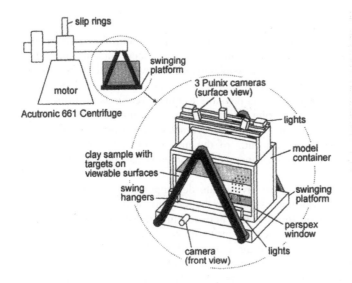

Figure 26. Arrangement of cameras for surface deformation measurement (Taylor et al. 1998)

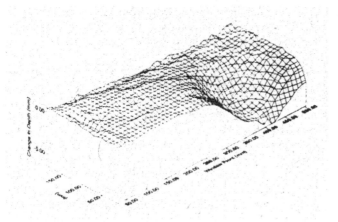

Figure 27. Surface movements caused by tunnel heading collapse (Taylor et al. 1998)

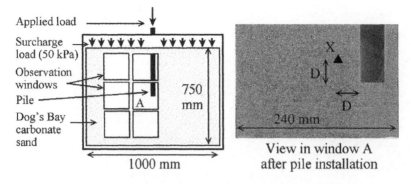

Figure 28. Arrangement of PIV measurement for a pile load test (White et al. 2001)

Attention must be given to the influence of friction on the perspex window on the image processing of a plane strain model. Different centrifuge centres developed different methods to reduce friction at the boundary wall in the two-dimensional tests. One method is to smear silicone oil between the perspex wall and a thin transparent plastic foil or latex rubber sheet, which becomes the contact interface to the soil. These methods reduce the friction significantly (a.g. adhesion angles of $2 - 5°$, Balachandran 1996), but zero friction is almost impossible to achieve, and thus the results do not represent the quantitative behaviour of the soil sample away from the wall. Nonetheless, it does allow identification and some quantification of the mechanism acting.

7.4 *Other transducers*

Other transducers are available for measuring all kinds of parameters. Thermistors can be used for measuring temperature, whereas constant acceleration transducers can be used to control the level of gravitation reached. Dynamic load influences are measured with a fast accelerometer (usually based on differential measurement compared with a steady field of acceleration). These acceleration transducers may also be used for determining the small strain soil stiffness inflight.

Most of the available measurement techniques in the non-centrifugal environments may be used in the centrifuge by adjusting the size and the measurement range. One such technique, now being transferred from use in site investigations and laboratory testing to measure deformations and to localise shear zones in centrifuge modelling, is the use of TDR probes as proposed by McAlister & Pierce (2001) and Fujiyasu & Pierce (2002). Other techniques include the use of computer topography and of X-rays, both of which are already used for post-processing samples after a test. Apart from these options, there are many ways of possibilities in improving existing and developing new inflight measurement techniques. As an example, many applications of biomedical research may be transferred to geotechnical modelling as the basic question – non-destructive and undisturbed insight into a body / sample – is similar.

8 MODELLING OF PROCESSES - SUMMARY

Much attention should be paid to the technique or the procedure of the modelling itself. This consists of a sequence of idealisations, by changing a prototype or a full-scale problem into a centrifuge model test. This includes a clear definition of possible achievements, difficulties and necessary simplifications with respect to the available machine, manpower and money as well as accuracy of the data and test results to be obtained.

In the previous chapters, principles of centrifuge modelling techniques have been discussed in terms of: the available centrifuge, the fabrication of soil samples, the loading, and instrumentation possibilities. In this chapter, the single points of interest will be summarised and extended to model test possible setups and the development of more advanced modelling techniques.

A centrifuge model test is always a composition of the single influencing parameters, as described before. Attention has to be paid to the effects of the modelling procedure itself, which will be discussed for three typical topics. These are achievements in relation to simulating the behaviour of the unsupported excavation of tunnels, the simulation of excavations adjacent to embedded walls and the influence of the installation methods on pile response.

In most of the pile tests, piles are installed in granular soils in the strongbox before the soil sample is prepared (e.g. Kotthaus et al. 1993). In clay, holes may be drilled into the consolidated sample, the piles inserted and the bounding annulus refilled with slurry. In both of these installation procedures, the respective prototype pile modelled will "grow" while spinning in the centrifuge at increasing accelerations. In the growth process, the pile may experience settlements,

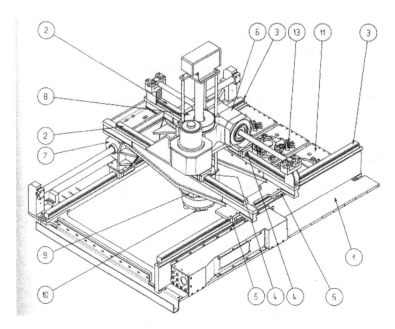

Figure 29. Drawing of the LCPC robot (Derkx et al. 1998). with 1) base beam 2) and 3) linear guide rails 4) and 5) skids 6) and 7) brushless motors driving a screw/nut assembly 8) screw/nut assembly for the z-axis 9) rotation arm 10) tool holder

Figure 30. Photo of the LCPC robot (Thorel 2001)

which have to be allowed for (e.g. Laue & Sonntag 1998). Even if movements are considered in respect of a preinstalled pile, differences in settlement of the pile and the surrounding soil stratum can occur due to differences in self weight. Real piles are most likely to be constructed by boring or jacking or driving e.g. by a falling weight. These procedures are more difficult to model, as they require two or more flights in most centrifuges. The use of a robot (Figs. 29 and 30) with tools such as a hammer or an actuator, which can apply axial load, can approximate the installation of a full scale pile. A load can be applied at a later stage with another (or the same) actuating tool.

The use of actuators with a long stroke allows the simulation of a single pile driven into the soil. The installation method for a pile is therefore crucial to the interpretation of pile data, since mechanisms representing the pile bearing behaviour will be influenced by this procedure (e.g. Dyson & Randolph 1998 among others, Fig. 31). Although centrifuge modelling allows useful information to be gathered and relationships to be explored, the transfer of the measured values to prototype behaviour needs to be analysed carefully.

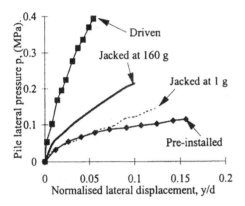

Figure 31. Influence of pile installation method on the lateral load-displacement curve (Dyson & Randolph 1998)

Modelling of excavations can be done step by step by removing soil on one side of an embedded retaining wall. Centrifuge start and stop cycles have to be conducted, causing an ongoing change of stress in the soil stratum (Jessberger & Güttler 1988). Further developments in the modelling of excavations are the use of water-filled bags on the excavation site (e.g. Schürmann & Jessberger 1994). As the centrifuge is accelerated, the starting state of the test is different to the prototype situation since the different self-weights of water and soil cause another stress field to develop. Allersma (1998b) and König (2001) introduce further developments of more realistic modelling procedures of the excavation process. Allersma (1998b) uses a linen bag construction filled with sand, which allows predefined variable layers of soil to be removed. König (2001) developed an inflight scraping system for layers of variable thickness.

Tunnelling excavation techniques, which have already been discussed in Sharma (2002), follow a similar development. To simulate the unsupported length, a pressure bag is placed in the tunnel to keep the unsupported area stable during the increase of gravity. The pressure in the bag has to be controlled by hand via a valve leading to possible over or "under" stressing around the tunnel area. The quality of the measurement is influenced detrimentally by this construction technique, as shown later by König (1998). He compared test results conducted simulating the excavation with a pressure bag with those obtained by using a tunnel construction machine consisting of a drilling unit, which allows a few centimetres of drilling without support of a tunnel liner.

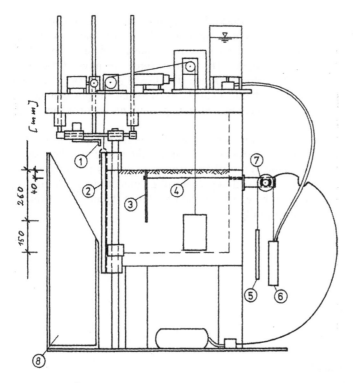

Figure 32. Modelling of the excavation process, with 1) lower edge of the scraper 2) front wall 3) sheet pile wall 4) anchor rod 5) counterweight for the tightening mechanism of the anchor rod 6) ballast tank for the anchor rod 7) brake mechanism for fixing the end of the anchor 8) ballast tank for material removed (König 2002)

Centrifuge technology offers a wide range of modelling possibilities, but care has to be taken in how the results are analysed and transferred to the real world. Centrifuge modelling data should therefore always be used in comparison with other analytical, numerical or site-specific studies. In respect of the interpretation and the further analysis, a designed test setup of a centrifuge model, which can never be a full representation of a field situation, should be as simple as possible to focus on the main influencing parameters. For further analysis, the influence of the modelling processes should be taken into account as well as location and installation procedures of transducers and the influence of the sidewalls of the strongboxes.

REFERENCES

Allersma, H.G.B. 1998a. Stress analysis on photoelastic particles in centrifuge tests. In T. Kimura et al. (eds), *Centrifuge 1998*: 61-66. Rotterdam: Balkema.
Allersma, H.G.B. 1998b. Development of cheap equipment for small centrifuges. In T. Kimura et al. (eds), *Centrifuge 1998*: 85-90. Rotterdam: Balkema.
Almeida, M.S.S. 1984. Stage constructed embankment on soft clay. PhD Thesis. Cambridge University.
Balachandran, S. 1996. Modelling of geosynthetic reinforced soil walls. PhD Thesis. Cambridge University.
Barker, H.R. 1998. Physical modelling of construction processes in the mini drum centrifuge. PhD Thesis, Cambridge University.
Bay Gress, C., Sieffert, J.G. & Laue, J. 1999. Modélisation de l'interaction sol - fondations superficielles. *Revue Française de Géotechnique* 59(3): 37-45.

Bolton, M.D., Gui, M.W., Garnier, J., Corte, J.F., Bagge, G., Laue, J. & Renzi, R. 1998. Centrifuge Modelling of Cone Penetration Tests. In T. Kimura et al. (eds), *Centrifuge 1998*: 155-160. Rotterdam: Balkema.

Bolton, M.D., Gui, M.W., Garnier, J., Corte, J.F., Bagge, G., Laue, J. & Renzi, R. 1999. Centrifuge cone penetration tests in sand. *Géotechnique* 49(3): 543-552.

Bucher, M. 1999. Setzungs- und Verformungsverhalten von Flachfundamenten aufgrund statischer und zyklischer Belastung sandiger homogener und geschichteter Böden. Diploma Thesis. ETH Zürich.

Burland, J.B. 1967. Deformation of soft clay. PhD Thesis. Cambridge University.

Corte, J.F. ed. 1988. *Centrifuge 1988.* Rotterdam: Balkema.

Crancon, P., Guy, C., Pili, E., Dutheil, S. & Gaudet, J.P. 2000. Modelling of capillary rise and water retention in centrifuge tests using a geotechnical centrifuge In J.Garnier et al. (ed.), *Int. Symp. on Physical Modelling and Testing in Environmental Geotechnics*: 199-206. La Chapelle sur Erdre: TOP Imprimerie.

Culligan-Hensley, P.J. & Savvidou, C. 1995. Environmental geomechanics and transport processes. In R.N. Taylor (ed.), *Geotechnical Centrifuge Technology*: 196-263. London: Blackie Academic and Professional.

Dean, E.T.R., James, R.G. & Schofield, A.N. 1990. Drum centrifuge studies for EEPUK. Contract No. EP-022R TASK ORDER 1-022. Phase 1. Draft Report. Cambridge.

Derkx, F., Merliot, E., Garnier, J. & Cottineau, L.M. 1998. On-board remote-controlled centrifuge robot In T. Kimura et al. (eds), *Centrifuge 1998*: 97-104. Rotterdam: Balkema.

Dupas, A., Cottineau, L.M., Thorel, L. & Garnier, J. 2000. Capacitive sensor for water content measurements in centrifuged porous media. In J.Garnier et al. (eds), *Int. Symp. on Physical Modelling and Testing in Environmental Geotechnics*: 11-17. La Chapelle sur Erdre: TOP Imprimerie.

Dyson, G.J. & Randolph, M.F. 1998. Installation effects on lateral load-transfer curves in calcareous sands. In T. Kimura et al. (eds), *Centrifuge 1998*: 545-550. Rotterdam: Balkema.

Egan, D. & Merrifield, C.M. 1998. The use of miniature earth pressure cells in a multi gravity environment. In T. Kimura et al. (eds), *Centrifuge 1998*: 55-80. Rotterdam: Balkema.

Ellis, E.A., Soga, K., Bransby, M.F. & Sato, M. 1998. Effect of pore fluid viscosity on the cyclic behaviour of sands. In T. Kimura et al. (eds), *Centrifuge 1998*: 217-222. Rotterdam: Balkema.

Fragaszy, R. & Cheney, J.A. 1981. Drum Centrifuge Studies of Overconsolidated Slopes. *Journal of Geotechnical Engineering Division ASCE.* 107 (GT7): 843-858.

Fujiyasu, Y. & Pierce, C.E. 2002. Development of TDR sensors for shear zone monitoring in centrifuge soil models. In Phillips et al. (eds), *Proc. of the International Conference of Physical Modelling in Geotechnics, St. John's*: 125-130. Rotterdam: Balkema.

Garnier, J., Ternet, O., Cottineau, L.-M. & Brown, C.J. 1999. Placement of embedded pressure cells. *Géotechnique,* 49 (3): 405-414.

Garnier, J., Thorel, L. & Haza, E. (eds) 2000. *Physical Modelling and Testing in Environmental Geotechnics.* La Chapelle sur Erdre: TOP Imprimerie.

Graemiger, E. 2001. Einbauverfahren und Bodenerkundung im Sand in der geotechnischen Zentrifuge. Diploma Thesis. ETH Zürich.

Gurung, S.B., Almeida, M.S.S. & Bicalho, K.V. 1998. Migration of zinc through sedimentary soil models. In T. Kimura et al. (eds), *Centrifuge 1998*: 598-594. Rotterdam: Balkema.

Iwasaki, T. & Tatsuoka, F. 1977. Effects of grain size and grading on dynamic shear moduli of sands. *Soils and Foundations*: 17 (3): 19-35.

Jessberger, H.L. & Güttler, U. 1988. Geotechnische Grosszentrifuge Bochum – Modellversuche im erhöhten Schwerefeld. *Geotechnik*: 11(2): 85–97.

Kimura, T., Kusakabe, O. & Takemura. J. (eds) 1998. *Centrifuge 98*. Rotterdam: Balkema.

Ko, H.Y. & McLean, F.G. (eds) 1991. *Centrifuge 91.* Rotterdam: Balkema.

König, D., Jessberger, H.L., Bolton, M., Phillips, R., Bagge, G., Renzi, R. & Garnier, J. 1994. Pore pressure measurement during centrifuge model tests: Experience of five laboratories. In C.F. Leung et al. (eds), *Centrifuge '94*: 101-108. Rotterdam: Balkema.

König, D. 1998. An inflight excavator to model a tunnelling process. In T. Kimura et al. (eds), *Centrifuge 1998*: 707-712. Rotterdam: Balkema.

König, D. 2002. Modeling of deep excavations. In Phillips et al. (eds), *Proc. of the International Conference of Physical Modelling in Geotechnics, St. John's*: 83-88. Rotterdam: Balkema.

Kotthaus, M., Laue, J. & Jessberger, H.L. 1993. Centrifuge Model Tests on the Interaction Behaviour of Deep and Shallow Foundations. In T. Moan et al. (eds), *Structural Dynamics. EURODYN 93*. Rotterdam: Balkema.

Kusakabe, O. & Gurung, S.B. 1997. Development of mini drum centrifuges and a few initial tests. In M. Almeida (ed.), *Int. Symp. on Recent Developments in Soil and Pavement Mechanics*. Rotterdam: Balkema.

Kusakabe, O., Hagiwara, T. & Kuroiwa, H. 1988. Design and operation of a drum centrifuge. In J.F. Corte (ed.), *Centrifuge 88*: 77-82. Rotterdam: Balkema.

Laue, J. 1996. Zur Setzung von Flachfundamenten auf Sand unter wiederholten Lastereignissen. *Schriftenreihe des Lehrstuhls für Grundbau und Bodenmechanik der Ruhr-Universität Bochum,* Heft 25, Bochum.

Laue, J. 1997. Transport Stability of Iron Ore Concentrate Heaps by means of Centrifuge Model Testing, In J.S. Chung (ed.), *Proc. of the Conference ISOPE 97*: Vol I 921-927. International Society of Offshore and Polar Engineers: Golden, CO.

Laue, J., Siemer,T., Grundhoff, T. & Jessberger, H.L. 1996. Experimental Methods in Soil Mechanics. In W. Krätzig & H.J. Niemann (eds), *Dynamic Behaviour of Material and Strucutural Components*: 597-623. Rotterdam: Balkema.

Laue, J. & Sonntag, T. 1998. Piles subjected to torsion. In T. Kimura et al. (eds), *Centrifuge 1998*: 187-192. Rotterdam: Balkema.

Laue, J., Nater, P., Springman, S. M. & Graemiger, E. 2002. Preparation of soil samples in drum centrifuges. In Phillips et al. (eds), *Proc. of the International Conference of Physical Modelling in Geotechnics, St. John's*: 143-148. Rotterdam: Balkema.

Laue, J., Nater, P., Chikatamarla, R. & Springman, S.M. 2002. Der Einsatz von „pressure pads" in geotechnischen Labor- und Modellversuchen. *Messen in der Geotechnik*, Schriftenreihe des Instituts für Grundbau, Braunschweig.

Leung, C.F., Lee, F.H. & Tan, E.T.S. eds. 1994. *Centrifuge 94*. Rotterdam: Balkema.

Lynch, R.J. & Bolton, M.D. 1996. A low cost fibre optic pH sensor for the in-situ measurement of groundwater. In Proc. of the International Society of Optical Engineering, Advanced Technologies for Environmental Monitoring and Remediation, Denver, Colorado: 221-227.

McAlister, C.M. & Pierce, C.E. 2001. Calibration of small diameter TDR cables for measuring displacement in physical soil models. In *TDR 2001 Symposium*. Evanston IL: Northwestern University.

Mitchell, R.J. 1998. A new geoenvironmental centrifuge at Queens University in Canada. In T. Kimura et al. (eds), *Centrifuge 1998*: 31-34. Rotterdam: Balkema.

Miyake, M. & Yanagihara, T. 1999. Heap Shape of Materials Dumped from Hopper Barges by Drum Centrifuges. In *Proc. of the ninth International Offshore and Polar Engineering Conference Brest*: 745-748. International Society of Offshore and Polar Engineers: Cupertino CA.

Muraleetharan, K.K. & Granger, K.K. 1999. The use of miniature pore pressure transducers in measuring matrix suction in unsaturated soils. *Geotechnical Testing Journal* 22(3): 226-234.

Paikowsky S.G. & Hajduk, E.L. 1997. Calibration and Use of Grid Based Tactile Pressure Sensors in Granular Material, *ASTM Geotechnical Testing Journal* 20(2): 218-241.

Renzi, R., Corté, J.F., Rault, G., Bagge, G., Gui, M.W. & Laue, J. 1994. Cone penetration tests in the centrifuge: Experience of five laboratories. In C.F. Leung et al. (eds), *Centrifuge '94*: 77-82. Rotterdam: Balkema.

Rossato, G., Ninis, N.L. & Jardine, R.J. 1992. Properties of some kaolin-based model clay soils. *ASTM Geotechnical Testing Journal* 15(2): 166-179.

Schofield, A.N. 1976. Use of Centrifugal Model Testing to Assess Slope Stability. Technical Report, Cambridge University Engineering Department, CUED/C, Soils TR 30. Cambridge.

Schofield, A.N. 1980. Cambridge University Geotechnical Centrifuge Operations. 20th Rankine lecture. *Géotechnique* 30(3): 227-268.

Schürmann, A. & Jessberger, H.L. 1994. Earth pressure distribution on sheet pile walls In C.F. Leung et al. (eds), *Centrifuge '94*: 95-100. Rotterdam: Balkema.

Sharma, J.S. 2002. Measurements of deformation - Trends or Numbers. In S.M. Springman (ed.) *Workshop on constitutive and centrifuge modelling: two extremes. Monte Verita*. Rotterdam: Balkema.

Shen, C.K., Li, X.S., Ng, C.W.W., Van Laak, P.A., Kutter, B.L., Cappel, K. & Tauscher, R.C. 1998. Development of a geotechnical centrifuge in Hong Kong. In T. Kimura et al. (eds), *Centrifuge 1998*: 13-18. Rotterdam: Balkema.

Siemer, T. 1996. Zentrifugenmodellversuche zur dynamischen Wechselwirkung zwischen Bauwerk und Boden infolge stossartiger Belastung, *Schriftenreihe des Lehrstuhls für Grundbau und Bodenmechanik der Ruhr-Universität Bochum,* Heft 27, Bochum.

Siemer, T. 2000. Centrifuge model investigations towards the dynamic shear modulus in the subsoil. In J.Garnier et al. (ed.), *Int. Symp. on Physical Modelling and Testing in Environmental Geotechnics*: 361-368. La Chapelle sur Erdre: TOP Imprimerie.

Smith, C.C. 1995. Cold regions engineering. In R.N. Taylor (ed.), *Geotechnical Centrifuge Technology*: 264-292. London: Blackie Academic and Professional.

Springman, S.M. 1993. Centrifuge Modelling in Clay: Marine Applications. In J. Clark et al. (eds), *Proc. of the 4th Canadian Conf. on Marine Geotechnical Engineering* 3: 853-896. St. Johns.

Springman, S.M. 2001. Lecture Notes for Modelling in Geotechnics. ETH Zurich.

Springman, S.M., Laue, J., Boyle, R., White, J. & Zweidler, A. 2001. The ETH Zurich Geotechnical drum centrifuge. *International Journal of Physical Modelling in Geotechnical Engineering*. 1(1): 59-70.

Springman, S.M., Nater, P., Chikatamarla, R. & Laue, J. 2002. Use of flexible tactile pressure sensors in geotechnical centrifuges. In Phillips et al. (eds), *Proc. of the International Conference of Physical Modelling in Geotechnics, St. John's*: 113-118. Rotterdam: Balkema.

Stewart, D.P. and Randolph, M.F. 1991. A new site investigation tool for the centrifuge. In H.Y. Ko & F.G. Mc Lean (eds), *Centrifuge '91*: 531-537. Rotterdam: Balkema.

Stewart, D.P., Boyle, R.S. & Randolph, M.F. 1998. Experience with a new drum centrifuge. In T. Kimura et al. (eds), *Centrifuge 1998*: 35-40. Rotterdam: Balkema.

Stroud, M.A. 1971. Sand at low stress levels in simple shear apparatus. PhD Thesis. Cambridge University.

Taylor, R.N. ed. 1995. *Geotechnical centrifuge technology.* London: Blackie Academic and Professional.

Taylor, R.N., Grant, R.J., Robson, S. & Kuwano, J. 1998. An image analysis system for determining plane and 3D displacements in soil models. In T. Kimura et al. (eds), *Centrifuge 1998*: 73-78. Rotterdam: Balkema.

Thorel, L., Garnier, J., Piau, S. & Haza, E. 2000. CLEOPATRE: a database on centrifuge physical modelling. In J. Garnier et al. (ed.), *Int. Symp. on Physical Modelling and Testing in Environmental Geotechnics*: 369-378. La Chapelle sur Erdre: TOP Imprimerie.

Thorel, L. 2001. Updated version of CLEOPATRE keywords and picture of the LCPC robot. Private communication.

White D. J., Take W.A, Bolton M.D. & Munachen S.E. 2001. A deformation measuring system for geotechnical testing based on digital imaging, close-range photogrammetry, and PIV image analysis. In *Proc. 15th ICSMGE, Istanbul, Turkey*: 539-542. Rotterdam: Balkema.

DISCUSSION

S.M. Springman & T. Weber
Institute for Geotechnical Engineering, Swiss Federal Institute of Technology, Zurich, Switzerland

Laurent Vulliet asked whether full saturation of various types of the soil sample had been achieved, particularly when using clay. Jan Laue that explained the procedure included mixing slurry made from powder or real soil slowly in the mixing unit, under vacuum, to remove all air from the system. He added that material was placed in the centrifuge using a nozzle mounted on the actuator, which was driven out below the water surface for soil placement underwater and subsequent consolidation. In this case, he was sure that full saturation was achieved. However he mentioned the possibility that, on stopping the drum (for a period), and depending on the permeability of the soil, a crust might develop (as the excess pore pressures dissipated), as has sometimes been done in a beam centrifuge (to achieve an overconsolidated top layer). Sarah Springman noted however that air might get into the slurry in pouring the slurry from the mixer into the hopper. She thought that as long as no air bubbles were trapped once the slurry reached the tube leading to the nozzle, then the soil would remain saturated, but that some samples should be taken from models made in this way to see whether the clay was nearly 100% saturated. Malcolm Bolton pointed out that the same principles would apply in making clay samples outside the centrifuge (in the laboratory) and then placing them in the centrifuge, in terms of mixing under vacuum and then consolidating. But he said, that if the clay was to remain saturated, that the final pore pressure in the sample near to the clay surface should be less than the air entry value of the clay at the time it was exposed to the air on removal from the consolidometer.

Jacques Garnier discussed the latest developments in using pressure cells for total stress measurement, which were thought to avoid (or at least minimise) the problems related to deflection of the diaphragm and redistribution of stress in granular material. He described them as a non-direct measurement system, using a circular diaphragm resting on a very thin oil layer, with the oil pressure measured by a very small transducer, all of which produced a very stiff response. He gave an example that the deflection of the transducer built by Askegaard from Denmark (Garnier et al. 1999) was less than 1 micron under a normal pressure of 100 kPa. He noted that the centrifuge had been used successfully to calibrate the cells, which were placed on a horizontal surface of sand and a given amount of sand was pluviated onto them to a known density. He explained that the g-level was increased to control the level of normal stress applied to the cell, while measuring the response. He felt that the main problems arose due to placement of the cell in the soil, because the load transfer was very dependent upon the relative stiffness of the soil and the cell. He commented that if the cell was relatively stiff, then more stress was concentrated onto it, whereas if the cell could settle, the stress was redistributed. He concluded that it was now possible to measure normal stress in sandy soil to an accuracy of more or less 5%, but it was vital to be very careful with the choice of cell and the placement technique.

Bolton was reminded that many of the challenges that have to be overcome in centrifuge testing were exactly the same as those that have to be faced when doing field tests or even laboratory tests. He called for input from non-centrifuge modellers to which Vulliet responded by mentioning their use of fibre optics for in situ measurement, mainly for strains and temperature. Bolton commented that the fibres may have a breaking strength of 20 Newtons in the field, and that by inserting these into a centrifuge model, it would be equivalent to multiplying this by the scale squared. He thought that the problems of relative stiffness and strength as well as potential for soil reinforcement could become an important issue, with which Laue agreed. Laue mentioned that a TDR cable would be a replacement for obtaining the deformation profile from insertion of spaghetti, which were very weak and flexible once saturated and would not influence the soil behaviour too much. But he thought that the TDR cables were thicker and that this could influence the measurement and should be taken into account. Bolton mentioned that Rod Lynch (Lynch & Bolton 1996) had used photometric transducers in Cambridge to deduce chemical concentration by sending light down optic fibres in order to measure the colour intensity in a little cell between these two fibres. He was aware that stress was not so important in many of these transport experiments and thought that no-one had had the courage to introduce optic fibres into a model in which stress and strain were very important, although the technology looked very promising.

David Muir Wood picked up on the comments made about the stress field in the centrifuge box in relation to work carried out by Kevin Stone in Bochum on the behaviour of very soft clay layers, which were prepared with a horizontal surface and no crust. He noted that the radial acceleration field on the beam centrifuge had caused slip surfaces to form with two sides of the hill sliding down towards the centre of the box, and although it had seemed obvious to make a flat surface, actually a curved surface would have been better. Laue agreed with this and pointed out that this was the advantage of the drum centrifuge! Bolton agreed and noted that the ponding of water and the shapes of slopes gave rise to exactly the same considerations in terms of hills in the gravity field or dales in the slopes.

Bill Craig asked about modelling on coarse granular material and wondered whether the grain size should be scaled down in a centrifuge model in view of the micromechanical views presented by Malcolm Bolton. He believed that provided the size of the grains was sensible compared to the system dimensions of the models, for example the diameter of a pile or a footing, or the width of a structural element, we should not reduce our grain sizes. Laue pointed out that often a very fine graded soil was used because of these restrictions in terms of the ratio of the diameter of foundations to the diameter of the soil particles. But he thought that some confusion existed, about the interpretation (by non-centrifuge modellers) of results obtained from centrifuge modelling, and recommended that care should be taken to improve communication of action taken and decisions made in relation to the grain size.

Ivo Herle was not convinced and thought that if the particles were scaled in any way, e.g. from sand to a very fine powder, there would be a variation of forces between grains and influences due to grain shape and roughness. He wondered how footing tests were modelled because he was aware that the grain size effect on a small footing could be pronounced. Bolton mentioned the existence of the 1500 papers listed in the Cleopatra database and noted that dozens of these had tried to address this question and that different conclusions had been reached, showing that there were still some big issues to be sorted out. He raised the example of a footing on granular soil and cited C.K. Lau's (1988) work on models at different scales, tested at the appropriate g-level with two soils, a silica flour and a silica sand, which had the same grading curve but at a factor of 50 times apart. He said that Lau had used a microscope to determine that the shape as well as the size distribution of the particles was very similar and he found that by scaling the particle size down, the material was actually slightly stronger in triaxial tests and in centrifuge tests. Bolton suggested that the smaller particles were less breakable and therefore caused more dilatancy and Lau had found that the footing on silica flour was stronger, when tested in the centrifuge, by exactly what he would have expected based on the triaxial tests. Bolton concluded that this had shown that there were no scale effects for this example that could not be understood. He thought that any boundary value problem could be studied in the centrifuge and if the data were understandable, then confidence would be raised in terms of facing the same situation in the field, even if the exact details would be different. He was not too worried about trying to replicate a particular field situation or soil, but thought that the main aim was simply to understand the mechanics of granular materials.

Herle was also interested to know whether some problems, in which partly saturated soil played a major role, had been modelled successfully in the centrifuge. He pointed out that this was also connected with grain size and grain size distribution and wondered whether the capillary forces scaled with the gravity. König responded that tests had been carried out to investigate the scaling of capillary rise under unsaturated soil conditions (Rezzoug et al. 2000). He confirmed that when the same grain size of soil had been used and tested at different scales, the capillary height and also the distribution of water content were scaled in the correct manner.

Garnier noted that the first problem to be faced when dealing with grain size effects was the ratio between the grain size and the model size and cited the first paper investigating this effect (Ovesen 1979), for which loading tests were performed on circular foundations at different scales. He confirmed that for bearing capacity problems, if the ratio between the diameter of the foundation and the grain size was larger than 30, there were no size effects, probably because the footing was resting on more than 1000 grains. He cited his recent work on size effects for shear interfaces (Garnier 2002), and declared that the critical scaling ratio increased to 100 for shear interfaces. Springman recalled that work by Phillips & Valsangkar (1987) on scaling effects related to the cone penetrometer had confirmed that a ratio of cone diameter to particle size should exceed 30 for the cone penetration event. She added another example of soil structure interaction for which this factor was even higher (Kutter et al. 1994), for the collapse of a cavity above a trapdoor. She had noted that the shape of the cavity varied quite significantly for lower ratios of trapdoor diameter to particle size and that it was only when the ratio exceeded 350, did the cavity look remotely the same. She was not sure whether the change in gravity with depth of the model, or any other factor, may have influenced the result, given the two extremes of centrifuge size available at Davis in California, where this work had been carried out.

Cino Viggiani was still not satisfied with the answers related to grain size and scaling with respect to the problem of failure. He returned to the issue of localised failure, which he acknowledged was only one of the possible modes of failure, and asked about extracting relevant information concerning the failure of such an experimental system if the internal length of the material could not be scaled. Bolton acknowledged that for problems of rupture, for which shear band formation dominated, the scaling of displacements was very complicated. He described recent tests at Cambridge for which a full scale pipe (diameter of 200 mm) was forced to displace upwards through a sand or gravel or even a clay layer to represent a real oil pipe bursting through soil cover. He explained that centrifuge models were also made at a 1/20th scale to model the same situation in the full-scale tank, and that these were tested at 20g. Despite checking three different soil particle sizes with respect to the pipe diameter, he said that the centrifuge model did not exactly reproduce the uplift data of normalised force against displacement. The small-scale models showed a higher initial peak followed by a reduction to the same normalised ultimate resistance. He noted that this was indicative of progressive failure with a length-scale that made small models less progressive than the full-scale. The displacement to generate peak uplift force could not be normalised by particle size; no simple scaling would permit the direct transfer from the small-scale model to the full-scale data. He admitted that just taking the scale ratio for displacements, as was usually done in centrifuge modelling, produced a clear error of a factor of 3 or more in the prediction of the movement of the full-scale pipe necessary to create peak resistance against uplift. He concluded that this was clearly an interesting research topic and that a centrifuge offered the best way of researching it, given the relationship to a boundary value problem and the local and global material response, which was almost impossible to study at full scale. He did not feel that a finite element analysis offered a solution either, until more data from centrifuge modelling and semi-full-scale testing could be provided (for constitutive model development).

Craig addressed two aspects of the problem, recalling work by Andrew Wright (1979), who had investigated the plane strain flow of material in silos, which was a challenge at the best of times! He looked at the difference between the grain size and the stress level effect in a model silo, which were sometimes confused or combined into a single effect. He had found that if he used a particular sand for a given silo model size (diameters of 100 to 150 mm), he would get a different flow pattern within the silo, depending upon the stress level at which he was modelling. He had also observed a radically different flow pattern, even for sand with a grain size of ~200 microns, depending upon whether it was nominally dry sand (i.e. with an atmospheric moisture content of ~1%) or it was oven dried immediately before the test. He had also scaled down the particle size, using a similar approach to Lau, and had replaced the silica sand with a rock flour, which had been milled down to a smaller grain size. He found a whole range of flow patterns at different stress levels, because this scaled down material was sufficiently fine to be almost cohesive when it was packed into a silo, rather like soil beneath the tip of a driven pile or pills, which were stamped in a pharmaceutical factory. He described the arching mechanism of the quasi-cohesive very fine material at low stress levels, which prevented any flow at all, as well as the rat hole flow that developed at a higher stress level and the mass flow regime that opened up at an even higher stress level.

Laue commented that the particle size effect *per se* was more of an interpretational problem in that the effects should be divided into three areas:

- relative size (as the difference between the size of the structure and the soil particles)
- roughness
- particle size (as the need for scaling of the soil particle size),

which were valid for any type of physical modelling including laboratory tests, and he called for a more specific description in future.

Diethard König explained how the size effect has been dealt with in terms of an element test conducted on a triaxial sample (Fig. 1a), for which different sizes of sample should produce results that would not depend upon the relationship between the grain size and the geometry of the sample (Fig. 1c). He noted that if the same soil had been used, the only variation should be the natural scattering from test to test and claimed similarity in the centrifuge in terms of the strong

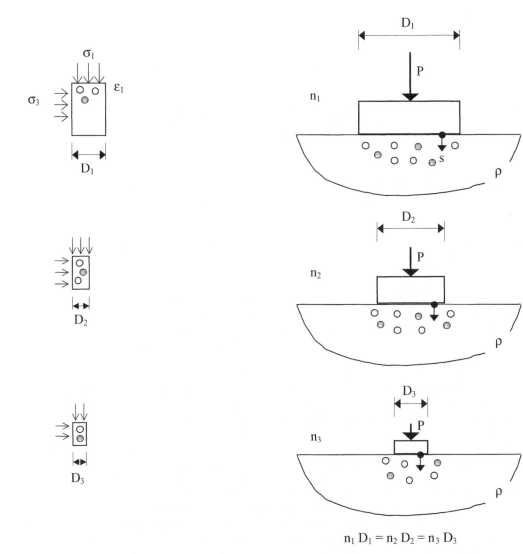

$$n_1\, D_1 = n_2\, D_2 = n_3\, D_3$$

(a) Triaxial samples of different diameters   (b) Three different geometries of footing tests

Figure 1. Verification of models (König & Laue 2001)

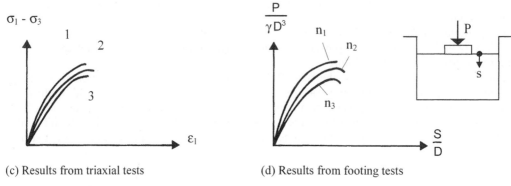

(c) Results from triaxial tests   (d) Results from footing tests

Figure 1. (Continued)

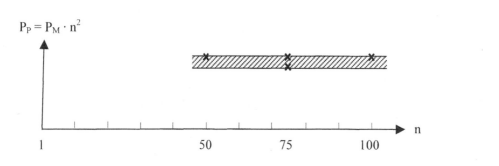

Figure 2. Results from footing tests for model cases converted to the prototype (König & Laue 2001)

box and a footing on the sand surface. He pointed out that varying the relationship between the geometry of the footing and the particle size could be investigated by using the same soil, but a different diameter of footing and the g-level, so that the prototype modelled remained the same (Fig. 1b). He expected that if the product of g-level multiplied by the diameter of the footing was always the same, the results should be plotted in terms of prototype dimensions or in dimensionless form (Fig. 1d), and if they were not identical (apart from some minor scatter), there would be a size effect (Fig. 2).

König also mentioned the example of a layer of soil above a trapdoor (Fig. 3, Stone & Muir Wood 1992, White et al. 1994, Viswanadham et al. 2001), which required the grain size to be scaled down because a discontinuity was introduced into the system as the trapdoor was lowered and shear bands formed in the soil layer. He said that pattern of shear bands was very dependent on g-level and grain size and that the grading curve (Fig. 3d) of the model soil, the g-level and the trapdoor diameters should all be scaled accordingly to observe the same pattern of shear bands (Figs 3a, c cf. b). These figures show the same prototype situation so that $n_1 < n_2$. He concluded that these verification of model tests were a strong measure of whether the test procedure was acceptable or not, because not only the grain size but also boundary conditions and interface problems would be integrated within the judgement of whether the modelling was effective. Laue added that most model tests using a sand are usually conducted with a fine graded fraction, which will still behave like sand when scaled up according to the scaling law relationships in terms of hardness of the particles, which is an important factor in terms of crushing and dilatancy. He concluded that the words *'particle size effect'* were a distraction and that this should be defined more specifically in each case and then dealt with accordingly.

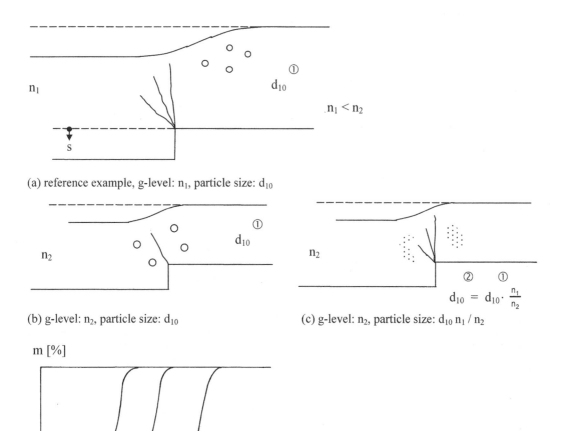

(a) reference example, g-level: n_1, particle size: d_{10}

(b) g-level: n_2, particle size: d_{10}

(c) g-level: n_2, particle size: $d_{10} n_1 / n_2$

(d) scaled grading curves

Figure 3. Trapdoor experiment (König & Laue 2001 after Stone & Muir Wood 1992, White et al. 1994 and others)

Sulem agreed and pointed out that there were two important lengths:
- the engineering structure length, which was the dominant length scale of the problem, and
- the fabric length of the material, i.e. for the relevant soil or the rock in terms of grain size or some dominant joint spacing.

He confirmed that when the ratio of these two dimensions was large, then the results would be acceptable, but that close to a localised failure mechanism there would be a change in the scale of the problem. He added that the thickness of the shear band should be compared to the material length and that this explained why some information about the fabric length was now being introduced into constitutive models to deal with localisation, even though the difficulty was to calibrate this internal length. He noted that one method of calibrating this internal length was through scale effect assessment and asked if there was a centrifuge test in which a precise measurement of scale effects was feasible in order to calibrate the relevant internal length. Laue thought that it was possible go in the opposite direction from trying to get the same results for the same problem, for example by using larger grains, although his focus was generally to have a really clearly defined system for a practical problem. Herle countered that this was only true for this one single particular model, the prototype, and that the scale effect was strongly dependent on the system configuration.

Muir Wood concluded that it was necessary to treat each case at face value and to think, from the micro-mechanics point of view, about what sources of possible size effects might exist in any modelling situation. He mentioned surface roughness, particle sizes or aspects to do with localisation or sizes of the object or sizes of the container, and remarked that there were obviously many ways in which dimensions would come into the modelling performed.

REFERENCES

Garnier, J. 2002. Size effects in shear interfaces. In S.M. Springman (ed.) *Workshop on Constitutive and Centrifuge Modelling: Two Extremes, Monte Verità.* Lisse: Swets & Zeitlinger.

Garnier, J., Ternet, O., Cottineau, L.-M. & Brown, C.J. 1999. Placement of embedded pressure cells. *Géotechnique* 49(3): 405-414.

König, D. & Laue, J. 2001. Discussion on "Particle Size". *Workshop on Constitutive and Centrifuge Modelling, Monte Verità: Two Extremes, Ticino, July 2001.*

Kutter, B.L., Chang, J.-D. & Davis, B.C. 1994. Collapse of cavities in sand and particle size effects. In Leung, C.F., Lee, F.H. & Tan, T.S. (eds) *International Conference Centrifuge 94, Singapore*: 809-816. Rotterdam: Balkema.

Lau, C.K. 1988. Scale effects in tests on footings. PhD thesis. Cambridge University.

Lynch, R.J. & Bolton, M.D. 1996. A low cost fibre optic pH sensor for the in-situ measurement of groundwater. *Proceedings of the International Society of Optical Engineering, (SPIE), 2835 Advanced Technologies for Environmental Monitoring and Remediation, Denver, Colorado*: 221-227.

Ovesen, N.K. 1979. The use of physical models in design. *VII ECSMFE, Brighton* (4): 319-323.

Phillips, R. & Valsangkar, A. 1987. An experimental investigation of factors affecting penetration resistance in granular soils in centrifuge modelling. *CUED/D TR210.*

Rezzoug, A., König, D. & Triantafyllidis, Th. 2000. Scaling laws in centrifuge modelling for capillary rise in soils. In J. Garnier, L. Thorel & E. Haza (eds) *International Symposium on Physical Modelling and Testing in Environmental Geotechnics, La Baule, France*: 217-224. Paris: LCPC.

Stone, K.J.L. & Muir Wood, D. 1992. Effects of dilatancy and particle size observed in model tests on sand. *Soils and Foundations* 32(4): 43-47.

Viswanadham, B.V.S., König, D. & Jessberger H.L. 2001. Discussion: Dimensional analysis for geotechnical engineers. *Géotechnique* 51(1): 91 – 93.

White, R.J., Stone, K.J.L. & Jewell, R.J. 1994. Effect of particle size on localisation development in model tests on sand. In C.F. Leung, F.H. Lee & E.T. Tan (eds) *Proc. Int. Conf. Centrifuge 94, Singapore*: 817-822.

Wright, A. 1979. Silos - model and field studies. PhD thesis. University of Manchester.

Constitutive and Centrifuge Modelling: Two Extremes, Springman (ed.)
© *2002 Swets & Zeitlinger, Lisse, ISBN 90 5809 361 1*

The philosophy of modelling versus testing

F.H. Lee
National University of Singapore, Singapore

ABSTRACT: This paper explores the philosophy of modelling and testing. It is argued herein that the objective of modelling is to obtain a prediction, whereas that of testing is to obtain a validation. Through case histories of model studies, it is demonstrated that the requirements of of a model are not always fully achievable. The important thing is for the modeller to know how the simplification he/she is making will affect his/her results. On the other hand, a test need not model a specific or even an idealised prototype; as long as it is definitive. Centrifuge and numerical modelling may be regarded as two complementary as well as competitive approaches to making predictions for geotechnical problems. Numerical modelling with good laboratory and in-situ testing are probably better in defining in-situ state and conditions of the ground, whereas the remoulding and reconsolidation processes involved in preparing centrifuge model may cause the in-situ state of the model to be rather different from that of the prototype. On the other hand, many problems involving very large deformation, break-up of material zones as well as moving and changing interfaces are likely to be much more readily solvable by a centrifuge modelling rather than a numerical modelling approach.

1 INTRODUCTION

Modelling and testing are commonly used in geotechnical research and engineering, and indeed in many scientific and engineering pursuits. They have quite different objectives and functions, and not surprisingly, philosophies. In this paper, the roles and philosophies of geotechnical centrifuging as a model and a test are first examined, and hopefully, clarified.

Figure 1 shows a construct which highlights the relationship and differences between a model and a test. For the purpose of our discussion, a model will be defined as a representation of an object of study, which we have often termed as a prototype. The latter may either be an idealised object or a real, specific object. The aim of constructing a model is to use it to make qualitative or quantitative predictions regarding modes of behaviour of the prototype. Many things that geotechnical engineers and researchers do, fall under this definition of model. In the simplest instance, a geotechnical engineer designing a foundation for a building sketches out an idealised representation of the foundation and the underlying soil. Through the application of some design formulae or charts, he/she then arrives at a prediction of the bearing capacity or settlement of the foundation. The idealised representation to which he applies the formulae or design charts is thus his/her model. A test, on the other hand, is a means of establishing the

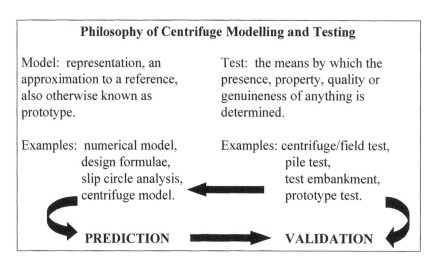

Philosophy of Centrifuge Modelling and Testing

Model: representation, an approximation to a reference, also otherwise known as prototype.

Test: the means by which the presence, property, quality or genuineness of anything is determined.

Examples: numerical model,
design formulae,
slip circle analysis,
centrifuge model.

Examples: centrifuge/field test,
pile test,
test embankment,
prototype test.

PREDICTION　　　　　**VALIDATION**

Figure 1. A construct of modelling and testing

property, validity or genuineness of a hypothesis, claim or prediction, the latter often arising from a model. For instance, during the construction of the foundation, the client or engineer would often check the validity of the design bearing capacity predicted by the engineer's model using a load test. Thus, a test can be considered the complement of a model; whereas the latter makes a prediction, the former validates or invalidates it. These two opposing ends of the construct set the framework for the subsequent discussion on their philosophies.

2 PHILOSOPHY OF MODELLING

We shall use the example of a simple geotechnical design problem of, say, a foundation, to illustrate some aspects of modelling philosophy. As mentioned earlier, we first construct an idealised representation of the problem, that is a model, in this case the foundation and the underlying soil. The fact that we need to idealise the problem is an acknowledgement that the model or representation will not reproduce all the features of the prototype, such as variable soil profiles and topography which may not be horizontal, local heterogeneity such as sand and clay lenses, effects of nearby foundations. In constructing the model, the engineer will naturally seek to reproduce all those prototype features that he/she can reproduce, particularly those which have significant effects on the prediction; this attribute will hereafter be referred to as accuracy.

Apart from fidelity, the engineer will often also want to ensure that the prediction of his/her model will err on the safe side. In other words, he/she will often try to get as close as possible to the prototype result, without exceeding it. This highlights another important attribute of modelling; that is boundedness.

The typical requirements of a model can be illustrated in a ground vibration research project which the Author was commissioned to undertake by the Ministry of Defence (MINDEF). The project was motivated by the need to build ammunition storage facilities underground in order to free land for other developments. However, all such facilities, whether above- or underground, can only be constructed if the land around it is not inhabited. Thus, an estimate needs to be made of the inhabited building distance (IBD) associated with building such facilities underground. In deep underground ammunition storage facilities, the IBD is often governed by the ground vibration that will be generated in the event of an accidental explosion, and the effect that such vibrations will have on structures and buildings. The engineers from MINDEF felt that the empirical design guidelines set by NATO (1992) and the US Department of Defense Explosive Safety Board (DDESB 1996) were overly conservative and there was room for further trimming of the IBD. In other words, an improvement in the model was felt to be necessary. The objective of the study was to set up an improved model, with the hope of reducing the IBD. Several modelling approaches were considered, including centrifuge modelling, 1g-modelling and numerical modelling. In the event, experimental modelling was abandoned as the underground chambers were to be constructed in heavily jointed granite rock, which cannot be modelled readily at a reduced scale. In addition, the extent of ground which needs to be covered would measure approximately 300 m x 300 m, with depth of at least 150 m. Even at the highest g-level achievable on the NUS Geotechnical Centrifuge, this would require a model which is too large to be accommodated on the swing platform. In the end, it was decided that numerical modelling would be used, with a reduced scale explosion in the field.

One major issue concerned the representation of the effect of the rock joints on the ground waves. The rock joints are too numerous to be modelled individually, so that their effects would have to be smeared out over the entire domain. A workshop was held in Singapore in 1998, with participation from MINDEF, Nanyang Technological University (NTU), NUS, Norwegian University of Science and Technology (NTNU) and the Norwegian Geotechnical Institute (NGI). The results of the presentations appeared to suggest that there is likely to be some reflection of waves occurring at the joints, and possibly some distortion to the tensile phase of the waves, if the joints go into tension. However, this seems unlikely to occur in the far-field region, where the interest lies, since the stresses will be far lower than the geostatic stress levels. Determining a set of representative properties is likely to be difficult, since rock material properties may not be representative of rock mass properties. To resolve this, it was decided that the properties of the rock mass be evaluated by back-calibration against a reduced scale explosion in a much smaller chamber.

Figure 2 shows the layout of the calibration test. The depth and dimensions of the charge chamber as well as the height of the compacted earth bund was scaled down from the depth and dimension of the prototype chamber and thickness of prototype soil cover, as the cube root of the charge weight ratios i.e.

$$\frac{l_m}{l_p} = \left(\frac{W_m}{W_p}\right)^{1/3} \tag{1}$$

in which the l is the linear dimension to be scaled, W the charge weight, and the subscripts m and p represents model and prototype quantities. The rationale of the exponent 1/3 is to preserve the energy density of the geological medium.

The back-calibration exercise involves finding (if possible) a set of material properties which best matches the ground response. Figure 3 shows the two-dimensional axisymmetric finite element mesh used for this purpose. A Mohr-Coulomb model was chosen for the rock and a non-linear, hysteretic model for the soil. Material damping was represented by a Rayleigh damping model. The first quantity that was matched was the peak particle velocity (PPV) in the rock, which is widely used in explosive and earthquake work. The primary rationale for choosing this quantity first is the notion that it is only dependent upon the maximum kinetic energy density which propagated through the soil medium. Figure 4 shows the measured and matched PPV at various locations. It turned out to be relatively easy to get a good match on this quantity; in fact, no reduction was required on the Young's modulus of the rock material, which suggests

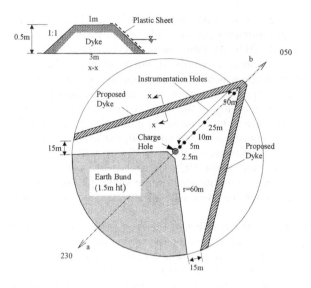

(a) Plan view and instrumentation layout

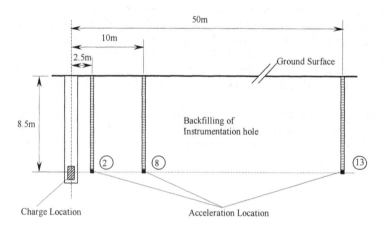

(b) Test and instrumentation layout in rock stratum along Section ab

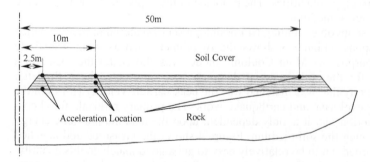

(c) Test and instrumentation layout in earth bund along Section ab

Figure 2. Ground vibration field test: underground ammunition storage facility

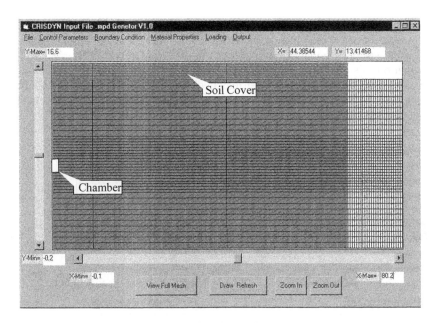

Figure 3. Finite element mesh for calibration analysis

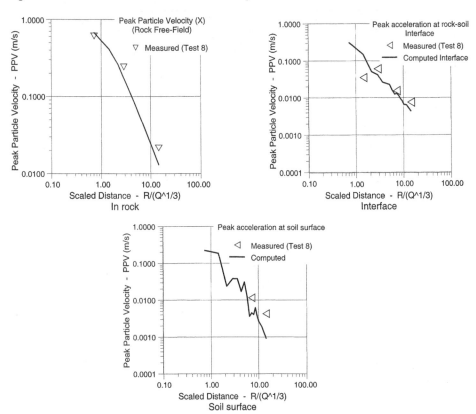

Figure 4. Plot of PPV against scaled distance (Charge weight 41.6 kg)

that the effect of the rock joints on the modulus may not be significant. The next quantity to be matched was the response spectrum, which is a measure of how the energy of the ground vibration was distributed over the various frequency components. As expected, it was much more difficult to get a good match on this quantity, as shown in Figure 5, and significant discrepancies still persist at some locations even with a set of "best matched" parameters. Three-dimensional finite element analyses using the mesh in Figure 6 were also conducted to obtain a corresponding calibration for subsequent prediction.

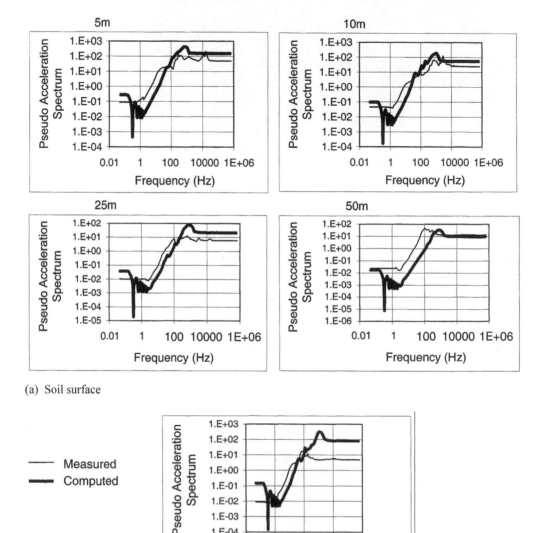

(a) Soil surface

(b) Mid-depth soil layer (10 m from charge)

Figure 5. Computed and measured acceleration response spectra at various distances from charge (Charge weight 41.6 kg ——— measured ■■■ computed)

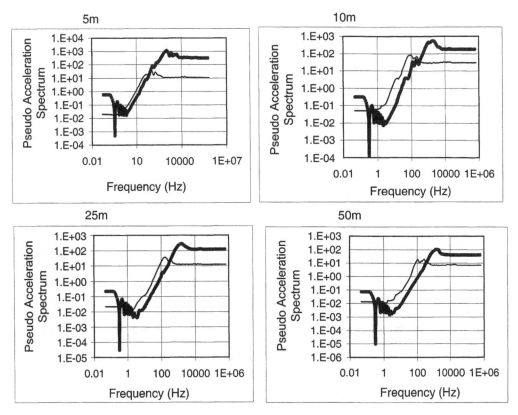

(c) Rock – soil interface

Figure 5. (Continued).

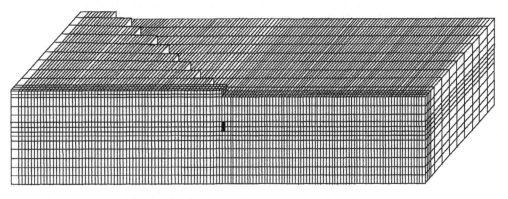

Figure 6. Three-dimensional finite element mesh for calibration test

In applying the analyses to the actual problem, a set of "safe" parameters were finally chosen. Each of the prototype storage chambers is roughly cylindrical in shape, with the longitudinal axis oriented horizontally. The length and radius of each chamber are about 100 m and 10 m, respectively. The cross-section of the chamber is more horse-shoe in shape than circular. Thus, the problem is actually three-dimensional in geometry. Most of the predictions were made using two-dimensional axisymmetric analyses, with some three-dimensional analyses for benchmarking purposes. Axisymmetric analysis was preferred over plane-strain analysis because of the long cylindrical shape of the chamber. A plane strain analysis would have precluded spreading of waves in the out-of-plane direction, that is, along the tunnel axis. This assumption would have been valid if the wavelengths of the dominant frequency components are much shorter than the length of the tunnel. However, the compression wave speed in the granitic rock at this site is about 5700 m/s. This implies that waves with a wavelength equal to the tunnel length will have a frequency of 57 Hz. All frequency components below this frequency will therefore have a wavelength longer than the tunnel length and thus diffract significantly around the ends of the tunnel. To account for this spreading effect, an axisymmetric analysis was preferred over a plane strain analysis. It will be noted that the axisymmetric analysis will over-estimate the amount of spreading for frequency components above 57 Hz, and thereby underestimate the PPV, which is quite dependent upon the high frequency components. This violates the attribute of lower-boundedness. However, the resulting ground motion is to be used as input into some dynamic structural analysis or response spectra analysis. Since structures do not generally respond strongly to frequencies higher than 57 Hz, even for vertical ground excitation, the underestimation of the higher frequency components was felt to be acceptable. In adapting the geometry from a horizontal cylinder to a vertical cylinder, the volume of the chamber was preserved. In order to convert the blast over-pressure on the walls of the chamber from the prototype geometry to model geometry, the peak horizontal and vertical force acting on each quadrant of the chamber was conserved.

Figure 7 compares the PPV attenuation plot from the three- and two-dimensional analyses. As expected, the two-dimensional model predicts slightly lower PPV values for given standoffs than the three-dimensional analyses. In particular, the three-dimensional analysis predicts a slight "hump" in the attenuation curve at a scaled distance of about 1.3. On the other hand, for the two-dimensional analysis, the "hump" was much less distinct and occurs at a smaller scaled distance, the latter being attributable to the fact that the idealised two-dimensional model chamber has a much smaller horizontal extent than the prototype. To investigate the effect of topography on the attenuation characteristics of the ground vibration, analyses were also conducted with several actual ground topographies.

Although the project is now completed, questions remain on the validity of using a small-scale event to back-calibrate rock mass properties which are to be used in much larger-scale event. The linear scaling ratio between the calibration and the designed event is about 15.9 and the energy scaling ratio is about 4000. Discussions are now underway with NATO to conduct a somewhat larger event (but still not full-scale) with an energy scaling ratio of about 25 and a linear scaling ratio of about 2.9. It is, however, unlikely that a full-scale prototype test will ever be conducted, deliberately.

This exercise in modelling a specific site problem illustrated several issues. The first issue is that some idealisation is almost inevitable in dealing with such complex problems. In the process of idealising the problem, trade-offs have to be made which will affect the accuracy and boundedness of the results. In our discussions with MINDEF engineers, it emerges that two sides do not exactly share the same perspective. As researchers, we were more concerned about the accuracy of the results and were perhaps more inclined to sacrifice the boundedness attribute of the results. On the other hand, MINDEF engineers appeared to be more willing to compromise on the accuracy of the results but were much more keen to ensure that the results err on the safe side. One of the lessons that we learnt here is that getting the correct answer is not necessarily always a paramount requirement. Many problems still exist, which are too complex to be modelled in their entirety and some idealisation is often inevitable. This applies to any modelling approach. The question is whether one can turn what is apparently a deficiency into an advantage.

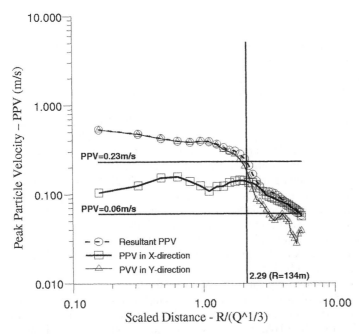

(a) 3D model

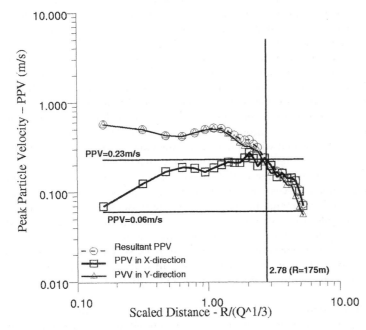

(b) 2D axisymmetric model

Figure 7. PPV against scaled distance for 250-tonne charge weight (Q denotes charge weight)

Another lesson that we learnt is the importance of sensitivity studies, which results directly from the fact that modelling of real problem almost invariably requires simplification. This may be more easily done in a numerical model than in an experimental model, but instances of such experimental sensitivity studies do exist. For instance, Rowe & Craig (1976), in their study of the Oosterschelde caisson, conducted sensitivity studies which showed that the presence of localized pockets of loose sands in the foundation has a very significant effect on the settlement of the caisson.

Modelling may also be used to predict the behaviour of an idealised prototype or prototype situation. This is commonly the approach employed to generate design formulae or charts. Some good examples of this approach include Holsapple & Schmidt's (1979, 1980) model study on explosive cratering and Neubecker & Randolph's (1994) study on drag anchors. For man-made events, it may still be feasible to validate model prediction by conducting large scale prototype tests. For natural events, such testing may be more difficult. Nonetheless, some indirect way of counterchecking the model prediction, if available, will help to increase confidence in the use of the results. One example of such a natural event is an earthquake. Although Singapore is located at least 350 km away from the nearest seismically active zone, it still occasionally experiences small earthquakes arising from the Sunda Arc. Anecdotal evidence suggests that such earthquakes are often most clearly felt by occupants of high-rise buildings constructed over soft marine clay strata. A study was initiated to investigate the amplification which results from soft soil foundations. Experience from the Mexican 1985 earthquake would suggest an amplification of about 3 to 4 on the peak horizontal acceleration (PHA). On the other hand, Sabetta & Pugliese's (1996) statistical study of field data concluded that the PHA, as well as peak spectra, amplification is about 1.7. However, it is difficult to compare the Mexican experience with Sabetta & Pugliese's (1996) findings since the geological setting and earthquake characteristics may be very different.

The study was conducted using centrifuge modelling, supported by field data and numerical analysis. Initial centrifuge models were tested using a laminar box seated on top of a shaking table actuated by a simple cocked-spring mechanism (Niu 1997). This mechanism uses the ringdown of four pre-compressed springs to generate the earthquake. Owing to the limitations of the cocked-spring mechanism, typically fairly large earthquakes of about 20% to 30% of model gravity are generated. Furthermore, the earthquake motion is highly harmonic, which translates into a very narrow frequency bandwidth, that is unsuitable for amplification studies. This was partially solved by increasing the friction at the sliding base of the shaking table to introduce some high-frequency components into the motion. However, the frequency spectrum remains highly uneven and some care was needed when interpreting the data. A series of tests were conducted using different thicknesses of remoulded marine clay, which allows response spectra amplification curves to be obtained for different thicknesses of soft clay layers. The divergence between the different amplification curves was significantly reduced by normalizing the actual amplification by the peak amplification and the frequency by the frequency of the peak. This allows the spectra amplification plot to be characterized by two parameters, the peak amplification and the frequency of this peak. Figures 8a and b show the variation of peak amplification and frequency with thickness of the marine clay stratum. The peak spectra amplification varies significantly with thickness of soil layer. Nonetheless, for soft clay layers with thickness exceeding 10 m, the amplification is roughly 2, which is in general agreement with Sabetta & Pugliese's (1996) findings.

However, it was felt that the findings of this initial study might not be applicable to the Singapore scenario since the PHA at the bedrock in Singapore is unlikely to exceed 2% in the worst case; this emerged from a study of ground motion data in Singapore. To enable much smaller earthquakes to be studied, a new shaking table, shown in Figures 9a and b, was developed based on electro-servohydraulic actuation, with a typical result given in Figure 10. As shown in Figure 11a, much higher peak spectral amplification factor of about 6 is now obtained.

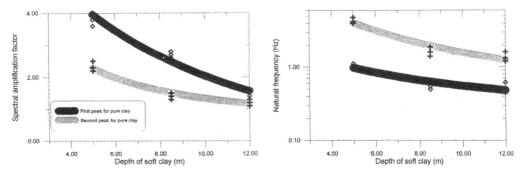

(a) Spectral amplification factor   (b) Frequency at peak amplification

Figure 8. Amplification characteristics for different soft clay layer thicknesses

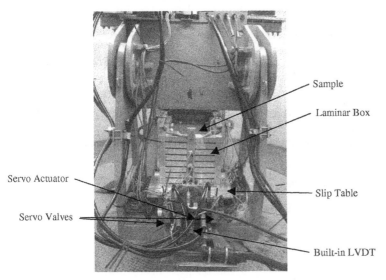

(a) Shaking table

(b) Earthquake accessories

Figure 9. Shaking table and accessories

123

Small Earthquake

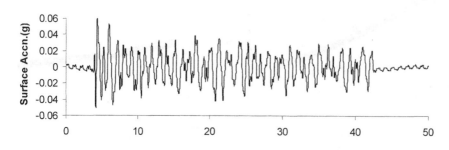

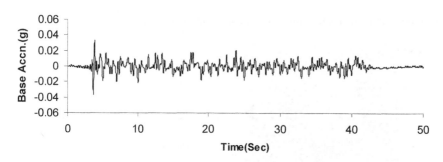

Figure 10. Acceleration in a small earthquake. Prototype thickness of soft clay layer: 12 m

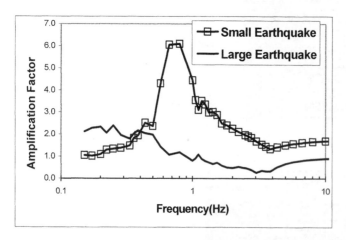

Figure 11. Response spectra amplification curves from model test. Prototype thickness of soft clay layer: 12 m

The next question is whether this result is applicable to the field situation, given the fact that we used remoulded marine clay in the centrifuge experiments, as opposed to the in-situ ground. However, Tan et al.'s (2001) comparison of the shear modulus of disturbed, re-consolidated Singapore marine clay samples and that of in-situ marine clay suggested that much of the in-situ shear modulus can be recovered if the disturbed samples are re-consolidated back to in-situ stress levels. This suggests that the errors arising from disturbance, and presumably remoulding, may not be so large as to render the experiments meaningless. We also counterchecked the centrifuge results with those obtained from the local seismological stations. The field studies were conducted and reported in an undergraduate BEng dissertation (Cheng 2001) after the centrifuge studies were completed. Figure 12 shows the seismological stations which were installed on Singapore island in November 1996. Of these stations, the Beatty School station (marked BES) and the Katong Park station (marked KAP) are located in 18 m and 28 m-thick marine clay strata, respectively. The remaining stations are located in rock or stiff soil.

Using ground motions from 43 small earthquake records, Cheng (2001) first tried to fit a common attenuation relation to all stations, without accounting for soft soil effects. He used a well-tried form, in which

$$Log(PHA) = a + bM + cR - d \log(R) \tag{2}$$

in which M is the magnitude, R the epicentral distance, and a, b, c and d are fitted coefficients. The best attempt in fitting data to Equation 2 yielded an R^2-value of only 0.51. More importantly, as shown in Figure 13, Equation 2 consistently over-estimates the PHA on the stiff soil sites and underestimates the PHA on the soft soil sites. When a binary site geology term S is included into Equation 2, the R^2 value increases to 0.8, and the resulting relationship predicts an amplification factor of about 6.3 for the PHA.

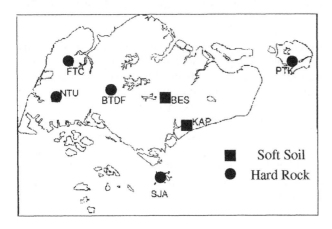

Figure 12. Seismological stations on Singapore island

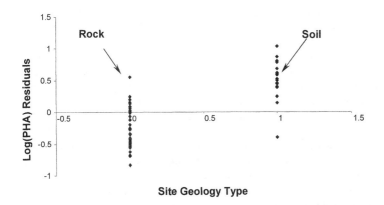

Figure 13. Residuals of Equation 2 with respect to site geology type

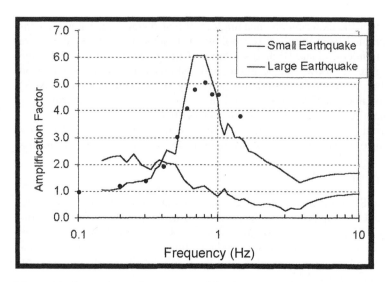

Figure 14. Comparison of spectra amplification curves from centrifuge model and field data (black circles)

Cheng (2001) also performed similar regression analyses for selected frequency components within the reliable signal band. His results are plotted together with the centrifuge data in Figure 14. The comparison is not definitive since the field data were obtained from soil strata thickness of 18 m and 28 m, respectively, whereas the centrifuge model data were obtained from a model which has a prototype thickness of about 12 m. Nonetheless, it is encouraging that the field data show a similar order of amplification as the centrifuge model. In this project, we were fortunate to have encountered a problem which does not present much difficulty in modelling. In many other instances, problems of boundary conditions, soil types and in-situ states could have easily arisen. This happened in another study in which the laminar box was used to house a model of a caisson-retained landfill. In this case, the boundary conditions imposed by the laminar box on the sides of the plane strain caisson model were evidently incorrect

3 PHILOSOPHY OF TESTING

Since the objective of a test is to validate rather than predict, the requirements of representing a prototype may be relaxed considerably. On the other hand, a test should be definitive in whatever it tests, whether it is a hypothesis (or conjecture), prediction or specified performance. One example of such a testing programme in which the Author is involved is the post-installation behaviour of sand compaction piles (SCPs) (e.g. Lee et al. 1996, Ng et al. 1998 and Lee et al. 2001). Sand compaction piles (SCPs), is the terminology commonly used to refer to a method of ground improvement in which sand columns are formed by forcibly injecting sand into the in-situ soft clay via a casing (Aboshi et al. 1979). In the field, SCPs are installed by driving a cylindrical casing containing sand into the soft clay to the desired depth, using a vertical vibratory hammer. The casing is then withdrawn for about 2 to 3 m and then redriven down by about 1 to 2 m as shown in Figure 15. By repeating the withdrawal-redriving sequence until the casing is fully withdrawn, a fully compacted sand pile, with a diameter that is often significantly larger than that of the casing, is obtained. Owing to the highly automated sequence of operations, SCPs have been used extensively in Japan and Singapore for rapid stabilization of the soft sea-bed soil in land reclamation works (e.g. Aboshi & Suematsu 1985).

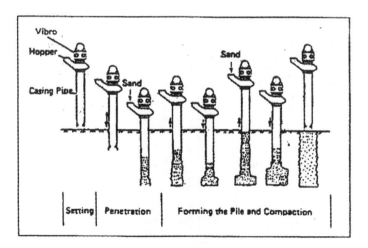

Figure 15. SCP installation in the field

Since soil is not removed from the ground prior to SCP installation, the insertion of the casing into the ground and the subsequent injection of sand will lead to a cavity-expansion type of displacement, and thereby some "set-up" in the strength of the soft soil. There has been some evidence that the set-up effects of SCP installation are significant. However, direct comparative evidence was not available and existing design is still based on the in-situ strength and state of soft clay and deposited sand. A centrifuge testing programme was therefore conducted to investigate the set-up and its effects on the behaviour of the improved ground.

The initial conjecture to be tested was "that set-up in the soft clay and sand piles has a significant effect on pile grid behaviour". To do this, two methods of installing SCPs were compared using centrifuge tests. The first is the "frozen pile" method (e.g. Kimura et al. 1985, Terashi et al. 1990 and Kitazume et al. 1996), in which SCPs are modelled by inserting miniature frozen saturated sand columns into pre-drilled holes in the model clay beds, under 1g. The second method involves using an apparatus to inject sand into the soft clay bed under high-g conditions, in order to represent the field installation procedures more closely (Ng et al. 1998). Sand is injected into the clay bed from a sand feed hopper via a steel casing by a hydraulically-rotated screw. The hopper-casing assembly is first injected into the clay bed. The casing is then withdrawn from the clay bed as the sand is injected. In order to allow pile grids to be installed in-flight, the casing, feed hopper and hydraulic cylinder are carried on a stepper-motor-driven carriage, which allows the casing to be moved into the desired location in-flight, Figure 16.

In order to study the post-installation state of the ground, the response of the ground to an idealised "undrained" loading was observed. This "undrained" loading was imposed by constructing a sand embankment using sand pluviation from a height of about 300 mm above the improved ground at 1g, and then rapidly increasing the g-level of the model to induce ground movement and failure. Evidently, such a build-up of g-level on an already constructed embankment model does not represent a realistic loading scenario. Its aim is to enable a definitive comparison of the behaviour of the two models with ground improvement, not to simulate a realistic load; this illustrates an important difference between modelling and testing. For this purpose, it has the advantage that the loading on the clay bed is only dependent upon the g-level, which can be accurately controlled. It would be more difficult to achieve the same degree of repeatability of loading increment with in-flight pluviation. The centrifugal acceleration was increased in steps of 10g. At each increment, the g-level was maintained for approximately 2½ minutes. As the build-up of g-level is very rapid in comparison to the time required for pore pressure to dissipate, the loading process is virtually undrained.

(a) Casing and sand hopper   (b) SCP model on centrifuge

Figure 16. In-flight SCP installer

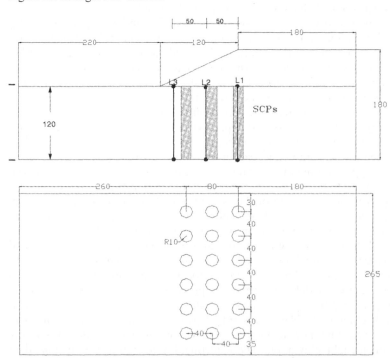

Figure 17. Side and plan views of an SCP model. Vertical lines L1, L2 and L3 are vertical grid lines pre-painted on the side of the model, next to the window. Dimensions in mm. SCP diameter 20 mm

128

The results of this comparison exercise have been presented and discussed by Lee et al. (2001). Figure 17 shows the side and plan views of one of the models studied, with locations of the SCPs and three vertical grid lines indicated. The variation of maximum lateral displacement of one of the grid lines, L2, with g-level is shown in Figure 18. The enhancement in stiffness of the improved ground with in-flight SCP installation seems fairly evident.

The next issue that was addressed was the increase in lateral stress in the ground during SCP installation. The conjecture here was "that build-up of lateral stress in the ground during SCP in-stallation is significant" (and therefore, presumably, accounts for the stiffness of the in-flight SCP model). Although Asaoka et al. (1994) tried to model this event numerically, certain salient features in the process were not well-represented. Most important of these is the fact that Asaoka et al. (1994) modelled the installation of the SCPs as a simultaneous expanding of a cavity in the soft clay. This neither replicates the penetration of the casing, which is akin to a pile being jacked into the ground, nor the sand injection process. To investigate this, total stress and pore pressure transducers were pre-embedded at pre-designated locations in model clay beds where SCPs were to be installed in-flight. This was done for a single SCP (P1) as well as for SCP groups. Figure 19a shows the transducer locations (TS) of one such test while Figure 19b shows the measured

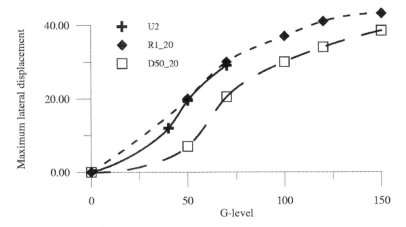

Figure 18. Maximum lateral displacement for three models. U2 is an unimproved model, R1_20 is a "fro-zen pile" model and D50_20 is an in-flight SCP model

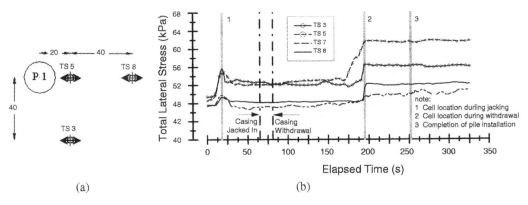

(a)   (b)

Figure 19. (a) Location of transducers and (b) changes in total lateral stresses during SCP installation in a 70g-centrifuge test. Depth of transducers: 70 mm

129

increase in total stresses. As can be seen, the build-up of lateral stress in the vicinity of a sand pile varies from about 8% of the in-situ stress at a standoff of 3 times the sand pile diameter to about 20% at a standoff of 1 times the sand pile diameter. In the field, SCPS are often installed at centre-to-centre spacings of less than 2 times the diameter. Thus, set-up effects in the clay are likely to be significant. Subsequent tests also showed that the increase in stress in the sand is much higher than that in the clay, thus indicating that the set-up also affects the properties and behaviour of the sand. Besides this, tests also showed that the set-up is further accentuated when piles are installed in a group, as is normally the case.

The above example illustrates some of the trade-offs which often occur in centrifuge experiments. The objective of the investigation is to validate (or disprove) a conjecture relating to set-up effects in SCPs in the field. For this reason, a method which can realistically replicate the process of SCP installation in the field was developed. On the other hand, realism is sacrificed in the loading method in favour of repeatability and control; so as to yield a definitive comparison.

4 CONCLUSIONS

Centrifuge modelling and numerical modelling may be regarded as complementary as well as competitive approaches to the prediction of prototype behaviour. Each has its own strength and weaknesses. Numerical modelling with good laboratory and in-situ testing are probably better in defining in-situ state and conditions of a ground whereas the remoulding and reconsolidation processes involved in preparing a centrifuge model may cause the in-situ state of the model to be rather different from that of the prototype. It might therefore be argued that, numerical modelling (with good laboratory and in-situ testing) may have some advantages over centrifuge modelling in such problems. On the other hand, many problems involving very large deformation, break-up of material zones as well as moving and changing interfaces are likely to be much more readily solvable by a centrifuge modelling rather than a numerical modelling approach. Examples of such problems are those involving penetration by projectiles, piles, footings and other objects as well as extrusion and intrusion of one material zone into another. The installation of sand compaction piles discussed above is one such example; there are no doubt many other similar problems. Therefore, a good strategy for the centrifuge modeller may be to identify the problems which are relevant to the local geotechnical community first and then to determine in which of these can the advantages of the centrifuge modelling be exploited.

REFERENCES

Aboshi, H., Ichimoto, E., Harada, K. & Emuki, E. 1979. A method to improve characteristics of soft clay on inclusion of large diameter sand columns. *Proc. Int. Conf. Soil Reinforcement – Reinforced Earth and other Techniques, Paris* 1: 211-216.
Aboshi, H. & Suematsu, N. 1985. The state of the art on sand compaction pile method. *Proc. 3rd Int. Geotech. Sem. Ground Improvement Methods*: 1-12. Singapore: Nanyang Technological Institute.
Asaoka, A., Kodaka, T. & Nozu M. 1994. Undrained shear strength of clay improved with sand compaction piles. *Soils and Foundations* 34(4): 23-32.
Cheng, K.W. 2001. *Spectral characteristics of earth tremors in Singapore.* BEng dissertation, National University of Singapore.
DDESB 1996. *Ammunition and explosives safety standards.* DOD 6055.9-STD, US Department of Defense Explosives Safety Board, Washington, D.C.
Holsapple, K.A. & Schmidt, R.M. 1979. A material strength model for apparent crater volume. *Proc. 10th Lunar and Planetary Science Conf.*: 2757-2777.

Holsapple, K.A. & Schmidt, R.M. 1980. On the scaling of crater dimensions 1. Explosive processes. *Jnl. Geophysical Research* 85: B12.

Kimura, T., Nakase, A., Kusakabe, O. & Saitoh, K. 1985. Behaviour of soil improved by sand compaction piles. *Proc. 11th Int. Conf. Soil Mech. Found. Engineering, San Francisco* 2: 1109-1112. Rotterdam: Balkema,

Kitazume, M., Miyajima, S. & Nishida, Y. 1996. Stability of revetment on soft clay improved by SCP. *Proc. 2nd Int. Conf. Soft Soil Engng, Nanjing, China* 2: 455-460.

Lee, F.H., Ng, Y.W. & Yong, K.Y. 1996. Centrifuge modelling of sand compaction piles in soft ground. *Proc. 2nd Int. Conf. Soft Soil Engineering, Nanjing, China* 2: 407-412.

Lee, F.H., Ng, Y.W. & Yong, K.Y. 2001. Effects of installation method on sand compaction piles in clay in the centrifuge. *Geotechnical Testing Journal, ASTM* 24(3): 314-323.

NATO 1992. Manual of safety principles for the storage of military ammunition and explosives. AASTP-1, AC/258, London.

Neubecker, S.R. & Randolph, M.F. 1994. Model testing and theoretical analysis of drag anchors in sand. In Leung, C.F., Lee, F.H. & Tan, T.S. (eds), *Proc. Int. Conf. Centrifuge 94, Singapore*: 765-770. Rotterdam: Balkema.

Ng, Y.W., Lee, F.H. & Yong, K.Y. 1998. Development of an in-flight sand compaction piles (SCPs) installer. In Kimura, T., Kusakabe, O. & Takemura, J. (eds), *Proc. Int. Conf. Centrifuge 98 Tokyo*: 837-843. Rotterdam: Balkema.

Niu, J. X. 1997. *Centrifuge modelling of seismic effects on sand and soft clay strata*. MEng dissertation, National University of Singapore.

Rowe, P.W. & Craig, W.H. 1976. Studies of offshore caissons founded on Oosterschelde Sand. *Proc. Conf. Design and Construction of Offshore Structures*, ICE, London: 49-55.

Sabetta, F. & Pugliese, A. 1996. Estimation of response spectra and simulation of non-stationary earthquake ground motions. *Bulletin of the Seismological Society of America* 86: 337-352.

Tan, T.S., Lee, F.H., Chong, P.T. & Tanaka, H. 2001. Effect of sampling disturbance on the strength and stiffness of Singapore marine clay. *Journal of Geotech. & Geoenv. Engng., ASCE* (submitted).

Terashi, M., Kitazume, M. & Minagawa, S. 1991. Bearing capacity of improved ground by sand compaction piles. *Deep Foundation Improvements: Design, Construction and Testing, ASTM STP* 1089: 47-61.

DISCUSSION

S.M. Springman & T. Weber

Institute for Geotechnical Engineering, Swiss Federal Institute of Technology, Zurich, Switzerland

Malcolm Bolton paraphrased David Muir Wood's statement from the first day that modelling engaged in intervention, observation, reflection and then decision, and he considered that Fook Hou Lee's contribution was very similar, but was concerned with centrifuge rather than constitutive modelling. He thought this showed why centrifuge modellers demanded technical sophistication of themselves as well as needing to aspire to such a philosophical and scholarly approach, so that a practical contribution could be made to geotechnical engineering.

Ivo Herle asked about the example on the installation of frozen piles to represent sand compaction piles. He wondered why this method had been used because it seemed inappropriate, given that fresh sand would be compacted into the soft layer in situ. He noted that this caused immediate drainage of the layer close to the sand pile, which would become stiffer, whereas an interface of remoulded soil with a very high water content would be created in this centrifuge model, which was opposite to what happened in situ. Lee agreed that this was true but said he had simply tried to compare two methods of installation, since the frozen pile method was normally adopted, despite the drawbacks. He wanted to develop a better method that modelled reality more closely, which permitted the casing to be pushed into the ground and then withdrawn repeatedly while a sand column was formed so that drainage started straightaway. He said that was precisely why it was necessary to adopt the inflight installation method and explained the technique in more detail in that an x-y-frame allowed the installer to be positioned at any point over the model. He described the hopper, which was connected to a casing, which could be filled with sand and driven hydraulically into the soft clay bed, or withdrawn from it. He pointed out that there was a screw inside the casing, which was actuated by a miniature hydraulic motor to force the sand out as the casing was withdrawn and explained that the rate of withdrawal of the casing and the rate of sand feed controlled the eventual diameter of the sand column.

Lee also clarified that the soft clay was allowed to reconsolidate at the higher g-level after pile installation before applying the embankment loading because it was important to simulate the reconsolidation process so that the excess pore water pressures were dissipated, as was also required in the field. He concluded that the reconsolidation under surcharge would be taken into account in the design procedure but not reconsolidation of the excess pore water pressure from the columns.

Cino Viggiani mentioned the first example in which a model was required to solve a very complex problem, as a simplification of reality. He supported the need for carrying out sensitivity studies in terms of looking at the influence of the numerical values of the different parameters of a given model, but he thought that this should be carried out in an even wider sense. He recalled Ivo Herle's excellent analogy of the chain in that the chosen constitutive model needed an initial state and a set of parameters, each of which had a numerical value, in addition to a numerical scheme for solving the boundary value problem. He was concerned that the importance of any single element of this chain was not always clear, for example, if a wonderful constitutive model, with reference to the specific problems under investigation, represented the essential ingredients of the problem, a poor prediction might still arise because the integration scheme (e.g. a first order one) was poor. He thought that not only the numerical values of the parameters of a given model should be varied in a sensitivity analysis, but that consideration of the integration scheme, the mesh and the elements should also be included. He had discovered that predictions were incredibly sensitive to such details and that the sum of these details had made a significant difference to the end result, the numerical prediction. He warned against underestimating the relevance of each component of the chain, i.e. the constitutive ingredient or the numerical ingredient. Lee agreed, having compared 3 to 4 constitutive models of various complexities and data obtained in collaboration with the US army engineers on limestone, which had totally different rate of attenuation from the rock in question. He had also chosen the size of the model in relation to their second order integration scheme and compared results (favourably!) with a rival group, who were only using a linear integration scheme. He also thought that the time integration or time stepping scheme was very important, in that a pacing scheme was acceptable in 2-dimensions but not for a 3-dimensional model. He explained that it had been impossible to use implicit integration on the Pentiums or workstations, and that an explicit integration scheme had been necessary, which had caused problems when incorporating full Rayleigh damping until an iterative method had been developed and adopted. He had been satisfied that internal consistency between the two schemes had been achieved. Lee also reiterated that the boundary conditions were very important for this 'explosive' problem in that the points of measurement should be located far enough from the boundary, which should be modelled as a transmitting boundary to try to absorb the waves arriving there so that they would not be reflected back into the domain. He said they had adopted a simple Lysmer boundary with an optimisation using dashpots at the boundaries to absorb the body waves.

Laurent Vulliet asked about how boundaries were modelled for such problems in physical modelling, and particularly in the centrifuge. Lee noted that the explosive problems had not been modelled by his team on the centrifuge, but that a laminar box was used to eliminate any reflection for the projects requiring simulation of an earthquake, so that the main focus of interest, e.g. a caisson could be investigated under free field conditions. He admitted, however, that the laminar box did not provide the perfect solution in that the sides would also be forced to move in a certain way as the ends moved, so that the soil was dragged along the side. He suggested that the best solution was to make the sample as wide as possible and scale the problem so that the main area of interest was sufficiently far enough away from the boundaries to be considered as a greenfield problem.

Andrew Schofield said that it was very important for shear problems to have complementary cyclical shear stresses (up and down) at the ends of the box, as well as having the shear applied at the base and wondered if there really was a serious problem if the sides moved with the ends. Lee thought that this was not problematic for uniform layers of soil, but that it would not be possible to guarantee that the side would move in unison with the ends for a backfilled caisson and a waterfront.

Schofield was optimistic about engineers adopting centrifuge model test data to improve the design of the foundation of a high-risk structure, such as a nuclear reactor, in an earthquake zone. Despite concern from many that it was necessary for the far field as well as the near field to be made to liquefy in a physical model, Schofield felt that it was only necessary to create local liquefaction in the region of the foundation provided that this represented the damage to the foundations from liquefaction below the nuclear reactor.

Sarah Springman returned to the issue of side friction, and whether the soil should be dragged along the inside the box during the cyclic loading, and recalled that when two latex rubber sheets had been greased to minimise friction between them, that the adhesion had been reduced to ~3° at best. She was not sure whether this could be adopted in a stacked ring box to which Lee responded that this was a big problem for dynamic experiments. He acknowledged that Peter Rowe had been the first to adopt this for static experiments but said that when tests were being carried out on soft clay, the rubber could be stiffer than the clay, so two very thin layers of polyethylene sheet were used instead with a layer of grease between them. Jacques Garnier warned about always trying to reduce the friction for some static applications of centrifuge modelling if some compressible material was introduced between the soil and the container. He noted that horizontal strain could develop during the spin up of the centrifuge, which might be enough to cause the earth pressure coefficient to drop from the at-rest value to the active one. He concluded that the wall aspect of the centrifuge models could be important, and that sometimes it was better not to try to reduce friction significantly so that the geostatic stress were not too different from the prototype stresses.

Constitutive and Centrifuge Modelling: Two Extremes, Springman (ed.)
© 2002 Swets & Zeitlinger, Lisse, ISBN 90 5809 361 1

Impact of centrifuge modelling on offshore foundation design

C.M. Martin
Department of Engineering Science, Oxford University, United Kingdom

ABSTRACT: This paper presents a survey of published work in which centrifuge testing has been used to model shallow offshore foundations. Separate sections are devoted to gravity bases, jack-up unit foundations and suction caissons. After a brief introduction to each foundation type, examples of both site-specific and 'general' centrifuge modelling studies are reviewed in roughly chronological order. The impact of centrifuge testing is assessed within the context of competing physical modelling options such as field tests and $1g$ model tests, which have traditionally played a very important role in the design of offshore foundations. Some remarks on likely future trends are given at the end of the paper.

1 INTRODUCTION

The emergence of centrifuge testing as a widely accessible and economically viable modelling technique has coincided with a period of rapid expansion in the field of offshore engineering. Given the very large size of most offshore foundations, coupled with the need to verify their performance under arduous regimes of cyclic storm-induced loading (quite unlike anything previously encountered onshore), centrifuge researchers in the mid 1970s must have been looking forward to a lifetime of lucrative consulting work. For various reasons, however, centrifuge testing "has not been widely used by [the offshore] industry even though the capability has existed for almost thirty years" (Murff 1996). Despite having a significant impact in certain areas, notably the modelling of jack-up unit foundations, centrifuge testing and its proponents have encountered significant resistance in an affluent industry that can often afford to fund large-scale field trials when testing innovative design proposals. For more routine foundation designs, the traditional approach of performing in situ and laboratory testing, followed by an appropriate suite of theoretical or numerical analyses, has served the industry well. In cases where field trials have not been performed, design techniques have typically been validated against small- or medium-scale laboratory tests at $1g$, rather than in the centrifuge. Nevertheless, there are signs that offshore foundation designers are increasingly willing to explore the use of centrifuge testing. This applies particularly to site-specific applications where there are no comparable case histories, so that modes of failure and deformation are uncertain.

It would be unrealistic to attempt, in a single paper, a totally comprehensive review of centrifuge modelling in offshore foundation design. The author's personal experience has mainly been with jack-up unit foundations, and to a lesser extent with gravity bases and suction caissons, so the survey is confined to these three types of shallow foundation. There has, of course, also been a great deal of centrifuge testing involving deep offshore foundations such as piles and drag embedment anchors. This could usefully form the subject of a separate article. The author would also like to point out that, having spent several years with the Centre for Offshore Foundation Systems at the University of Western Australia, his reporting will inevitably be somewhat biased towards the centrifuge work that has been conducted there. A fairly wide-ranging search of the journal and conference literature has been conducted, but there have no doubt been many other important centrifuge studies (in particular those sponsored by the offshore industry), which remain confidential.

2 GRAVITY PLATFORM FOUNDATIONS

The development of offshore hydrocarbon fields began in the Gulf of Mexico, where the seabed generally consists of soft, normally consolidated clays. In these conditions, the familiar steel 'jacket' platform, anchored by driven piles, quickly established itself as the most economical solution for the shallow to intermediate water depths then under development. Quite different soil conditions were encountered when exploration began in the North Sea; in many areas strong clays (overconsolidated by glaciation) were interbedded with dense sand deposits. This meant that large concrete gravity structures could be placed directly on the seabed, although the many benefits of providing skirts below base level soon became apparent (Watt 1976). Concrete gravity platforms began to supplement conventional jacket structures in the North Sea from the early 1970s onwards.

Foundation designers for these early gravity structures faced a number of significant, unprecedented challenges. In particular, there was concern about the ability of a shallow foundation to resist large, cyclically applied horizontal loads and overturning moments (large, that is, by comparison with the concurrent vertical load). To investigate this problem, a wide-ranging programme of centrifuge modelling was initiated at Manchester University in 1973. Both site-specific and general studies were conducted; sponsors included Shell and the UK Building Research Establishment. Early work concentrated on the performance of gravity structures on sand, but most of the later tests used overconsolidated clay. These gave valuable insights into the modes of failure under various combinations of vertical dead load and cyclic horizontal and moment loading. Figure 1 (Craig & Al-Saoudi 1981) illustrates one of the key findings, that the failure mechanism under cyclic lateral loading is much shallower than that caused by monotonic loading. Soil elements just beneath the structure experience the highest cyclic shear strains, with consequent localisation of the failure. As pointed out by Rowe (1981), this means that "[offshore] foundation analysis on clays is not a question of introducing a reduced strength into a conventional analysis of a foundation subjected to an inclined eccentric load". Craig & Al-Saoudi (1981) also found that the beneficial effect of skirts could be greatly diminished under cyclic lateral loading, mainly due to the formation of gaps adjacent to the skirts. Numerous other publications by P.W. Rowe, W.H. Craig and co-workers emerged from their centrifuge tests on model gravity platforms. Concise reviews are given by Rowe (1981), Craig (1983) and Tani (1990).

French researchers Le Tirant et al. (1977) were also engaged in some early centrifuge testing of model gravity structures, focusing on the response to cyclic lateral loading on dense sand. Other work, funded by Chevron, was conducted at the California Institute of Technology (Prevost et al. 1981). These experiments were similar to the Manchester ones in that cyclic horizontal and moment loads were applied, and the soil used was an overconsolidated silt. However only one level of vertical load was considered, and the skirt configuration was identical in all tests. It appears that the primary objective was to generate calibration data for a parallel numerical modelling study, and as such there is little qualitative description of the observed foundation behaviour. Around the same time, another research effort involving centrifuge modelling of gravity platforms on clay was being sponsored by Exxon. Murff (1996) describes the main objectives and accomplishments of this work, commenting that it was "focused both on understanding mechanisms and in calibrating numerical procedures", but the results appear to be unpublished.

As discussed by Poulos (1988), the Norwegian Geotechnical Institute has played a dominant role in the development of rational design procedures for offshore gravity structures. The review article by Eide & Andersen (1984) provides a comprehensive summary of the NGI methods and their evolution up to that point. Considering that the article is 48 pages long, and that NGI had previously collaborated with Prevost, Rowe and Craig (see Andersen et al. 1978, Andersen et al. 1979), it is perhaps surprising that centrifuge testing receives only one brief mention: a reference to the paper by Rowe & Craig (1979). Nevertheless it seems likely that the Manchester centrifuge work did play a significant role in directing NGI's subsequent decision to base their stability checks on relatively simple limit equilibrium and finite element analyses, but with a strong focus on the use of soil strength and (secant) stiffness properties that were appropriate to the cyclic loading regimes experienced at different locations underneath the platform. To support this strategy, NGI has compiled unrivalled databases of (i) cyclic triaxial and simple shear tests on various soils (ii) field monitoring measurements covering both installation and in-place performance of large offshore platforms.

In the area of physical modelling, NGI researchers conducted a range of medium-scale field tests and small-scale 1g studies to help develop and validate their design methods (Eide & Andersen 1984, Andersen et al. 1989). In general, the agreement between predicted and observed behaviour was excellent, which no doubt partly explains why NGI did not become more heavily involved with centrifuge modelling in the 1970s and 80s. It should be borne in mind, however, that nearly all of the North Sea gravity platforms installed to date have been founded on clay. Experience with large shallow foundations on sand has been much more limited, and this provided the impetus for a major centrifuge modelling study in the early 1990s sponsored by NGI, Delft Geotechnics, Shell and Norwegian Contractors (Allard et al. 1994, Andersen et al. 1994).

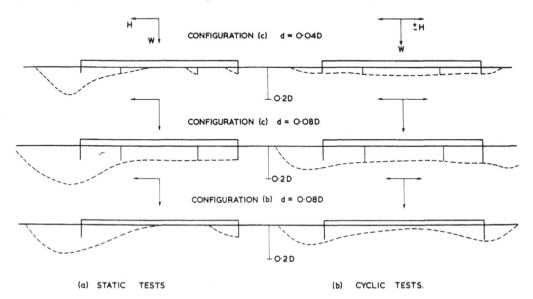

Figure 1. Modes of failure for skirted gravity foundation on clay (Craig & Al-Saoudi 1981)

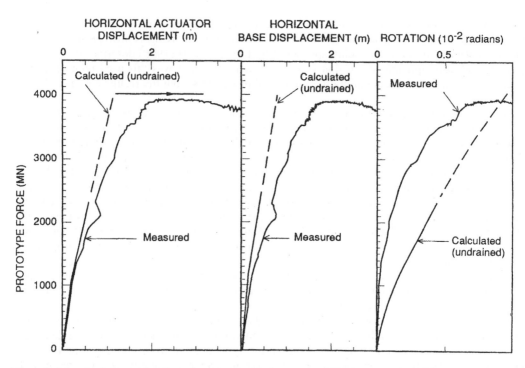

Figure 2. Monotonic lateral loading of gravity foundation on sand (Andersen et al. 1994)

The prototype structure chosen for testing was similar to the existing Ekofisk tank in the North Sea, with a diameter of 97 m and 1.5 m skirts. Tests involving monotonic lateral loading and simulated storm loading were conducted at 300g on very dense sand, with highly detailed instrumentation of the interface pore pressures and total stresses. A special pore fluid was used to ensure that both drainage and cavitation were modelled correctly. The strongly dilatant behaviour of the dense sand during (virtually) undrained shearing meant that very large horizontal loads were developed; for example a monotonic horizontal load of 1.95 times the weight of the structure could be sustained for several prototype minutes. As shown in Figure 2, however, the horizontal displacements needed to mobilise this full capacity would be unacceptably large for a real offshore structure. Limit equilibrium and finite element calculations using properties derived from soil element tests generally showed reasonable agreement with the observed pore pressure accumulation and load–displacement responses. The actual Ekofisk tank was installed in 1973, and was one of the first gravity structures to be modelled in the centrifuge at Manchester. Allard et al. (1994) and Andersen et al. (1994) make no reference to this earlier work, or to the paper by Le Tirant et al. (1977). A detailed comparison between the 'pioneering' and 'mature' centrifuge test results would certainly be an interesting exercise.

Several major concrete gravity structures have been installed offshore within the last few years, but none of the designs seems to have involved any centrifuge model testing. Geotechnical design for the Wandoo platform off Western Australia is described by Humpheson (1998). The soil profile consisted of a thin (approximately 1 m) layer of dense calcareous sand overlying strong, well-cemented caprock material. The plan dimensions of the platform were 114×69 m and the water depth was 54 m, so it was clear that the stability design would be governed by sliding; no model testing or numerical analysis was needed to determine the critical failure mode. This allowed the detailed geotechnical design to focus on the accumulation of excess pore pressures in the surficial sand layer, using soil responses observed in cyclic triaxial tests and cyclic simple shear tests. A centrifuge model test was apparently not considered necessary. Foundation design for the Malampaya gravity platform off the Philippines, installed in June 2000, was similar to that for Wandoo. The seabed was again composed of calcareous sand, and although there was no cemented material present it was once more evident (from the large plan area of the platform and the shallow water depth) that sliding failure under cyclic storm loading would govern the design. No centrifuge testing was performed. Other major gravity platforms to be installed in recent years include Hibernia (off Newfoundland) and South Arne (North Sea). Again neither project seems to have made any use of geotechnical centrifuge testing – at least none that is reported in the public domain.

3 JACK-UP UNIT FOUNDATIONS

The mobile drilling platforms known as jack-up units play a vital role in the offshore industry, being used for both work-over operations adjacent to fixed structures and, increasingly, as production drilling platforms in their own right. Typical modern-day jack-ups have three independent legs that can be raised and lowered by jacking equipment on the hull (which is floatable). Each leg is equipped with a large pad footing known as a spudcan. Because they are mobile rigs, decisions on whether a jack-up is capable of operating safely at a given site must be made on a case-by-case basis, often with very sketchy information about the soil conditions likely to be encountered. Several aspects of the predicted foundation performance have an important bearing on these site-specific assessments. First, it is necessary to make an accurate estimate of the depth to which the spudcans will penetrate during the preloading (proof testing) phase of installation, in which the hull is temporarily filled with ballast water to force the footings deeper into the seabed. Second, the ability of the preloaded foundations to resist various combinations of vertical, moment and horizontal (V, M, H) load during extreme storm events must be verified. Third, estimates of foundation stiffness under different levels of (V, M, H) loading must be provided for use in the dynamic structural analysis of the platform.

Ever since small self-elevating 'spud barges' were developed in the 1950s for use in the Mississippi Delta, there has been a constant trend towards bigger jack-ups with bigger foundations, capable of operating for extended periods in deeper water (and hence harsher environmental conditions). About 20 years ago the industry started to become concerned about perceived over-conservatisms in the approaches being used for spudcan analysis. In particular it was realised that if a significant rotational fixity could be shown to exist at foundation level, this would lead to a reduction in the critical leg bending moments at hull level (compared with the moments obtained using a traditional 'pinned footing' structural analysis). A Joint Industry Study, funded by 18 companies and institutions involved in the offshore industry, was set up in 1984 to investigate this issue (final report: Noble Denton & Associates 1987). The three main strands of research were 1g model testing of spudcans on clay at Oxford University, centrifuge model testing of spudcans on sand at Cambridge University, and finite element analysis by consultants Fugro.

The centrifuge work was performed by R.G. James and M.F. Randolph, with assistance from a student already working on a closely related project (Shi 1987). Model spudcans with various underside profiles were tested on dense sand under various combinations of vertical, moment and horizontal load. The main focus of the study was on monotonic loading, and the framework of work-hardening plasticity theory was found to be very helpful when interpreting the results. For example there was evidence of a well-defined yield surface in (V, M, H) load space, with a size controlled by the vertical spudcan penetration and a cross-sectional shape that could be approximated using simple empirical expressions (see Fig. 3). James and Randolph also found "definite evidence of 'elastic' behaviour inside the yield surface indicating that some moment restraint is available, however at the yield surface this restraint drops dramatically". As requested by the project sponsors, the centrifuge test results were used to develop a simplified, practical method for calculating the moment–rotation response of a spudcan foundation on sand, for a general (V, M, H) load path (Osborne et al. 1991). The Oxford workers developed a similar method for spudcans on clay.

Numerous references to centrifuge modelling studies can be found in the industry-standard assessment procedures for jack-up units (latest edition: SNAME 1997). The current rotational fixity calculation for spudcans on sand closely resembles the one originally developed for the Joint Industry Study; it is slightly simpler and has been modified to take account of additional industry-funded centrifuge tests performed at Cambridge since 1987. The (V, M, H) yield surface, for example, is based on that of Dean et al. (1992). The SNAME procedures for calculating vertical spudcan penetration in sand acknowledge the importance of the so-called scale effect, referencing (among others) centrifuge tests performed by Kimura et al. (1985) at the Tokyo Institute of Technology. In layered soils, particularly when sand overlies soft clay, there is a risk of catastrophic spudcan 'punch-through' during both preloading and storms. Historically this has been a major cause of jack-up foundation failures and it is addressed at length in the SNAME procedures. Approximate analytical methods for dealing with punch-through are discussed, and the findings of centrifuge tests at Manchester (Craig & Higham 1985, Craig & Chua 1990) are summarized. A final example of centrifuge modelling referenced in SNAME (1997) is the study of spudcan–pile interaction conducted at Cambridge (Siciliano et al. 1990, Murff 1996). This problem is important when (as is frequently the case) a jack-up unit needs to be installed adjacent to a piled offshore platform on soft soil. It is interesting to note that the centrifuge studies of punch-through and spudcan–pile interaction were both motivated by the need to shed light on a complex geotechnical problem for which the available analytical and numerical modelling techniques had failed to provide a satisfactory basis for design. Furthermore, both pieces of work were funded by industrial sponsors (Craig & Chua by Shell and others; Siciliano et al. by Exxon), confirming that the advantages of centrifuge testing were well-known and accepted within the jack-up community as early as the mid 1980s.

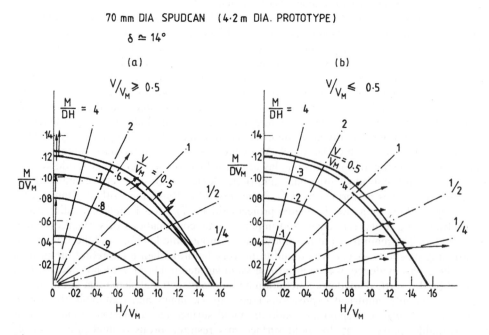

Figure 3. Yield loci for spudcan foundation on sand (Noble Denton & Associates 1987)

Significant programmes of industry-sponsored research into jack-up foundation behaviour continued into the 1990s, and a number of these studies involved centrifuge testing. As well as continuing their investigations of single spudcan footing behaviour on sand (e.g. Dean et al. 1997, Tsukamoto 1999), the Cambridge group embarked on an ambitious and intricate series of centrifuge model tests using complete three-legged jack-up rigs. Largely funded by Exxon, these tests involved monotonic and cyclic loading experiments on both sand and clay. Results of the tests, and of parallel attempts to model the observed soil–structure interaction numerically, are reported in a series of conference papers spanning 1991-95 (see Dean et al. 1995 for references) and by Dean et al. (1998). This work was extremely important, mainly because it verified the general soundness and conservatism of the simplified rotational fixity calculations that had been adopted for the SNAME guidelines. Field monitoring of jack-up performance has now become a major research focus for the industry, and this should allow further validation and calibration of the fixity calculation procedures. Other recent centrifuge work at Cambridge has investigated factors such as cyclic loading, the behaviour of skirted spudcans, and the availability of tensile pullout capacity for spudcans embedded in clay.

Another example of centrifuge modelling related to jack-up units is the work of Springman & Schofield (1998), again conducted in Cambridge with funding from Exxon. In very soft clays such as those in the Gulf of Mexico it is well known that spudcans penetrate to considerable depth during preloading, though usually less than 2.5 diameters (Endley et al. 1981). Backflow into the hole created by a penetrating spudcan is virtually complete, and the lattice framework of the jack-up leg becomes completely enveloped by clay below mudline level. This clearly gives rise to a foundation with increased V, M and H load capacities and enhanced small-displacement stiffnesses compared with those provided by the spudcan footing alone. In fact the Springman & Schofield centrifuge tests (Fig. 4) showed that p-y type interaction between the leg and the infill soil totally dominated the contribution of the spudcan. The circumstances being modelled were admittedly quite extreme: a spudcan penetration in excess of 4.5 diameters, and a lattice leg having a horizontal presented area ratio of 42% (which seems high). Nevertheless, the results make interesting reading for the author, whose $1g$ spudcan tests on soft clay (Martin 1994) never suffered from the 'inconvenience' of significant infill above the footing that was being studied in such detail. Note that in stronger clays spudcan penetration is limited, with minimal backflow (Craig & Chua 1990). In these cases the ability to understand and model the behaviour of an isolated spudcan is essential.

Researchers at the University of Western Australia (UWA) have also performed centrifuge modelling of spudcan foundations for various clients connected with the jack-up industry. Concerns about the performance of shallow foundations on calcareous soils, and the perceived inadequacy of existing design methods (Poulos & Chua 1985, Dutt & Ingram 1988) led to a major series of centrifuge tests partly sponsored by Woodside (Finnie 1993, Finnie & Randolph 1994a, 1994b). Flat circular footings with prototype diameters ranging from 1 m to 15 m were subjected to both vertical and combined (V, H) loading on typical calcareous soil profiles, some involving artificially cemented crusts so that punch-through mechanisms could be studied (Fig. 5). The tests showed that conventional bearing capacity theories and numerical modelling approaches were not directly applicable to the highly frictional yet highly compressible calcareous soils, and appropriate modifications were proposed. Realistically scaled cyclic loading tests confirmed fears that liquefaction failure was indeed a major design issue, particularly if there was no time for significant consolidation to occur prior to storm loading. Other centrifuge studies relevant to jack-up operation on calcareous soil include the work by Nauroy & Golightly (1991) at the Laboratoire Central des Ponts et Chaussées (LCPC), where the bearing response of a flat circular footing (prototype diameter 3.6 m) was examined on both calcareous sand and silica sand. The greater compressibility of the calcareous sand was to some extent offset by its higher angle of friction, and the observed bearing capacities were similar.

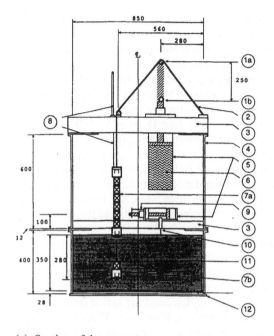

1a & b	Piston mounted pulley wheel
2	Drive chain
3	Beam
4	Extension
5	Actuator
6	Water
7a & b	Lattice leg
8	Vertical PT
9	Horizontal LVDT
10	Settlement LVDT
11	Kaolin clay
12	Drainage layer. Dimensions in millimetres

(a) Section of the apparatus

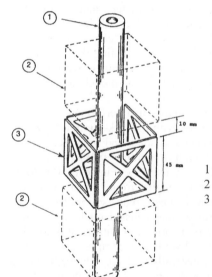

1	Central tube, 18mm OD, 7mm ID
2	Outline of adjacent sets of 4 baffle plates
3	1mm thick Dural baffle plate. Note: Connecting collars and stubs (cruciform) omitted for clarity. Dimensions in millimetres

(b) Detailed view of the lattice leg

Figure 4. Apparatus for lateral loading of jack-up leg in soft clay (Springman & Schofield 1998)

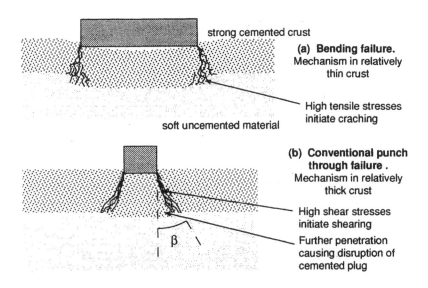

Figure 5. Footing penetration into calcareous soil with cemented crust (Finnie & Randolph 1994b)

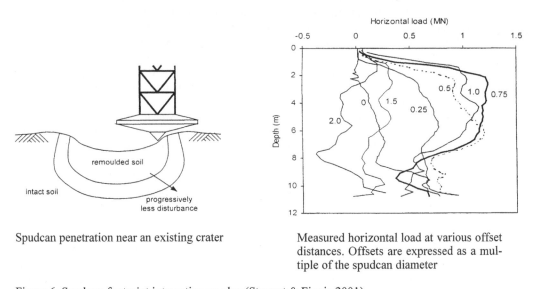

Spudcan penetration near an existing crater

Measured horizontal load at various offset distances. Offsets are expressed as a multiple of the spudcan diameter

Figure 6. Spudcan-footprint interaction on clay (Stewart & Finnie 2001)

Recently the new drum centrifuge at UWA was used to study a problem which is of growing concern to jack-up operators, namely interaction between penetrating spudcan footings and 'footprints' created by previous jack-up activity at the same site. The problem is most severe on clays of medium strength such as those in the Gulf of Thailand, where open footprints more than 10 m wide and 10 m deep can be encountered. Installation of a different jack-up (or inaccurate re-installation of the original unit) may result in uncontrolled vertical and lateral spudcan displacements. This can lead to overstressing of the legs or even a collision between the jack-up and the adjacent fixed structure. The drum centrifuge tests, performed by Stewart & Finnie (2001), used a 12 m diameter prototype spudcan attached to a rigid leg. As shown in Figure 6, the critical offset from footprint centre to spudcan centre is about 0.75 diameters, while offsets greater than about 1.5 diameters show negligible interaction. In reality, of course, the flexibility of the jack-up leg creates a soil–structure interaction problem, for which Stewart & Finnie developed an approximate calculation method. This is yet another example of an awkward jack-up foundation problem where centrifuge testing has provided rapid, cost-effective guidance to practising engineers. Advanced three-dimensional finite element analyses may well have yielded similar results, but the cost would surely have been prohibitive.

4 SUCTION CAISSON FOUNDATIONS

Over the last two decades, skirted caisson foundations installed with the assistance of suction have emerged as a versatile and economical solution for a wide range of offshore anchoring applications. During installation, the foundation is first vented and allowed to penetrate under its own weight. A differential pressure is then created by pumping water out of the skirt compartment, the additional driving force being used to complete the penetration. In coarse-grained soils, this process induces significant seepage flows through the soil; end-bearing resistance on the skirt tips and internal skin friction can be virtually eliminated by (controlled) liquefaction of the soil plug. By contrast in clays and silts the seepage flows are negligible, and installation is essentially a brute force operation. As well as the active suction applied during installation, the passive suction developed during undrained or partially drained pullout loading of the in-place foundation can be used to advantage. Suction caissons are generally circular in shape, and are used both as stand-alone foundation elements and in clusters. Skirt length to diameter ratios L/D generally fall in the range 0.3 to 4, depending on the soil conditions and the nature of the applied loading. Typical diameter to wall thickness ratios D/t are 30 to 40 for concrete skirts and 300 to 500 for steel skirts. In the latter case, ring stiffeners are commonly used to provide local reinforcement (e.g. at anchor chain attachment points) and to prevent compression buckling of the thin-walled skirt during the suction phase of installation.

The first commercial application of the suction caisson concept was a set of 12 catenary chain anchors installed at Gorm in the North Sea in 1980. Preparatory research and development by Shell involved $1g$ model testing together with field experiments at medium and full scale (Hogervorst 1980, Senpere & Auvergne 1982). The tests successfully demonstrated that (i) suction-assisted installation in medium dense sand could be achieved (ii) simple analytical methods could provide reasonably accurate predictions of lateral holding capacity. Some years later, the suction-assisted installation of Gullfaks C (a North Sea gravity platform with 22 m skirts) was also preceded by a large-scale field penetration test (Tjelta 1986). The next major field application came in 1991, when clusters of concrete suction caissons were used to anchor the floating tension leg platform at Snorre, a soft clay site in the North Sea. Yet again the industry showed its preference for proving the soundness of a novel foundation concept through medium-scale field trials (Fig. 7), though as stressed by Dyvik et al. (1993), "the primary intention of the project was not to extrapolate model test results directly to the prototype, but instead to provide results that could be compared to the predictions made for these tests by analytical foundation design procedures". Another important development was decision to use suction caissons rather than piles to support the Europipe and Sleipner SLT jacket structures, installed in 1994 and 1996 respectively. Both platforms are on very dense sand in the North Sea. Numerical simulation of large-scale field trials and of $1g$ model tests (some conducted at NGI, some at Oxford) played a crucial role in validating the foundation design method (Bye et al. 1995). The NGI laboratory tests were performed at a moderate scale, about 1/25, which perhaps explains why no centrifuge testing was considered necessary.

In the early 1990s, a number of general experimental studies of suction caisson behaviour were also conducted, funded by various government research agencies, oil and gas companies and offshore engineering consultants. Many of these studies were performed at 1g, but centrifuge researchers were also involved. Fuglsang & Steensen-Bach (1991) at the Danish Geotechnical Institute investigated the pullout behaviour of model caissons on two different types of clay. At the ISMES centrifuge in Italy, Renzi et al. (1991) performed installation and (compressive) vertical bearing capacity tests on suction caissons in clay, focusing on the 'set-up' phenomenon. Hjortnaes-Pedersen & Bezuijen (1992) at Delft Geotechnics studied the fundamental mechanisms of skirt penetration in clay, though their model caisson (perhaps deliberately) had a very stocky D/t ratio of around 10. By far the most ambitious of these general investigations was a series of Exxon-funded centrifuge tests at LCPC which examined the feasibility of using suction caissons to anchor tension-leg platforms in typical Gulf of Mexico clays (Clukey & Morrison 1993, Morrison et al. 1994, Clukey & Morrison 1995, Murff 1996). Tests at 100g were used to model a steel caisson with prototype dimensions $D = 50$ ft and $L/D = 2$. Initial tests focusing on monotonic pullout loading were followed by an evaluation of performance under realistic sequences of cyclic (V, H) loading. The monotonic results showed excellent consistency with finite element simulations (Fig. 8a), while the cyclic tests were used to develop simple empirical design approaches. Another interesting aspect of this work was its successful replication of the Dyvik et al. (1993) field tests mentioned above. The four-caisson cluster in soft clay was modelled at 1/10 scale and tested at 10g, first using Speswhite kaolin and then using reconstituted soil from the test site at Lysaker in Norway. Figure 8b shows an example of the impressive agreement obtained.

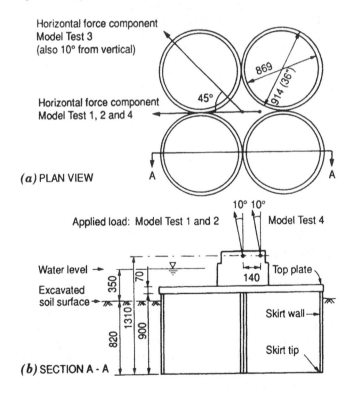

Figure 7. Field trials of suction caisson cluster on clay (Dyvik et al. 1993)

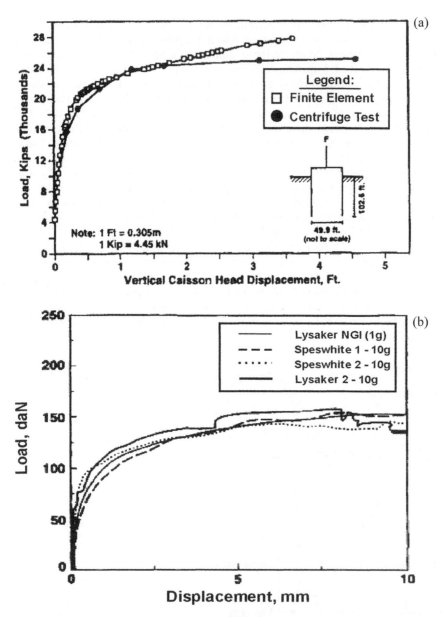

Figure 8. Centrifuge modelling of suction caissons on clay (Murff 1996)

More recently, M.F. Randolph and co-workers at UWA have conducted numerous centrifuge studies of suction caisson behaviour. Several of these have involved site-specific modelling of caissons and caisson groups being considered for offshore developments on the calcareous soils of Australia's North-West Shelf (Watson & Randolph 1998). Early offshore developments in this region were designed with driven piles, and problems with inadequate shaft friction were encountered. This, coupled with the ongoing absence of rational, well-tested design methods and constitutive models for carbonate soils, has led to "a key role for physical modelling in validating potential foundation designs". Watson & Randolph describe various sets of centrifuge tests performed for their industry clients: an individual suction caisson foundation for a compliant tower; a set of four independent suction caissons for a jacket structure; a skirted raft foundation for a jacket structure. Problematical aspects of soil behaviour such as volumetric collapse and possible liquefaction under cyclic loading were of concern in all cases, meaning that accurate reproduction of the in situ strength and consolidation properties was essential. Watson & Randolph conclude that "centrifuge modelling provided a key role in demonstrating the nature of the foundation response, and in quantifying magnitudes of cumulative settlement and overall capacity of foundations subjected to combined vertical, horizontal and moment loading".

Centrifuge testing also played an important part in the design of suction caisson foundations for the Laminaria development in the Timor Sea (Randolph et al. 1998, Fahey & Bruno 1999). The soil at the site was a calcareous silty clay with a mudline strength of about 10 kPa, increasing with depth at about 2 kPa/m. Steel suction caissons ($D = 5.5$ m, $L/D = 2.3$) were designed to anchor the catenary mooring chains for a large floating production vessel. The load to be applied to the caissons was predominantly horizontal, though there was also a component of pullout (the chain was attached about 35% of the way up the skirt in order to optimise the lateral capacity). Centrifuge modelling at 120g was carried out on behalf of the joint venture partners Woodside, Shell and BHP Petroleum. The primary goal of the tests was to verify the monotonic lateral load capacity that had been predicted using upper bound plasticity analysis, and to investigate the effects of cyclic loading. Figure 9 shows the apparatus and some results obtained in one of the tests. The measured capacity under a monotonic chain pull is 9.2 MN. One-way cyclic loading to 5.5 MN and to 7.1 MN (100 cycles each) results in negligible lateral displacement, but failure begins to set in after ten or so cycles to 9.2 MN. The predicted monotonic capacity was 10.6 MN, and even better agreement was obtained by taking account of strength anisotropy in the upper bound calculation.

Other recent centrifuge tests at UWA have been devoted to fundamental studies of suction caisson behaviour. Typically these tests have been performed in conjunction with analytical or numerical modelling, examining aspects such as compressive vertical bearing capacity (Hu et al. 1999) or performance under general (V, M, H) loading conditions (Watson et al. 2000). Most of the tests to date have involved low aspect ratio caissons with $L/D = 0.4$ to 0.5. Although some cyclic loading tests have been performed (Watson 1999), the focus of the research so far has been on monotonic undrained loading in soils such as kaolin clay and calcareous silt. Centrifuge work currently in progress at UWA includes a detailed investigation of the installation, set-up and pullout behaviour of thin-walled suction caissons in clay (House & Randolph 2002). An interesting feature of this work is its meticulous scale modelling of internal ring stiffeners. In stronger clays it is thought that these stiffeners may cause a column of soil to be extruded during installation, rather than flowing around the stiffeners and regaining contact with the inside of the skirt. The consequences for penetration resistance and pullout capacity are considerable (Randolph et al. 1998).

There is now a large body of field experience and published experimental data relating to both the installation and in-place performance of suction caissons in clay. The use of these foundations in sand remains relatively rare, and detailed results from the field trials and 1g tests conducted for the Europipe and Sleipner SLT projects are confidential. Many important aspects of suction caisson performance on sand (such as drained and partially drained cyclic loading) remain poorly understood, particularly where combinations of vertical, moment and horizontal load are involved. This is a major obstacle to the potential use of suction caissons as 'all-purpose' foundations for small offshore structures such as minimal facility platforms, anemometer masts and wind turbines.

147

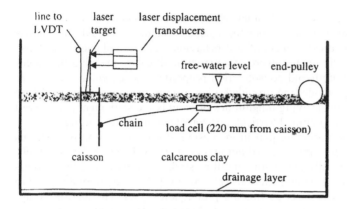

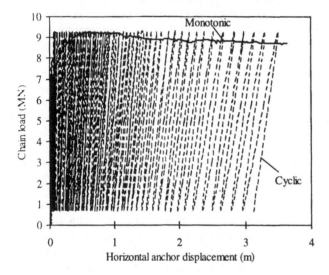

Figure 9. Anchor chain loading of suction caisson on calcareous clay (Randolph et al. 1998)

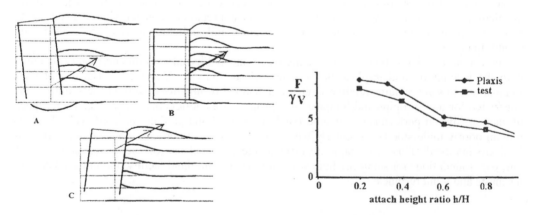

Figure 10. Anchor chain loading of suction caisson on sand (Allersma et al. 1999)

Byrne (2000) has performed extensive 1g testing of suction caissons on fine sand under monotonic and cyclic (V, M, H) loading, including some partially drained tests using silicon oil as the pore fluid. Published centrifuge tests involving suction caissons on sand have mostly been conducted by H.G.B. Allersma and co-workers at Delft University of Technology, though Watson (1999) did perform a limited number of caisson tests on calcareous sand. Examples of the Delft work include Allersma et al.'s (1997) study of installation behaviour, where the penetration resistance during controlled suction installation was up to eight times smaller than that occurring during direct mechanical installation. Allersma et al. (1999) studied the monotonic lateral capacity of chain-loaded suction caissons in sand, using small models ($D = 30$ mm, $L/D = 1$ to 2.3) tested at 150g. Various chain attachment points and load inclinations were considered, and some typical results are shown in Figure 10. Interestingly the lowest chain attachment point, which leads to a backward rotation failure, gives the highest capacity (this contrasts with experience in clay, where the highest capacity is generally associated with an attachment point giving pure lateral translation). The finite element results in Figure 10 show a remarkably similar trend. Previous work at Delft has also involved some industry-sponsored investigations of other novel foundation concepts, for example the deployable 'suction embedded anchor' developed by Suction Pile Technology (Riemers et al. 1999). Even though the installation process could not be simulated in detail, centrifuge tests were invaluable in confirming analytical and finite element predictions of pullout capacity in sand. Other recent work at Delft has considered cyclic loading of suction caissons on sand, but the centrifuge tests have so far only involved pure vertical loading and fully drained conditions.

5 CONCLUSIONS

This paper has given a historical overview of the role of geotechnical centrifuge modelling in the design of offshore foundations. Attention has been confined to shallow foundations, the three main types being gravity platforms, jack-up unit 'spudcans' and suction caissons. Key design issues in all cases are (i) the very large horizontal loads and overturning moments resulting from wave and wind loading (ii) the cyclically applied nature of these loads.

In the days of the first North Sea gravity platforms, centrifuge technology was still in its infancy. Although the centrifuge made some useful early contributions to the development of design procedures, medium-scale field trials and 1g laboratory tests were generally preferred for the validation of emerging design methods. It is important to note, however, that the majority of existing North Sea gravity structures are founded on clay soils. The need for improved understanding of large gravity foundations on sand is widely acknowledged, and centrifuge modelling has already been used in some important fundamental research on this topic. Future offshore developments involving gravity structures on sand, particularly calcareous sand, may well create opportunities for site-specific centrifuge modelling studies.

The jack-up industry has been a keen supporter of centrifuge modelling since the mid 1980s, taking a particular interest in the behaviour of spudcan foundations on sand. Detailed centrifuge studies of bearing capacity and stiffness under combined vertical, moment and horizontal loading have been carried out, and the results have formed the basis for industry-standard calculation procedures. Centrifuge modelling of complete jack-up units (both on sand and on clay) has also been funded by industry, in an attempt to develop insights into complex soil–structure interaction processes. Other difficult aspects of spudcan behaviour that have been elucidated by centrifuge modelling include interaction with previous spudcan 'footprints' in soft clay, and punch-through behaviour in layered soils. There is every reason to suppose that demand for similar centrifuge studies will continue.

Suction caisson foundations have now been used for a number of major offshore projects in the North Sea and elsewhere, but centrifuge testing has played only a small rôle in the development of this innovative concept. The preferred approach for physical modelling has generally involved 1g tests at small and medium scale, in conjunction with field trials. An exception is the recent Laminaria project in the Timor Sea, where centrifuge modelling was used to confirm analytical predictions of suction caisson holding capacity, and to assess the effects of cyclic loading. As with gravity platforms, the in-place performance of suction caissons on sand remains poorly understood, particularly with regard to partially drained cyclic loading. This will be a prime area for both general and project-specific centrifuge modelling in coming years.

REFERENCES

Allard, M.A., Andersen, K.H. & Hermstad, J. 1994. Centrifuge model tests of a gravity platform on dense sand; I: testing technique and results. *Proc. BOSS '94, MIT* 1: 231-254. Oxford: Pergamon.

Allersma, H.G.B., Kirstein, A.A., Brinkgreve R.B.J. & Simon, T. 1999. Centrifuge and numerical modelling of horizontally loaded suction piles. *Proc. 9th ISOPE Conf., Brest* 1: 711-717.

Allersma, H.G.B., Plenevaux, F.J.A. & Wintgens, J.-F.P.C.M.E. 1997. Simulation of suction pile installation in a geocentrifuge. *Proc. 7th ISOPE Conf., Honolulu* 1: 761-766.

Andersen, K.H., Allard, M.A. & Hermstad, J. 1994. Centrifuge model tests of a gravity platform on dense sand; II: interpretation. *Proc. BOSS '94, MIT* 1: 255-282. Oxford: Pergamon.

Andersen, K.H., Dyvik, R., Lauritzsen, R., Heien, D., Harvik, L. & Amundsen, T. 1989. Model tests of gravity platforms. II: interpretation. *J. Geotech. Eng. Div. ASCE* 115(11): 1550-1568.

Andersen, K.H., Hansteen, O.E., Hoeg, K. & Prevost, J.H. 1978. Soil deformations due to cyclic loads on offshore structures. In O.C. Zienkiewicz, R.W. Lewis & K.G. Stagg (eds), *Numerical methods in offshore engineering*: Ch. 13. Chichester: Wiley.

Andersen, K.H., Selnes, P.B., Rowe, P.W. & Craig, W.H. 1979. Prediction and observation of a model gravity platform on Drammen clay. *Proc. BOSS '79, London* 1: 427-446.

Bye, A., Erbrich, C., Rognlien, B. & Tjelta, T.I. 1995. Geotechnical design of bucket foundations. *Proc. 27th Offshore Technology Conf., Houston*, Paper No. OTC 7793.

Byrne, B.W. 2000. Investigations of suction caissons in dense sand. DPhil thesis, University of Oxford.

Clukey, E.C. & Morrison, M.J. 1993. A centrifuge and analytical study to evaluate suction caissons for TLP applications in the Gulf of Mexico. *ASCE Geotechnical Special Publication No. 38*: 141-156.

Clukey, E.C. & Morrison, M.J. 1995. The response of suction caissons in normally consolidated clays to cyclic TLP loading conditions. *Proc. 27th Offshore Technology Conf., Houston*, Paper No. OTC 7796.

Craig, W.H. 1983. Simulation of foundations for offshore structures using centrifuge modelling. In P.K. Banerjee & R. Butterfield (eds), *Developments in soil mechanics and foundation engineering – 1. Model Studies*: Ch. 1. London: Applied Science.

Craig, W.H. & Al-Saoudi, N.K.S. 1981. The behaviour of some model offshore structures. *Proc. 10th ICSMFE, Stockholm* 2: 83-88.

Craig, W. H. & Chua, K. 1990. Deep penetration of spud-can foundations on sand and clay. *Géotechnique* 40(4): 541-556.

Craig, W.H. & Higham, M.D. 1985. The applications of centrifugal modelling to the design of jack-up rig foundations. *Proc. Int. Conf on Offshore Site Investigation, London*: 293-305. London: Graham & Trotman.

Dean, E.T.R., Hsu, Y.S., James, R.G., Schofield, A.N., Murff, J.D. & Wong, P.C. 1995. Centrifuge modelling of 3-leg jackups with non-skirted and skirted spuds on partially drained sand. *Proc. 27th Offshore Technology Conf., Houston*, Paper No. OTC 7839.

Dean, E.T.R., James, R.G., Schofield, A.N., Tan, F.S.C. & Tsukamoto, Y. 1992. The bearing capacity of conical footings on sand in relation to the behaviour of spudcan footings of jack-ups. *Proc. Wroth Memorial Symp. 'Predictive Soil Mechanics', Oxford*: 230-253. London: Thomas Telford.

Dean, E.T.R., James, R.G., Schofield, A.N. & Tsukamoto, Y. 1997. Theoretical modelling of spudcan behaviour under combined load. *Soils and Foundations* 37(2): 1-15.

Dean, E.T.R., James, R.G., Schofield, A.N. & Tsukamoto, Y. 1998. Drum centrifuge study of three-leg jackup models on clay. *Géotechnique* 48(6): 761-785.

Dutt, R.N. & Ingram, W.B. 1988. Bearing capacity of jack-up footings in carbonate granular sediments. *Proc. Int. Conf. on Engineering of Calcareous Sediments, Perth* 1: 291-296. Rotterdam: Balkema.

Dyvik, R., Andersen, K.H., Hansen, S.B. & Christophersen, H.P. 1993. Field tests of anchors in clay. I: description. *J. Geotech. Eng. Div. ASCE* 119(10): 1515-1531.

Eide, O. & Andersen, K.H. 1984. Foundation engineering for gravity structures in the northern North Sea. *Publication No. 154*, Oslo: Norwegian Geotech. Inst.

Endley, S.N., Rapoport, V., Thompson, P.J. & Baglioni, V.P. 1981. Prediction of jack-up rig footing penetration. *Proc. 13th Offshore Technology Conf., Houston*, Paper No. OTC 4144.

Fahey, M. & Bruno, D. 1999. Model testing applied to various geotechnical problems. *Proc. Int. Conf. Offshore and Nearshore Geotech. Eng., Navi Mumbai*: 207-219.

Finnie, I.M.S. 1993. *Performance of shallow foundations in calcareous soil*. PhD thesis, University of Western Australia.

Finnie, I.M.S. & Randolph, M.F. 1994a. Bearing response of shallow foundations on uncemented calcareous soil. *Proc. Conf. Centrifuge 94, Singapore*: 535-540. Rotterdam: Balkema.

Finnie, I.M.S. & Randolph, M.F. 1994b. Punch through and liquefaction induced failure of shallow foundations on calcareous sediments. *Proc. BOSS '94, MIT* 1: 217-230. Oxford: Pergamon.

Fuglsang, L.D. & Steensen-Bach, J.O. 1991. Breakout resistance of suction piles in clay. *Proc. Conf. Centrifuge 91, Boulder*: 153-159. Rotterdam: Balkema.

Hjortnaes-Pedersen, A.G.I. & Bezuijen, A. 1992. Offshore skirt penetration in the geo-centrifuge. *Proc. BOSS '92, London* 1: 528-542.

Hogervorst, J.R. 1980. Field trials with large diameter suction piles. *Proc. 12th Offshore Technology Conf., Houston*, Paper No. OTC 3817.

House, A.R. & Randolph, M.F. 2002. Centrifuge modelling of caisson capacity under uplift loading. In S.M. Springman (ed.), *Constitutive and centrifuge modelling: two extremes. Workshop, Ascona, Switzerland, 8-13 July 2001*, Rotterdam: Balkema.

Hu, Y., Randolph, M.F. & Watson, P.G. 1999. Bearing response of skirted foundations on non-homogeneous soil. *J. Geotech. Eng. Div. ASCE* 125(11): 924-935.

Humpheson, C. 1998. Foundation design of Wandoo B concrete gravity structure. *Proc. Int. Conf on Offshore Site Investigation and Foundation Behaviour 'New Frontiers', London*: 353-382. London: Soc. Underwater Technology.

Kimura, T., Kusakabe, O. & Saitoh, K. 1985. Geotechnical model tests of bearing capacity problems in a centrifuge. *Géotechnique* 35(1): 33-45.

Le Tirant, P., Luong, M.P., Habib, P. & Gary, G. 1977. Simulation en centrifugeuse de fondations marines. *Proc. 9th ICSMFE, Tokyo* 2: 277-280.

Martin, C.M. 1994. Physical and numerical modelling of offshore foundations under combined loads. DPhil thesis, University of Oxford.

Morrison, M.J., Clukey, E.C. & Garnier J. 1994. Behaviour of suction caissons under static uplift loading conditions. *Proc. Conf. Centrifuge 94, Singapore*: 823-828. Rotterdam: Balkema.

Murff, J.D. 1996. The geotechnical centrifuge in offshore engineering. *Proc. 28th Offshore Technology Conf., Houston*, Paper No. OTC 8265.

Nauroy, J.-F. & Golightly, C. Bearing capacity of a shallow foundation on calcareous sand. In H.Y. Ko (ed.), *Proc. Conf. Centrifuge 91, Boulder*: 187-192. Rotterdam: Balkema.

Noble Denton & Associates 1987. Foundation fixity of jack-up units: a joint industry study. London: Noble Denton & Associates.

Osborne, J.J., Trickey, J.C., Houlsby, G.T. & James, R.G. 1991. Findings from a joint industry study on foundation fixity of jackup units. *Proc. 23rd Offshore Technology Conf., Houston*, Paper No. OTC 6615.

Poulos, H.G. 1988. *Marine geotechnics*. London: Unwin Hyman.

Poulos, H.G. & Chua, E.W. 1985. Bearing capacity of foundations on calcareous sand. *Proc. 11th ICSMFE, San Francisco* 3: 1619-1622.

Prevost, J.H., Cuny, B. & Scott, F. 1981. Offshore gravity structures: centrifugal modeling. *J. Geotech. Eng. Div. ASCE* 107(2): 125-141.

Randolph, M.F., O'Neill, M.P. & Stewart, D.P. 1998. Performance of suction anchors in fine-grained calcareous soils. *Proc. 30th Offshore Technology Conf., Houston*, Paper No. OTC 8831.

Renzi, R., Maggioni, W., Smits, F. & Manes, V. 1991. A centrifugal study of the behaviour of suction piles. In H.Y. Ko (ed.), *Proc. Conf. Centrifuge 91, Boulder*: 169-176. Rotterdam: Balkema.

Riemers, M.E., Kirstein, A.A. & Allersma, H.G.B. 1999. Development and capabilities of the suction embedded anchor (SEA). *Proc. IGB Seabed Geotechnics Conf., London.*

Rowe, P.W. 1981. Use of large centrifugal models for offshore and nearshore works. *Proc. Symp. Geotech. Aspects of Offshore and Nearshore Structures, Bangkok*: 1-18.

Rowe, P.W. & Craig, W.H. 1979. Applications of models to the prediction of offshore gravity platform foundation performance. *Proc. Int. Conf. on Offshore Site Investigation, London*: 269-281. London: Graham & Trotman.

Senpere, D. & Auvergne, G.A. 1982. Suction anchor piles – a proven alternative to driving or drilling. *Proc. 14th Offshore Technology Conf., Houston*, Paper No. OTC 4206.

Shi, Q. 1987. Centrifugal modelling of surface footings subject to combined loading. PhD thesis, University of Cambridge.

Siciliano, R.J., Hamilton, J.M., Murff, J.D. & Phillips, R. 1990. Effect of jackup spudcans on piles. *Proc. 22nd Offshore Technology Conf., Houston*, Paper No. OTC 6467.

SNAME 1997. *Recommended practice for site specific assessment of mobile jack-up units*, Rev. 1. Jersey City, NJ: Soc. Naval Architects and Marine Engineers.

Springman, S.M. & Schofield, A.N. 1998. Monotonic lateral load transfer from a jack-up platform lattice leg to a soft clay deposit. In T. Kimura et al. (eds), *Proc. Conf. Centrifuge 98, Tokyo*: 563-568. Rotterdam: Balkema.

Stewart, D.P. & Finnie, I.M.S. 2001. Spudcan-footprint interaction during jack-up workovers. *Proc. 11th ISOPE Conf., Stavanger.*

Tani, K. 1990. Stability of skirted gravity foundations on very soft clay. PhD thesis, University of Manchester.

Tjelta, T.I. 1986. Large scale penetration test at a deepwater site. *Proc. 18th Offshore Technology Conf., Houston*, Paper No. OTC 5103.

Tsukamoto, Y. 1999. Some aspects of the behaviour of offshore footings on sand. *Géotechnique* 49(4): 503-514.

Watson, P.G. 1999. Performance of skirted foundations for offshore structures. PhD thesis, University of Western Australia.

Watson, P.G. & Randolph, M.F. 1998. Skirted foundations in calcareous soil. *Proc. ICE Geotech. Eng.* 131, July: 171-179.

Watson, P.G., Randolph, M.F. & Bransby, M.F. 2000. Combined lateral and vertical loading of caisson foundations. *Proc. 32nd Offshore Technology Conf., Houston*, Paper No. OTC 12195.

Watt, B.J. 1976. General practice for gravity structures. In D.M. Wood & P.J. George (eds), *Offshore soil mechanics*: 272-284. Cambridge University Engineering Dept.

DISCUSSION

S.M. Springman & T. Weber
Institute for Geotechnical Engineering, Swiss Federal Institute of Technology, Zurich, Switzerland

Fouk Hou Lee asked about software capable of simulating cyclic loading response of offshore foundations, and mentioned a code from the Norwegian Geotechnical Institute. Martin thought that there was some cyclic loading software for a total stress approach but thought that effective stress programs were not used in routine design because it was too expensive to do cycle by cycle analysis of the response. Muir Wood commented that NGI's approach to cyclic design of offshore platforms seemed very similar to the shake analysis for earthquake engineering, for which a quasi-linear but variable stiffness was adopted, with stiffness changing according to number of cycles and build up in pore pressure. He referred to papers by Suzanne Lacasse, which described the success of their predictions, and this was confirmed by Martin, who thought they had done some impressive class A predictions on some of their 1g testing on Drammen clay. He noted however, that it was easier if the same material was always used.

Constitutive and Centrifuge Modelling: Two Extremes, Springman (ed.)
© *2002 Swets & Zeitlinger, Lisse, ISBN 90 5809 361 1*

Construction of models in the ETH Zurich drum centrifuge

Ph. Nater
Swiss Federal Institute of Technology, ETH Zurich, Switzerland

ABSTRACT: The significance of a centrifuge test strongly relies on the quality of the soil model. Nearly as many model building procedures exist, as there are different soils. This is due to technical limitations and requirements on creating a specific soil structure. Building layered soil model implies the use of different soils in different consistencies and states and therefore the adoption of a variety of procedures and apparatus within the same test process. This causes additional effort and influences the soil stress history, especially compared to natural formation conditions. With the aim of building layered soil models containing both sand and clay layers, several model building procedures are described in this contribution, together with associated aspects of technical limitations. Some of them have been recently used in the drum centrifuge device of the ETH Zurich.

1 MODEL BUILDING PROCEDURES FOR DIFFERENT SOILS IN A DRUM CENTRIFUGE

1.1 *Special challenge*

Since there is a rigid hollow cylindrical container in a drum centrifuge, the model does not follow the resultant direction of the g-field (in a vertical plane) as it does in a beam centrifuge supplied with a swing platform. The resultant of the acting forces rotates nearer to the horizontal with increasing rotational speed, but the orientation of the container remains unchanged. Therefore the stability of structures and soil in the initial state at 1g is challenged because the horizontal model surface under an artificial g-field is vertical in the stationary drum. In consequence, some models have to be built in flight while the drum is rotating, which means that material has to be transferred from a stationary to be placed within a rotating system.

1.2 *Sand*

In dry sand no true cohesion occurs. The steepness of a slope follows the plane normal to the acceleration vector plus the angle of internal friction ϕ'_{crit}. Thus a vertical surface sculpted in dry sand is not stable without a supporting horizontal g component. When there is no possibility to run a whole centrifuge test without stopping the drum, additional strength has to stabilise the material. Therefore the material is partially saturated so that capillary suction is allowed to develop to increase locally the mean effective stress acting between the sand particles. This is sometimes referred to as an apparent cohesion.

Suction will be increased when a sand contains more fines, however dust becomes an issue in terms of health. Most kinds of technical instrumentation are also badly affected by dust. Decreasing the void size also improves the capacity for capillary suction. Therefore increasing the density of a soil sample has a major effect on the temporary additional strength delivered due to suction. Subsequently, for building models which require further treatment under 1g conditions, only medium dense to dense sands are recommended so that the model surface remains stable. Roughening the drum wall surface also has a major impact on the stability of a deposit with a vertical face without a stabilising horizontal g component. The importance of this in terms of changing the principal stress directions becomes less significant at g-levels under normal operation.

Wind effects in the rotating drum may also cause the development of dune patterns on dry sand surfaces once the centrifuge speed increases above a critical level. Therefore by developing suction in the soil near the surface the resistance against erosion caused by wind will be improved.

1.2.1 *Model building by hand*
Fine-medium sand is mixed with water outside the centrifuge and manually placed inside the stationary system on the drum wall manually with masonry tools. Suction can be maximised according to the material type through adding an optimal amount of water, so a reasonable shaping of the model surface has nearly no limitation. On the other hand no control over the density or the homogeneity of the model is possible during the model making process.

1.2.2 *Spraying material by air pressure*
Dry or slightly moist sand is accelerated in a tube by air pressure. The material will be sprayed directly from the stationary mould into the rotary system of the drum onto the wall or onto the already built in model surface from a nozzle.

To use capillary suction to stabilise and shape the surface after stopping the centrifuge, the following procedure can be applied: the dry sand can be wetted via flooding the base drain in flight using the water supply system of the centrifuge or through raining from a centrally mounted actuator. After saturation, a drain at the base of the drum channel can be opened so that free water will be pressed out again, which means that only water bound between the meniscus of grain contact zones remains (Dean et al. 1990). These menisci are responsible for the negative pore pressure which increase the effective stress and hence the strength available.

1.2.3 *Spraying material by a spinning disc*
Sand falls through a tube onto a rotating disc (on a vertical axis), on which four radial blades are mounted at 90° to each other. The blades guide the sand as it is accelerated in angular and radial directions. Particles are ejected from the spinning disc with radial and angular components of acceleration (and a vertical component out of the 1g free space) and propelled onto the drum wall. The same system is in use for spraying seed and fertiliser in agriculture or salt against ice-formation on roads in infrastructure maintenance in the winter. Figure 1 in paragraph 2.1 shows a sand spreader. For inducing capillary suction, the same procedure as mentioned in 1.2.2 can be applied.

1.3 *Clay*

Natural deposits of clay contain inhomogeneities which affect results obtained from physical modelling. To improve homogeneity, the material can be remoulded prior to the model building procedure. A common homogenisation method is to mix up the clay with water so that the water content is above the liquid limit, sieving out stones and other bigger particles and then keeping it in suspension in a stirrer. This slurry can be placed into a container and consolidated to achieve a specific strength profile.

1.3.1 *Model building in the laboratory*

Clay cakes can be produced in a consolidation device such as an Ödometer of suitable large diameter which is fixed within a load frame. These cakes can then be placed inside the drum by hand. The gaps can be caulked with other clay and the whole model can then be subjected to additional times of consolidation inside the drum. To avoid boundary effects because of the platen ends, tests are to be conducted in the centre of the cakes: However disturbance caused by the transfer of these cakes is unavoidable and the exact stress history will be unclear.

1.3.2 *Placing material in flight by nozzle*

Comparable to model building with sand using compressed air, liquid clay can also be passed through a tube from the stationary to the rotating world. After mixing to wetter than the liquid limit outside the drum, the clay is transported via a hopper and funnel system into the tube which is fitted to a t-shaped outlet mounted on the actuator (Fig. 5), which has 3 degrees of freedom of movement. To pass from the stationary system outside the drum to the rotating system inside the drum, a rotary coupling system is required. An even distribution of the slurry will be achieved by placing the t-outlet underneath the water table. The intended stress history will be achieved after consolidation under increased g-level.

2 SYSTEMS APPLIED IN THE DRUM CENTRIFUGE OF THE ETH ZURICH

At the moment, the priority lies in building layered soil models for the investigation of the soil structure interaction of circular footings. The properties of the clay layers should be in accordance with their modelled depth in nature. The groundwater level will be at or close to the model surface and each layer should be fully saturated. Specific properties of the ETH Zürich geotechnical drum centrifuge can be found in Springman et al. (2001).

2.1 *Sand*

Sand is placed in the drum using a rotating disc as described in 1.2.3 (Fig. 1). Dry material is to be prepared in a container outside the centrifuge and coupled to a tube guiding it onto the spinning disc. There the grains are accelerated radially and circumferentially so that they leave the disk with enough speed to traverse to the drum wall or other material, which already has been placed and is under the influence of the applied artificial g-field.

The whole tool can be moved up and down over the 700 mm height of the drum by moving the safety shield. The sand has to be saturated from the bottom to the surface through the plumbing system built into the drum or by pluviation from a tool fixed to the centrally mounted actuator (Fig. 2 shows actuator with another tool). Then the water will be drained again to develop enough capillary suction to shape the surface under low or no horizontal acceleration. Sand is scraped off to form a flat vertical surface (Fig. 2) with a blade mounted on the actuator which rotates relative to the drum. Other shapes of scrapers could also be used in the future, as required to create different surfaces e.g. slopes and trenches. The material that has been scraped off has to be removed by hand in order not to affect the drum or the actuator system during the main test. Therefore the drum has to be turned off at this stage, which means not only that the model has to be stable under vertical 1g, but also that one „load cycle" has already been performed.

The parameters influencing the model density, which governs the main properties of the sand, are the flow rate of the sand exiting the storage container, the rotation speeds of the spinning disc and the drum. The flow rate has been varied during the tests conducted in the centrifuge and a linear correlation was observed between relative density achieved and the flow rate (Fig. 3, Grämiger 2001).

a)   b)

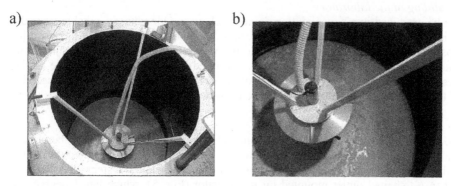

Figure 1. (a) Sand spreader mounted on the safety shield, (b) close up to the spinning disc

a)   b)

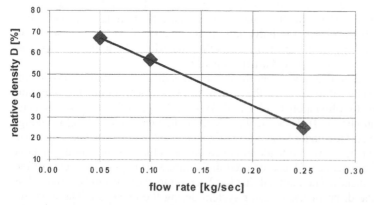

Figure 2. Scraper in action from outside the drum (a), and as a „screen shot" by video transmission from the onboard infrared camera (b)

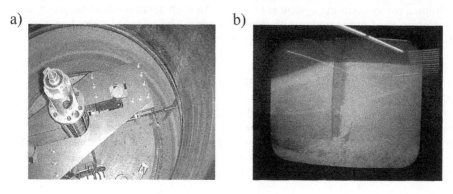

Figure 3. Relative density of a sand model as a function of the material flow rate out of the storage container for given disk rotation speed (Grämiger 2001)

2.2 Clay

The clay used for the recent tests is a lacustrine silty clay from the area of Birmensdorf. Common properties have been derived in laboratory tests and in a previous consolidation test in the centrifuge (Fauchère 2000, Züst 2000).

Lumps of natural clay are broken up and mixed with water to a water content of 90% - 100% and remoulded using a standard paddle stirrer (Fig. 4a). From there the liquid is transferred into a 200 litre vacuum mixing device (Fig. 4b) for further homogenisation and deairing which will require between 2-5 hours per charge. The outlet of the vacuum stirrer leads directly to a nozzle (Fig. 5) through tubing and a rotary coupling.

To prevent further air access to the deaired clay, the nozzle is placed initially under water, which means that the free water table in the model has to be set to about 10 cm over the soil surface. The actuator moves at slow speed relative to the drum to disperse the clay all over the model container. The shaping of the surface is similar to the procedure used for sand surfaces. Therefore the drum has to be stopped for cleaning and safety inspections before other equipment is mounted on the tool platform. The control over the clay properties is given by the initial water content for stirring and subsequent post consolidation values, which are dependent upon centrifugal acceleration in the drum, consolidation time and the ratio of overconsolidation compared to the later use of the soil model.

a)

b)

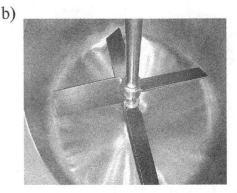

Figure 4. Stirring devices: (a) standard stirrer (b) vacuum mixer for homogenisation / deairing (inside view)

Figure 5. Nozzle and tube of the clay placing system mounted on the actuator

2.3 First experience in model building in the drum centrifuge of the ETH Zurich

Homogeneous soil models have been built with the methods described in paragraph 2.1 and 2.2. The procedures are easy to handle and the quality concerning the homogeneity of the soils is good. The quality has been proven by optical impression, cone penetration (CPT) (Fig. 6) and T-bar tests in flight and laboratory examination of soil samples taken from the models.

Meanwhile a layered soil model has been built in two tubs of 400 mm diameter and 200 mm height, which can be mounted diametrically opposite each other via the use of fitting plates fixed to the drum wall. These samples contain three sand layers with two clay layers in between. A displacement controlled load test with a circular footing of 0.028 m diameter under centrifugal acceleration of 50g has been conducted together with two CPT tests at the same g-level. The models were excavated after the tests (Fig. 7), inspected visually with measurement of the layer thickness and samples of the clay deposit have been tested in the laboratory for undrained shear strength s_u.

The surfaces of each layer are quite clean and sharp, with nearly no transition zone between two neighbouring layers of sand and clay. The undrained shear strength s_u of the two clay layers derived from the laboratory results after the test corresponds well to the current vertical effective stress at 1g and the overconsolidation ratio for the modelled prototype.

a)  b)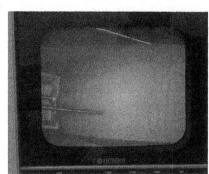

Figure 6. CPT-device for the inflight examination of soils in the drum: (a) mounted on the actuator, (b) CPT-test in flight

Figure 7. Cut through layered soil model in the tub after testing

3 PERSPECTIVE

With the proposed solutions and the available tools, the construction of layered soil models is manageable in the drum centrifuge. No immediate changes of the model building systems are scheduled at this stage, but evolution of the techniques is to be expected. For the further use of the existing equipment, the following problems occur and must be resolved:

- What additional measures should be taken into account to provide fully saturated and consolidated soil conditions and how can these conditions be controlled within the model building procedure?
- How can these procedures be improved in terms of achieving high quality of the end product, in larger quantity (i.e. full drum c.f. the tubs) within a manageable time frame?
- Are there other problems which will influence the significance of the results in terms of the soil model?

ACKNOWLEDGEMENTS

Thanks are due to everyone involved in the centrifuge team of the IGT: Sarah Springman, Jan Laue, Bernhard Sperl, Markus Iten, Alfredo Privitello and with him the whole workshop, Ernst Bleiker, Adi Zweidler, Ravikiran Chikatamarla and also Josh White and Richard Boyle for all their work during commissioning.

REFERENCES

Dean, E.T.R., James, R.G. & Schofield, A.N. 1990. Drum centrifuge studies for EEPUK. Contract No. EP-022R TASK ORDER 1-022. Phase 1. Draft Report, Cambridge.

Fauchère, A. 2001. Preliminary Studies for the Simulation of a Spread Base Bridge Abutment on Lacustrine Clay in the ETHZ Geotechnical Centrifuge. Diploma project. IGT, ETH Zurich.

Grämiger, E. 2001. Einbauverfahren und Bodenerkundung im Sand in der geotechnischen Zentrifuge Zürich. Diploma project. IGT, ETH Zurich.

Springman, S., Laue, J., Boyle, R., White & J. Zweidler, A. 2001. The ETH Zurich Geotechnical Drum Centrifuge. *IJPMG* V. 1: 59-70.

Springman, S., Trausch-Giudici, J., Heil, M. & Heim, R. 1999. Strength of a soft Swiss lacustrine clay: cone penetration and triaxial data. *Transportation Research Record 1675*: 1-9.

Züst, Y. 2001. Vorstudien zur Simulation einer Dammschüttung als Strassendamm auf Seebodenlehm in der geotechnischen Zentrifuge Zürich. Diploma project. IGT, ETH Zurich.

DISCUSSION

S.M. Springman & T. Weber
Institute for Geotechnical Engineering, Swiss Federal Institute of Technology, Zurich, Switzerland

Introduction

The speaker presented aspects of his research on circular footings on layered soils, focussing on the creation of the soil models. He described the drum and actuator system, in which various tools were clamped and moved vertically up and down, rotated at various speeds, or driven radially. The intention was to create several layers of sand and clay, and because of the different characteristics of the two materials, different techniques were needed to place them in the drum and to process the surface using a grader. The pluviation of sand, which was carried out with the spinning disk system, was described together with the successive cycles of saturation and desaturation carried out to create suction in the sand to maintain a stable model in the drum when the centrifuge had stopped rotating. He explained how this was controlled by the water supply and drainage system.

The initial test series concentrated on creating homogeneous sand layers in the drum. Clay layers were placed by lowering the nozzle of the clay inlet system below the water table to keep the clay fully saturated and to allow the slurry to flow across the sand surface prior to permitting consolidation to take place. Once consolidation was well advanced, it was possible to stop the centrifuge to process the surface, using the same grading system as for the sand (or to change the tool while the centrifuge was still running by lowering the safety shield).

The challenge of combining these two systems was discussed, particular the creation of a 'clean' interface between the sand and the clay layers (which had already been achieved in the axisymmetric strongbox), and with a well-defined initial state (and stress history). In particular the clay should be fully saturated and in a state of normal consolidation. This was important given that numerical back analysis should be carried out subsequently. He asked for the opinion of those present on the feasibility of the current plans for creating layered models in the drum and wondered how this could be improved without changing too many aspects.

Discussion

Malcolm Bolton embraced the sentiments in the heartfelt question from the speaker as he had been trying to respond to the challenges raised about creating more realistic soil profiles. David Muir Wood was impressed by the technologies introduced in the centrifuge model preparation but queried where the sources of clay or sand were located and how a sand embankment would be built. Philippe Nater replied that he used sand and clay containers (hoppers), which were suspended above the rotating centrifuge, and that sand placement could be controlled locally by using a hopper and nozzle system and moving the actuator up and down. He confirmed that CPT testing had been used to check consistency from which it was found that the initial heaving had no significant impact on the homogeneity of the soil. Muir Wood recalled that the behaviour of random material in hoppers, in terms of the structures in the granular material deposited, was extremely sensitive to the way in which the material was poured into the hoppers or silos. He thought that this would not lead to uniform (1D) deposition over the whole area of the model, and even though the cone tests had shown a uniform profile, he was unsure what this meant in terms of the real stress history of the sand. Nater agreed that this was a concern and referred back to Lis Bowman's comments on how the direction of the grains developed following deposition. Muir Wood pointed out that there were two important aspects: the direction of the grains and the nature of the force alignment between the grains, which were not necessarily the same things and that the force alignment could not be seen. Sand heap formation has a self-organised criticality and the soil is constantly flowing out to the sides from the deposition region rather being dumped over the whole area.

Bolton asked if it would be possible to use the spinning disk to distribute the clay slurry, bearing in mind an equivalent piece of agricultural machinery! Muir Wood countered with a suggestion to use a whole series of spinning disks at the same time as a multi-point distribution system rather than just one. Nater commented that the mess would be significant (especially during starting and stopping the slurry flow). Jacques Garnier thought that sample preparation was not a specific problem that afflicted solely centrifuge modellers but a problem for the world soil mechanics community. Muir Wood maintained however that the preferred solution for individual element tests would be to try to pluviate over the whole section of the sample simultaneously rather than from a point source. Michael Davies suggested that samples could be impregnated with resin and cut to form sections so that the structure could be inspected at different locations. Nater reported his observation that the density profile of the sand bed varied over the height of the sample, with a higher density at the bottom than at the top.

König questioned the stability of the soft clay surface after it had consolidated under self-weight, when the centrifuge had been stopped. Nater agreed that the surface was really soft and explained that the top 2 cm of soil just slid away when the centrifuge was stopped, while the rest remained stable. He said that the transition zone was about 1 mm thick when sand was sprayed on top and that the sand particles had not penetrated the clay. König asked if there was a problem with air entry after unloading (giving suction values greater than...) so that the clay was not fully saturated. Nater acknowledged that this was a challenge because of the free surface, but that water would be supplied to saturate the overlying sand layer, while additional consolidation time was allowed. He thought this would help to refill the pores again although he was unsure whether the clay would really fully resaturate.

Andrew Schofield commented on self-weight consolidation, referring to the tendency for the strength of the (normally consolidated) clay to increase with depth. Despite the consistency of the soft paste at the top approaching that of face cream, he said that consolidation at above 30g would always deliver enough strength on each layer, even to hold the very soft paste at the top and that this was supported by the calculations. He explained that a very thin layer face cream would be stable whereas an inch or a cm of face cream would creep down. Nater agreed that this was true and thought that the surface instability was caused by the surface water draining off as the centrifuge stopped, and as the water flowed out, it took some material with it. Schofield commented that it was important to be very careful with any change of speed whenever the centrifuge was decelerated, otherwise erosion would occur. He thought that the best solution would be to lower the water level until it was 1 mm above the soil so that the change of speed would occur without setting up a big wave.

Constitutive and Centrifuge Modelling: Two Extremes, Springman (ed.)
© *2002 Swets & Zeitlinger, Lisse, ISBN 90 5809 361 1*

The Seven Ages of Centrifuge Modelling

W.H. Craig
University of Manchester, United Kingdom

The paragraphs below arise from my contribution to a debate held at the Workshop at Monte Verita in which I was asked to present the case that

> *Centrifuge tests are a sufficient means of solution to geotechnical design problems.*

I have fought for a variety of lost causes in my time and this one was lost before we took the floor in the knowledge that the word '*sufficient*' would be interpreted in the mathematical sense as meaning necessary and sufficient for a particular purpose. I think there is not now, and never has been, an engineer to whom the centrifuge modelling technique has seemed so versatile and ubiquitous that no other tools or approaches to design are necessary. However there are some who consider that the centrifuge is one of the array of tools that can contribute directly to project-specific design and a rather wider group by whom the centrifuge is accepted in the development of more generic design approaches and as a means for determining the effects of parametric variations in design and in identifying mechanisms of behaviour in serviceability and limit states.

I took as a theme the speech of the 'melancholy' Jaques in William Shakespeare's play As You Like It (1599), that has become known as The Seven Ages of Man.

> *All the world's a stage,*
> *And all the men and women merely players.*

I tried to enliven and enlighten the debate by looking at seven individuals who had, or might have had, an influence on the development of centrifuge technology and its possible application to design over the last 150 years. In the note below I have expanded this a little but kept to the seven stages of my initial talk.

1 EDOUARD PHILLIPS

Born in Paris in 1821, the son of an English father and a French mother, Edouard Phillips became a naturalized French citizen during his period of study at the Ecole Polytechnique. Details of his life and work were reported briefly by Craig (1989) on the centenary of his death, drawing heavily on the obituary notice written by Sauvage (1891).

Phillips (Fig. 1) made his career principally in the railway and mining industries and he also held a number of teaching appointments. He published extensively from 1845 until his death, but in his obituary only a single paper is cited at length – the seminal work dealing with the possible use of centrifuge models in engineering that was published in full in the 1869 Memoires of the Academie des Sciences and summarized in the Comptes Rendus of 11 January of that year (Phillips 1869a). The paper recognized the limitations of contemporary elastic theory in the analysis of complex structures and considered the use of models. He introduced scaling relationships and examined several different scenarios, concluding with the case where self-weight body forces are significant. He proposed the exploitation of centrifugal acceleration to generate increased body forces on reduced size models, taking as examples the Britannia Bridge over the Menai Strait in Wales, designed 20 years earlier, the tubular bridge in Conway, also in Wales, and a possible bridge across the English Channel where he discussed briefly the foundation problems.

Figure 1. Edouard Phillips

In the first paper he considered only quasi-static problems, of analysis and design, but later in the same year (Phillips 1869b) he extended his thoughts to dynamic effects. His conclusion that, in the centrifuge, inertial time scaling is the same as linear scaling, is today commonly accepted by centrifuge modellers.

It appears that the idea was ahead of its time and there is no record of anyone in the nineteenth century using a centrifuge for modelling elastic or other structures.

2 PHILIP BUCKY

The first mention of applied centrifuge modelling in the literature appears to be that of Philip Bucky (1931), working at the University of Columbia in New York. Bucky (Fig. 2) worked on mining problems and first refers to the centrifuge after noting that

Models may be made the best means of designing works of great magnitude.

It is not clear whether Bucky knew of Phillips' work, written in French, but as an educated American may well have read the lines of Walt Whitman (1855):

One of the centripetal and centrifugal gang
I turn and talk like a man leaving charges before a journey.

Figure 2. Philip Bucky

Little did he know of the journey ahead. Bucky's work in mining continued for several years and was applied specifically to design – see Bucky & Fentress (1934). It was followed later by that of Louis Panek at the US Bureau of Mines in College Park Maryland – see for example Panek (1952), also working on design problems and by others - see Clark (1988), before the explosive growth of the technique in North America in the 1980s and 1990s.

3 G.Y. POKROVSKY

At about the same time as a centrifuge was first used in the USA, two pioneers were developing the technique, apparently independently in the USSR. N. N. Davidenkov and Georgi Yosifovitch Pokrovsky both published papers in Russian in 1933 and the first paper was published in mainstream geotechnical literature, from the same source, at the First International Conference of Soil Mechanics and Foundation Engineering (ICSMFE) - Pokrovsky & Fedorov (1936).

Figure 3. G.Y. Pokrovsky

Pokrovsky came to dominate the field in the USSR and the scale and impact of his work only became apparent at the 8th ICSMFE, held in Moscow in 1973 when he appeared (Fig. 3) and indicated the scope of his work over 40 years at a special seminar. His valuable books were subsequently translated into English (Pokrovsky & Fedorov 1975) and enjoyed a limited circulation along with a volume of work by Malushitsky (1975), also translated from Russian.

Pokrovsky had a prime technical interest in explosives and their use for civilian and military purposes. He is described in a review of Soviet science that I bought and read as a schoolboy (Vassiliev & Gouschev 1961), as a man of culture who painted, composed music and wrote poetry – it is tempting to speculate whether he also might have read Whitman.

With changes in the political and economic situation in the former Soviet Union the visible output from centrifuge groups in this area has been limited in recent years. Nonetheless there can be little argument with the belief that Pokrovsky was a giant in the field and the significance of his work should not be underestimated. There is no doubt that his experiments were utilized within the design process on a grand scale.

4 KARL TERZAGHI

While Terzaghi (Fig. 4) is widely considered the founding father of Soil Mechanics his background was that of a geologist (Goodman 1998), whose early career was in Europe. In 1931, while writing a book on Foundation Engineering, which would be a step forward from *Erdbaumechanik,* he took a vacation in Ticino, at Lugano – only a short distance from Monte Verita. He sought the tranquility to concentrate.

Figure 4. Karl Terzaghi

Faced with political turmoil in Europe at this time, he was offered a post at Columbia University but chose not to pursue the opportunity. A later possibility of a sabbatical there in 1934-35 also failed to materialize. However he was aware of the activity of Bucky in that institution and acknowledged his work in the period 31-34 in Article 153 in Theoretical Soil Mechanics (Terzaghi 1943).

Not usually keen on physical models, though he did perform his own trap-door experiments, Terzaghi was well aware of their potential and corresponded in the 1950s with Peter Rowe in relation to the latter's classic work on large 1g models of anchored bulkheads and sheet-piled walls. In this context it is interesting to look at the presentation of Herle (2002) to this Workshop and to make the suggestion that someone armed with Rowe's flexibility method of analysis and little more than geometric data and stratification would have fared quite well in the prediction game.

Terzaghi was heavily involved in site-specific design problems throughout his working life. He sought peace to write by coming to Ticino and might well have worked in close proximity to Bucky had he so chosen. He did recognize the role of model testing in certain circumstances but elected not to become personally involved in any major way.

5 PETER ROWE

Peter Rowe worked in Manchester from the early 1950s and built his own centrifuge there in the Department of Engineering of the University of Manchester shortly after Andrew Schofield built one a kilometre or so away at UMIST – an associated institution at that time and now a separate University. It was my own good fortune to work there with Rowe on Schofield's new machine in 1969-70 and subsequently to become involved with the bigger machine built in 1970-71 in the laboratory that now bears Rowe's name.

Figure 5. Peter Rowe

Rowe (Fig. 5) was an academic and a major geotechnical consultant. The machine he built was capable of carrying models 2.0 x 1.0 x 0.6 m deep to 120 g from the outset (and structurally capable of reaching 200 g, though the necessary drive power has never been installed). This specification was a direct result of his appreciation of the need for a major piece of hardware to assist the design studies for major projects. The first large models related to studies for two major (in UK terms) water retaining embankments up to 60 m high, incorporating quite elaborate geometrical details and, in one case, large undisturbed blocks of site clay with fabric that controlled foundation drainage. By chance, soon after machine installation, the development of oil and gas reservoirs beneath the North Sea began and the Manchester laboratory was kept busy with a series of modelling programmes related to huge offshore gravity platforms and later piled and jack-up structures. These have been followed, since Rowe's retirement, by studies of drag anchor behaviour and of upheaval buckling of pipelines when the combination of 100g+ accelerations and a 2 m model length have been particularly valuable.

Rowe's perception of the role of the centrifuge is best illustrated by major programmes in the 70s and 80s relating to the design of the Oosterschelde Storm Surge Barrier (Rowe & Craig 1976, 1978) and for the concrete gravity platform being developed by Norske Shell for the Troll field (Craig 1993). In both cases, model studies over several years were major contributors to the selection of overall structural form and subsequently to performance prediction studies for structures with no close antecedents. These instances may be among the closest that the centrifuge has come to being 'necessary' for design, though no claim is made for 'sufficiency'.

Like Terzaghi, Rowe spent vacations in the Swiss mountains, though his desire was to walk and his region of choice, as I recall, was the Bernese Oberland rather than Ticino.

6 ANDREW SCHOFIELD

In the debate at Monte Verita I singled out Professor Schofield to represent the sixth age. In truth I should have linked his name with that of Professor Mikasa in Japan. Both men developed centrifuge technology in their own countries in the mid 1960s starting from the twin considerations of consolidation of soil and slope stability problems.

Andrew Schofield (Fig. 6) was present at the Workshop and was surrounded there by his former students and colleagues. Neither he nor his work needed introduction at the debate and it was inappropriate to dwell there on his contribution. He introduced Peter Rowe to the idea of centrifuge modelling and he and Mikasa should be considered ahead of Rowe in strict chronological order. Departure from that order reflects the fact that Rowe and others above are now deceased while Schofield and those below continue to contribute in the twenty-first century. It was also tempting to link Rowe with Terzaghi and to bring Schofield closer to his former Cambridge student Springman.

Figure 6. Andrew Schofield

Schofield's range of work has been vast and much addresses directly or indirectly the broader concepts of identifying mechanisms of behaviour and developing strategies for avoiding problems – all of which are embraced by the broader concept of engineering design. I do not attempt to enlarge on Schofield the man and only make a personal selection of two examples of his work – others might choose differently.

London in the 1970s was threatened by flooding in the event of storm surges driving water down the North Sea and into the Thames estuary – essentially the same problem as that at Oosterschelde. The Thames Barrier was built, before the Dutch structure, and downstream riverside embankments were raised. Schofield worked with centrifuge models, during a period at UMIST in Manchester, probing the stress paths that might lead to the bursting of river waters into the marsh areas outside the embankments – as it were on the 'dry' side (Hird et al. 1978). He later worked at Cambridge on centrifuge model tests that elucidated the causes of collapse of some of the Mississippi flood levees towards the river – as it were on the 'wet' side (Schofield 1980). In both cases, by the time the centrifuge modelling was over, the problem identification and conceptual design needed to overcome those problems was complete and only detailed design and implementation of solutions remained.

It is appropriate to say more of Professor Mikasa. Working exclusively in Japan he developed theories of consolidation for soft clay and utilized the centrifuge in these studies. His paper (Mikasa et al. 1969) at the 7th ICSMFE coincides with that of Avgherinos & Schofield (1969) at the same meeting – both were related to slope stability. His work continued for many years, though with a lower international profile than that of Schofield, who has travelled the world preaching the gospel of the centrifuge. Like Rowe, Mikasa had an involvement in site-specific design of large dams. The widespread use of the centrifuge today in the Japanese construction industry has its roots in the early work of Mikasa and his successors, notably Professor Kimura. The different national culture of Japan may be partly responsible for the substantially different centrifuge infrastructure that has developed there. Elsewhere in the world, centrifuges remain predominantly in academic and government agency hands. In Japan, a large part of the centrifuge activity lies much closer to the principal designers and constructors of new works.

7 SARAH SPRINGMAN

Reversing the order of Jaques' seven, the last Age is typified by the youngest of the men and women I have chosen. In the nineteenth century the seed of an idea for the centrifuge was sown and in the twentieth the technique was born and came of age. At the start of the twenty-first century there is a relatively mature technology in place, though nothing stands still. The giants above have brought many advances and the centrifuge has demonstrated capabilities that Phillips could never imagine. Sarah Springman (Fig. 7) has brought the centrifuge reality to Switzerland and in particular to ETH. She has told of the historical links of her own family with Monte Verita and all who attended the Workshop enjoyed the location and the atmosphere and the science that was discussed.

The ETH in Zurich, and its geotechnical group will develop solutions to problems that are peculiarly Swiss as well as those that are of more general concern. Sarah Springman is better

Figure 7. Sarah Springman

known as an athlete than as an artist. All who attended the Workshop will wish her, and the group of young researchers she has assembled, well – their output as evidenced in the papers to the Workshop is increasing in volume and importance.

8 CONCLUSION

That the centrifuge will play a part in design in its broadest sense is not in doubt. This will not be sufficient for full design solutions, but will nonetheless be an integral part of many specific projects as well as of generic solutions presented is growing in volume and significance.

ACKNOWLEDGEMENT

Some of the material above has been published before and the opinions expressed are my own. Professors Schofield and Springman provided their own images for which I am grateful. I am further indebted to the staff of the Library of the University of Columbia who dredged through their archives to unearth the photograph of Bucky and to J. Garnier of LCPC in Nantes who provided the photograph of Phillips some years ago. The photo of Professor Terzaghi has been taken from Terzaghis memorial internet site at Oklahoma State University (http://geo-tech.civen.okstate.edu/People/terzaghi/terzaghi.htm).

REFERENCES

Avgherinos P. J. & Schofield A. N. 1969. Drawdown failures of centrifuged models. *Proc. 7th Int. Conf. Soil Mechanics & Foundation Engineering* 2: 497-505. México: Sociedad Mexicana de Mecánica de Suelos.

Bucky P. B. 1931. The use of models for the study of mining problems. *Technical Publication 425*. New York: Am. Inst. Of Min. & Met. Engng.

Bucky P. B. & Fentress A. L. 1934. Application of principles of similitude to design of mine workings. *Trans. Am. Inst. Min. & Met. Eng.* 1: 25-50.

Clark G.B. 1988. Centrifugal testing in rock mechanics. In W.H. Craig, R.G. James & A.N. Schofield (eds), *Centrifuges in Soil Mechanics*: 187-198. Rotterdam: Balkema.

Craig W. H. 1989. Edouard Phillips (1821-89) and the idea of centrifuge modelling. *Géotechnique* 39(4): 697-700.

Craig W. H. 1993. Partial similarity in centrifuge models of offshore platforms. In J. Clark et al. (eds) *Proc. 4th Canadian Conf. on Marine Geotechnical Engineering* 3: 1044-1061. Newfoundland: Memorial University of Newfoundland, St. Johns.

Davidenkov N. N. 1933 The new method of the application of models to the study of equilibrium of soils. *J. Tech. Physics* 3: 131-136, Moscow (In Russian).

Goodman R. E. 1998. Karl Terzaghi - The Engineer as Artist. *ASCE Press*. Virginia: Reston.

Herle I. 2002. Difficulties related to numerical prediction of deformations. In S.M. Springman (ed.) *Workshop on Constitutive and Centrifuge Geotechnical Modelling: Two Extremes, Monte Verita, Ticino, Switzerland, July 2001*. Rotterdam: Balkema.

Hird C. C., Marsland A. & Schofield A. N. 1978. The development of centrifugal models to study the influence of uplift pressure on the stability of a floodbank. *Géotechnique* 28(1): 85-106.

Malushitsky Y. H. 1975. The centrifugal testing of waste-heap embankments. Cambridge: Cambridge University Press.

Mikasa M., Takada N. & Yamada K. 1969. Centrifugal model test of a rockfill dam. *Proc. 7th Int. Conf. Soil Mechanics & Foundation Engineering* 2: 325-333. México: Sociedad Mexicana de Mecánica de Suelos.

Panek L. A. 1952. Centrifugal testing applied to the design of mine structures with special reference to roof control. *Proc. 7ʰ Int. Conf. Directors of Safety in Mines Research.*

Phillips E. 1869a. De l'equilibre des solides elastiques semblables. *C. R. Acad. Sci., Paris* 68: 75-79.

Phillips E. 1869b. Du mouvement des corps solides elastiques semblables. *C. R. Acad. Sci., Paris* 69: 911-912.

Pokrovsky G. Y. 1933. On the application of centrifugal force for modelling earth works in clay. *J. Tech. Physics* 3: 537-539, Moscow (In Russian).

Pokrovsky G. Y. & Fedorov I. S. 1936. Studies of soil pressures and soil deformations by means of a centrifuge. In A. Casagrande, P.C. Rutledge & J.D. Watson (eds) *Proc. 1ˢᵗ Int. Conf. On Soil Mechanics & Foundation Engineering* 1: 70. Cambridge, Massachusetts: Harvard University.

Pokrovsky G. Y. & Fedorov I. S. 1975. *Centrifugal testing in the construction industry.* Vols. 1 & 2. English translation by Building Research Establishment Library Translation Service of monographs originally published in Russian, Watford.

Rowe P. W. & Craig W. H. 1976. Studies of offshore caissons founded on Oosterschelde sand. *Design and Construction of Offshore Structure*: 49-55. London: Institution of Civil Engineers.

Rowe P. W. & Craig W. H. 1978. Prediction of caisson and pier performance by dynamically loaded centrifuge models. *Symp. On Foundation Aspects of Coastal Structures* 2, Paper IV.3, 16p, Delft.

Sauvage E. 1891. Notice Necrologique sur Edouard Phillips. *Ann. Mines, Series 8* 19: 343-378.

Schofield A. N. 1980. Cambridge geotechnical centrifuge operations. *Géotechnique* 30(3): 227-268.

Shakespeare W. 1599. *As You Like It.*

Terzaghi K. 1943. *Theoretical Soil Mechanics.* New York: Wiley.

Vassiliev M. & Gouschev S. (eds) 1961. *Life in the twenty-first century.* Harmondsworth, Middlesex: Penguin Books.

Whitman W. 1855. *The Leaves of Grass.* New York: Brooklyn.

Problems governed by failure

Constitutive and Centrifuge Modelling: Two Extremes, Springman (ed.)
© 2002 Swets & Zeitlinger, Lisse, ISBN 90 5809 361 1

Observed failures of laboratory sand samples and constitutive modelling

A. Gajo
Department of Mechanical and Structural Engineering, University of Trento, Trento, Italy

ABSTRACT: Failure of a quasi-statically loaded sample may be viewed as the ultimate manifestation of a complex mechanical process in which different mechanisms emerge and subsequently interact in leading towards global collapse. One of these mechanisms, however, often prevails over the others and dominates the global response. These different mechanisms may be described as follows:
- a quasi-static deformation path can degenerate into a dynamic motion, thus yielding a global collapse;
- an essentially homogeneous deformation field may become 'weakly' non-uniform, due to the emergence of inhomogeneous deformation modes of bifurcation;
- strongly localised deformation patterns may emerge in an essentially homogeneous deformation field.

In granular materials, all the above instabilities have been observed experimentally. In particular Begemann et al. (1977) and Lindemberg & Koning (1981) were the first to recognise that samples of saturated loose sand may undergo a sudden collapse under drained, static, stress controlled conditions. On the other hand, strain localisation of granular materials has been widely studied experimentally (Arthur et al. 1977, Vardoulakis 1980, Desrues et al. 1985). Concerning the second type of instability, it is worth recalling that excessively slender samples are generally avoided in experimental testing, in order to retard the appearance of inhomogeneous deformation modes.

Herein the attention will be focused on the onset of mechanical instability and the occurrence of strain localisation in granular media. Particular attention will be paid to the 'ingredients' of constitutive models which enable a reliable description of the failure mechanisms.

The analyses of failure of laboratory samples will be performed by using the Severn-Trent model (Gajo & Muir Wood 1999a, 1999b), which can be considered representative of the modern class of constitutive models that incorporate several aspects of mechanical behaviour of granular materials within a unique framework. In particular, the Severn-Trent model simulates the existence of an elevated stiffness at small strains, the progressive degradation of the elasticplastic stiffness and the pressure-dependence and density-dependence of the mechanical behaviour of granular media. It is a kinematic hardening, bounding surface plasticity model, based on well accepted concepts such as a Mohr-Coulomb-like strength criterion, critical state, a hyperbolic relationship for representing plastic stress-strain behaviour, dependence of strength on a state parameter and a flow rule derived from the Cam-Clay Model. The yield surface and the bounding surface are assumed to be smooth non-circular cones with the apex at the origin of the

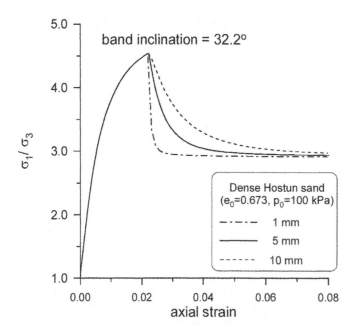

Figure 1. Effects of band thickness in simulated drained biaxial tests on dense sand (Gajo et al. 2001)

axes; the yield surface can freely rotate inside the bounding surface and can touch it only at the critical state, i.e. in the limit of infinite strains.

In its original form, the proposed model can deal only with the anisotropy of plastic behaviour, but, for the simulation of failure, particular attention must be paid also to the elastic anisotropy (Gajo et al. 2001). This is why the Severn-Trent model has been enhanced with the linear, anisotropic elasticity model proposed by Valanis (1990) and Zysset & Curnier (1995).

The problem of occurrence of strain localisation can be reduced to the well-known condition of vanishing of the determinant of the elastic-plastic acoustic tensor (Rudnicki & Rice 1975), which was solved for elastic anisotropy by Bigoni & Loret (1999) and Bigoni et al. (2000). The analysis of the post-localisation regime was performed, as described by Gajo et al. (2001), according to Hutchinson & Tvergaard (1980), namely assuming the existence of two homogeneous states of deformation, one inside and the other outside the shear band. These must satisfy the stress and the strain compatibility conditions along the two interfaces between the shear band and the remaining sample. Moreover, the kinematic conditions related to the boundary conditions must be taken into account. For the simulation of drained tests, this requirement leads to just one condition concerning the total vertical strain; whereas for the simulation of the undrained tests (i.e. globally isochoric tests), a further kinematic condition on the volumetric strains must be introduced (Gajo et al. 2001). In the latter case, the effects induced by membrane and system compliances were neglected and the pore pressure was implicitly assumed to be equilibrated everywhere.

Comparisons of theoretical simulations with experimental results obtained on dense Hostun sand by Desrues & Hammad (1989), for biaxial tests at different confining pressures, and by Desrues et al. (1996), for triaxial tests, show that in drained biaxial and triaxial tests the occurrence of strain localisation and the post-localisation behaviour are completely dominated by the stress induced elastic anisotropy. In particular, in the simulation of drained biaxial tests on dense sand, the assumption of elastic anisotropy leads the material outside the band to reload plastically, a phenomenon described as 'band saturation' by Tvergaard (1982). On the other hand, the

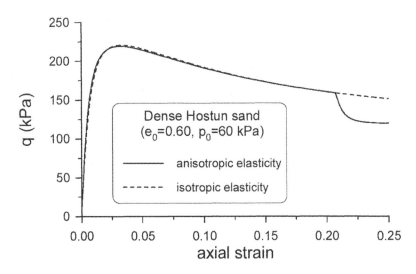

Figure 2. Typical simulated conventional stress-strain behaviour of drained triaxial tests on dense sand (Gajo et al. 2001)

assumption of elastic anisotropy leads to a fully developed post-localisation behaviour, which is consistent with experimental observations by Desrues & Hammad (1989), as shown in Figure 1, where the effects induced by band thickness on the simulated post-localisation behaviour of a dense Hostun sand sample are presented. In the simulation of drained triaxial tests performed on dense sand (Fig. 2), the introduction of elastic anisotropy leads strain localisation to occur at a conventional axial strain of 20 %, consistent with experimental observation by Desrues et al. (1996).

The fact that a saturated loose or medium dense sand sample may undergo a sudden collapse within the steady state envelope is a clear indication of the occurrence of instability, in the sense that a small increment in some variables leads to a sudden, large increment in other quantities. By using a specially built triaxial apparatus with no friction on the loading ram, and enabling a 100 mm diameter sample to be loaded by both dead weights and a pneumatic piston (which was practically frictionless due to a rolling diaphragm), Gajo et al. (2000) performed an accurate analysis of many factors affecting the onset of collapse, namely the loading method (dead weight or pneumatic piston), the loading path, the loading rate, the pre-shearing and the initial density.

The stability of the sample can be analysed using Hill's (1958) stability condition, which expresses that the second order variation of the work of deformation is strictly positive for all smooth velocity fields, taking null values where displacements are prescribed. As pointed out by Petryk (1982, 1985), when the loading device has a potential energy, then the stability criterion must be corrected by adding the potential energy of the loading device.

The analysis of collapse behaviour was performed by comparing the experimental results with the theoretical existence of different, energetically possible, departure paths involving homogeneous, small strains and mobilising, at different levels, the various sources of compliance of the loading apparatus. The second order variation of the energy of the various loading devices was evaluated by considering different assumptions about the meaningful strain rate at collapse, which can be sufficiently slow to allow air to flow freely through air pressure regulators, or sufficiently fast to prevent air flow through air pressure regulators and to induce an adiabatic compression of the compressed air inside the pneumatic piston and the volume change measurement device, or, finally, extremely fast to prevent pore water flow through drainage lines.

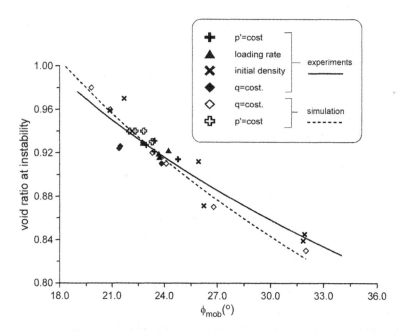

Figure 3. Comparison between void ratio and mobilised friction angle at collapse, observed experimentally and deduced theoretically (Gajo 2002)

Theoretical predictions are compared with experimental results in Figure 3. The excellent agreement shown is obtained only if the compressibility of the apparatus is taken into account. In particular, the samples behave as if the meaningful strain rate at collapse is large enough to induce an adiabatic compression of the air contained in the loading devices. This is confirmed also by the experiments performed by loading the sample by the pneumatic piston. Moreover some new experiments performed by using a smaller (and consequently less compressible) volume change measuring device confirm the theoretical expectations about the role of the compliance of the loading apparatuses (Gajo 2002).

REFERENCES

Arthur, J.R.F., Dunstan, T., Al-Ani, Q.A.J.L. & Assadi, A. 1977. Plastic deformation and failure in granular media. *Géotechnique* 27(1): 53-74.

Begemann, H.K., Koning, H.L. & Lindemberg, J. 1977. Critical density of sand. In Japanese Society of Soil Mechanics and Foundation Engineering (eds), *Proc. IX ICSMFE, Tokyo, Japan* 1: 43-46.

Bigoni, D. & Loret, B. 1999. Effects of elastic anisotropy on strain localisation and flutter instability. *J. Mech. Phys. Solids* 47: 1409-1436.

Bigoni, D., Loret, B. & Radi, E. 2000. Localisation of deformation in plane elastic-plastic solids with anisotropic elasticity. *J. Mech. Phys. Solids* 48: 1441-1466.

Desrues, J., Lanier, J. & Stutz, P. 1985. Localisation of the deformation in tests on sand sample. *Engng. Fracture Mech.* 21: 909-921.

Desrues, J. & Hammad, W. 1989. Shear band dependency on mean stress level. In E. Dembicki, G. Gudehus & Z. Sikora (eds) *Int. Workshop Num. Meth in Localisation and Bifurcation of Granular Materials, Gdansk, Poland*: 57-67.

Desrues, J. Chambon, R. Mokni, M. & Mazerolle, F. 1996. Void ratio evolution inside shear bands in triaxial sand specimens studied by computed tomography. *Géotechnique* 46(3): 529-546.

Gajo, A. 2002. The influence of system compliance on collapse of triaxial sand samples. *Submitted.*

Gajo, A. & Muir Wood, D. 1999a. Severn-Trent sand: a kinematic hardening constitutive model for sands: the *q-p* formulation. *Géotechnique* 49(5): 595-614.

Gajo, A. & Muir Wood, D. 1999b. A kinematic hardening constitutive model for sands: the multiaxial formulation. *Int. J. Numer. Anal. Meth. Geomech.* 23: 925-965.

Gajo, A., Piffer, L. & De Polo, F. 2000. Analysis of certain factors affecting the unstable behaviour of saturated loose sand. *Mech. of Cohes. Frict. Mater.* 5: 215-237.

Gajo, A., Bigoni, D. & Muir Wood, D. 2001. Stress induced elastic anisotropy and strain localisation in sand. In Muehlhaus H. (ed.) *Proc. Int. Workshop on Bifurcation and Localisation Theory in Geomechanics*: 37-44. Lisse: Swets & Zeitlinger.

Hill, R. 1958. A general theory of uniqueness and stability in elasto-plastic solids. *J. Mech. Phys. Solids* 6: 236-249.

Hutchinson, J.W. & Tvergaard, V. 1980. Shear band formation in plane strain. *Int. J. Solids Structures* 17: 451-470.

Lindemberg, J. & Koning, H. L. 1981. Critical density of sand. *Géotechnique* 31(2): 231-245.

Petryk, H. 1982. A consistent energy approach to defining stability of plastic deformation processes. *Stability in the Mechanics of Continua, Proc. IUTAM Symp., Numbrecht (1981)*: 262-272. Springer.

Petryk, H. 1985. On energy criteria of plastic instability. Plastic Instability, Proc. Considere Memorial, Ecole Nat. Ponts Chauss., Paris: 215-226.

Rudnicki, J.W. & Rice J.R. 1975. Conditions for the localisation of deformation. *J. Mech. Phys. Solids* 23: 371-394.

Tvergaard, V. 1982. Influence of void nucleation on ductile shear fracture at a free surface. *J. Mech. Phys. Solids* 30: 399-425.

Valanis, K. C. 1990. A theory of damage in brittle materials. *Eng. Fract. Mech.* 36: 403-416.

Vardoulakis, I. 1980. Shear band inclination and shear modulus of sand in biaxial tests. *Int. J. Numer. Anal. Meth. Geomech.* 4: 103-119.

Zysset, P. K. & Curnier, A. 1995. An alternative model for anisotropic elasticity based on fabric tensors. *Mech. Materials* 21: 243-250.

Constitutive and Centrifuge Modelling: Two Extremes, Springman (ed.)
© *2002 Swets & Zeitlinger, Lisse, ISBN 90 5809 361 1*

Centrifuge modelling of failures and safe structures

R. Phillips
C-CORE, St. John's, Newfoundland, Canada

ABSTRACT: Geotechnical engineering constructions are typically redundant systems comprising compliant natural deposits and stiffer man-made elements. Failure of these systems may comprise excessive deformation or exceedance of acceptable load levels. The impact of centrifuge modelling on a series of typical geotechnical engineering construction problems, that are driven by failure, are discussed. The importance of differentiating element failure from system failure will be demonstrated from models involving liquefaction phenomena. Various system failure definitions will be reviewed during consideration of various ring foundations under different combined loading conditions. A range of geotechnical problems will then be discussed involving modelling for the design of safe structures through failing 'soil' masses. Examples will include pipelines across landslides, ice scour and ice-structure interaction. The importance of validating the model test results will be emphasised.

1 CANLEX FLOWSLIDES

The Canadian Liquefaction Experiment (CANLEX), reported by Robertson et al. (2000), investigated the liquefaction potential of loose sand deposits under monotonic shear stress increments. The constitutive behaviour of the different sands was very well defined. A full-scale field event was designed at one site to cause a flowslide and this was based on numerical analyses incorporating the measured constitutive response of oil sand tailings. Alternative geotechnical construction scenarios, which could be adapted to increase shear stress ratios rapidly in the liquefaction susceptible foundation sand, were also considered, including heightening of a sand embankment and impoundment of fluidised tailings behind a clay dyke. Three numerical analyst groups modelled the measured constitutive response. These constitutive models were integrated spatially and temporally over the proposed construction with appropriate boundary conditions. The finite element and finite difference analyses considered compatibility and equilibrium and predicted the physical response of the field event.

Centrifuge model tests were added to the CANLEX program to simulate the physical response of the field event. Four tests (II-1 to II-4) of the first scenario (Fig. 1), and one test of the second scenario were undertaken. Model tests II-1 to II-3 were homogenous and tested at 29%, 40% and 50% relative density respectively. The geometry was similar for model test II-4 but the embankment was about 49% relative density and the sand layer was about 25%. Liquefaction was observed in 3 of the 4 first tests. Liquefaction was sufficiently contained in test II-2, that

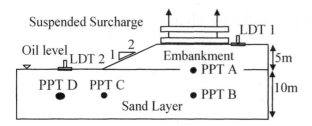

Figure 1. Model configuration for scenario 1

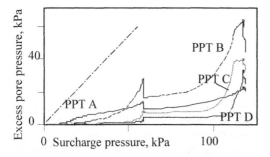

Figure 2. Excess pore pressure response, test II-2

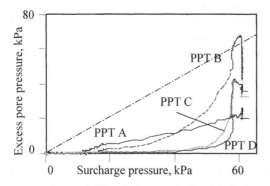

Figure 3. Excess pore pressure response, test II-4

despite local failure of the sand at points B & C (Fig. 2), a flowslide did not occur. In tests II-1 and II-4, the liquefaction was more extensive and flowslides ensued, (Figs 3 & 4), causing significant deformation of the embankment.

The simulations of one analyst and the centrifuge test data were mutually validated as the work proceeded. Both of these validated techniques predicted that the second scenario would not result in a flowslide. The other 2 numerical analysts predicted potential for a flowslide for this scenario. There was no evidence of significant liquefaction or a flowslide in the subsequent field event using scenario 2.

This example demonstrates the necessity of incorporating the essential element of constitutive behaviour into a complete analytical framework to predict the response of redundant engineering constructions. Further, the centrifuge model simulations provide a reasonable approximation of the geotechnical construction response in this case, as validated by a numerical analysis and field data.

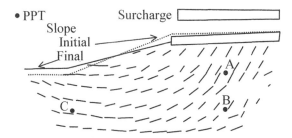

Figure 4. Deformation vectors, test II-1

2 RING FOUNDATIONS

Ring foundations have been used for offshore gravity based structures, grain silos and towers, including the Leaning Tower of Pisa. Failure modes for these foundations include bearing capacity, sliding and overturning from loads and excessive settlement or differential movement in respect of deformations. A differential movement limit of 1:1000 may be appropriate for a grain silo, but a similar limit would remove the appeal of the Leaning Tower of Pisa. Physical model tests presented by Phillips (1993) (Figs 5 & 6) and Viggiani (at this workshop) show that differential settlement problems can be simulated successfully in the centrifuge. Herle and Sharma both develop the theme of modelling deformation also at this workshop. The remainder of this paper will focus on failures from load.

Figure 5. Pisa Tower model

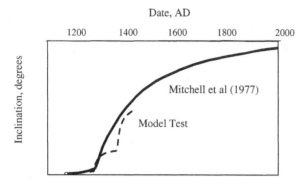

Figure 6. Tower inclination reconstructions

The Confederation Bridge has about 60 piers with ring foundations founded on sedimentary jointed rock (Fig. 7). These offshore foundations are subject to unusual loading conditions from high overturning moments from environmental loads such as first year ice sheets. The LRFD design used for these foundations included project specific calibration of the load and resistance factors. The factored load exceeded the factored resistance in the initial design at about 10 of the pier foundations. A series of centrifuge model tests of the design conditions were used to identify the mechanisms associated with the 3 different load failure modes considered (Phillips et al. 1998). The model loading paths were not biased by differences in the partial load factors. The observed bearing capacity failure mechanism (Fig. 8) provided additional capacity over that assumed in the initial design. This mechanism was validated by additional limit equilibrium analyses, and then used to design the remaining ring foundations.

Zhu (1998) and Zhu et al. (2001) extended this project to provide generic design guidelines for ring foundations on dense sand under eccentric vertical loading. Centrifuge model tests conducted on a series of foundations of constant area (Fig. 9) showed failure mechanisms changing from punching to general shear with increasing ring ratio (Fig. 10). The bearing capacity of a

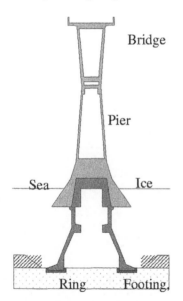

Figure 7. Typical pier configuration

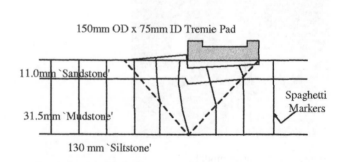

Figure 8. Typical failure mechanism

Figure 9. Ring foundation models

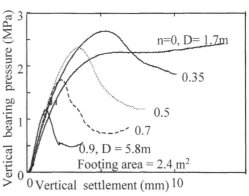

Figure 10. Ring load displacement response

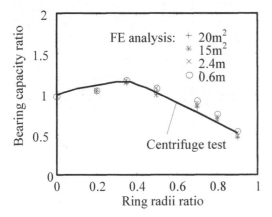

Figure 11. Ring foundations bearing capacity

ring foundation is bounded by the values for circular and strip footings. Test results for these bounds were validated against the work of others and by characteristic and finite element analyses. The finite element analyses also validated the measured variation of bearing capacity factor with ring ratio for foundation areas equivalent to 0.6 to 20 m² (Fig. 11).

3 LARGE DEFORMATION DESIGNS

Centrifuge modelling has also proven important in designing structures to withstand large relative soil (or ice) deformations. Paulin et al. (1996) simulated pipelines located in active slope movement areas in overconsolidated clay. They determined that the pipeline loads under drained conditions (Fig. 12b) were higher than those under undrained conditions (Fig. 12a), and that these loads were consistent with the ASCE guidelines for frictional and cohesive soils respectively. The soil-pipe interaction mechanism changed from a series of Coulomb wedge failures to plastic flow.

Ice gouges are caused as deep floating ice, such as icebergs and pressure ridges, touch and then move through the seafloor. Submarine pipelines and other seafloor structures must be designed to withstand these events. Centrifuge model tests have provided the necessary engineering data of subgouge soil deformations and associated loads in a range of soils (Clark et al. 1998). Figure 13 shows a deformation field in sand over clay with shear strains of order of tens of percent. A series of failure surfaces are observed. The underlying clay impinging on (or flapping over onto) the overlying sand confirmed field observations from excavations through relict gouge features. In a test with clay over sand, sand boils confirmed observations from video taken in a recent seafloor gouge feature. Large strain, large deformation finite element analyses have validated the centrifuge measurements for gouges in soft clay. Similar analyses have been unsuccessful in sands. The centrifuge measurements have now provided the basis for engineering designs for about 8 offshore pipelines in ice-gouged seabeds.

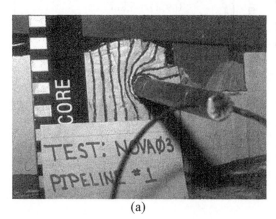
(a)

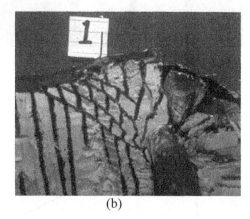
(b)

Figure 12. Pipe-soil failure mechanisms associated with undrained and drained interactions in stiff clay

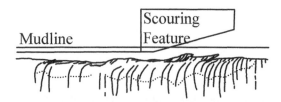

Figure 13. Subgouge deformation field

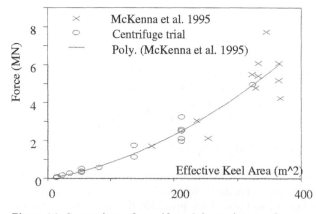

Figure 14. Comparison of centrifuge & ice tank tests of unconsolidated ice keel-structure interaction

Large scale 1g ice tank model test data are widely accepted as the basis for design for pack ice interaction with offshore structures. An ongoing study has evaluated the potential for centrifuge model tests to complement the larger ice tank tests. Barrette et al. (2000) have shown a good correlation between ice sheet - conical structure interaction loads measured in the centrifuge and predicted by accepted ice load models. A few direct comparisons of ice tank and centrifuge tests have also shown good load correlations, and, perhaps more importantly highlighted transitions between different interaction mechanisms. Figure 14 shows the loads required to fail a structure in an unconsolidated ice keel of varying cross-sectional area. The polynomial fit to the ice tank data also provides a good fit to more recent centrifuge model test data. Centrifuge tests have potential for addressing some ice structure interaction problems.

REFERENCES

Barrette, P.D, Lau, M., Phillips, R., McKenna, R.F. & Jones, S.J. 2000. Interaction Between Level Ice and Conical Structures: Centrifuge simulations Phase II. OMAE paper P&A-1004.

Clark, J.L., Phillips, R. & Paulin, M. 1998. Ice scour research for the safe design of pipelines: 1975-1997. In R. Phillips (ed.), *Proceedings of Ice Scour & Arctic Marine Pipelines Workshop. 13th International Symposium on Okhotsk Sea & Sea Ice, Mombetsu, Japan,* St. John's, Canada: C-CORE.

Herle, I. 2002. Constitutive modelling. Difficulties related to numerical predictions of deformation. In S.M. Springman (ed.), *Constitutive and centrifuge modelling: two extremes. Workshop, Ascona, Switzerland, 8-13 July 2001,* Rotterdam: Balkema.

Mitchell, J.K., Vivatrat, V. & Lambe, T.W. 1977. Foundation performance of Tower of Pisa. *ASCE Journal of the Geotechnical Engineering Division* 103(2): 227-249.

Paulin, M.J., Phillips, R. & Boivin, R. 1996. An experimental investigation into lateral pipeline interaction. *OMAE Florence,* Vol. V: 313-323.

Phillips, R. 1993. Tower construction on layered ground. Cambridge University Engineering Department Technical Report CUED/D-Soils TR261.

Phillips, R., Kosar, K.M. & Walter, D. 1998. Physical modelling of foundations for Confederation Bridge, Canada. In Kimura et al. (eds), *Centrifuge '98:* 447-452, Rotterdam: Balkema.

Robertson, P.K., Phillips, R. et al. (34 authors). 2000. The Canadian Liquefaction Experiment: An Overview. *Canadian Geotechnical Journal* 37(3): 499-504.

Sharma, J.S. 2002. Centrifuge modelling. Measurements of deformation – Trends or numbers? In S.M. Springman (ed.), *Constitutive and centrifuge modelling: two extremes. Workshop, Ascona, Switzerland, 8-13 July 2001,* Rotterdam: Balkema.

Viggiani, C. 2002. Public lecture: The leaning tower of Pisa: back to the future. In S.M. Springman (ed.), *Constitutive and centrifuge modelling: two extremes. Workshop, Ascona, Switzerland, 8-13 July 2001,* Rotterdam: Balkema.

Zhu, F. 1998. Centrifuge Modelling and Numerical Analysis of Bearing Capacity of Ring Foundations on Sand. Ph.D. Thesis, Memorial University of Newfoundland, St. John's, Canada.

Zhu, F, Clark, J.I. & Phillips, R. 2001. Scale effect of strip and circular footings resting on dense sand. *ASCE Journal of Geotechnical Engineering* 127(7): 613-621.

Constitutive and Centrifuge Modelling: Two Extremes, Springman (ed.)
© 2002 Swets & Zeitlinger, Lisse, ISBN 90 5809 361 1

The use of centrifuge modelling to investigate progressive failure of overconsolidated clay embankments

W.A. Take & M.D. Bolton
Cambridge University Engineering Department, United Kingdom

ABSTRACT: Three technologies have been applied to the centrifuge modelling of clay embankments including matric suction measurement, an environmental chamber and image-based deformation measurement. Using these technologies, a "rubblisation" failure mode has been observed in response to rainfall infiltration. The failure mode is described within the framework of presenting the challenges involved in modelling the observed behaviour numerically.

1 INTRODUCTION

Delayed failure of embankments and cut slopes has the potential to pose risk to human life, and cause great disruption to transportation infrastructure. Many of the reported failures occur after periods of wet weather, indicating an interaction between the environment, the resulting transient pore pressures and slope stability.

Research into the failure of clay embankments and cut-slopes has largely taken the form of limit equilibrium and finite element (e.g. Potts et al. 1990) back-analyses of slope failures, and numerical analysis assuming the significance of certain parameters governing rate effects (e.g. Alonso et al. 1995, Potts et al. 1997). The difficulty associated with measuring soil suctions, and the limitations associated with conventional "target-based" methods of deformation measurement have restricted the full application of physical modelling techniques to observe rainfall-induced instability under controlled laboratory conditions.

2 EXPERIMENTAL METHODOLOGY

Three technologies have been developed to model the stability of overconsolidated clay embankments more accurately during rainfall infiltration in the centrifuge. These technologies improve the ability to measure matric suctions, to control the environmental boundary conditions surrounding the slope, and to observe soil deformation without the recourse to embedded target markers.

2.1 Matric suction probe

Effective stress analysis of any observed embankment behaviour requires the knowledge of the pore water pressures within the model embankment. Recent advances in the technology of direct measurements of negative pore water pressures have indicated that the key design features of such a device are the elimination of spaces likely to trap cavitation nuclei, a small water reservoir, a high air-entry porous element, and an effective saturation procedure (Ridley 1993, Guan & Fredlund 1997). A miniature matric suction probe based on these design criteria has been developed for use in centrifuge models. With a nominal 3 bar air-entry ceramic filter, and a rigorous programme of ceramic saturation, the probe is capable of measuring negative pore water pressure to a value of -400kPa.

2.2 Environmental chamber

Results from centrifuge testing, like all modelling exercises, are susceptible to errors due to incorrect boundary conditions. For the modelling of clay embankments, one often-overlooked boundary condition is that of the relative humidity above the embankment surface. If the air above the model embankment is allowed to be exchanged with dry air from the centrifuge chamber, the embankment will tend to lose moisture over time. Such a moisture loss will lead to the development of matric suctions and associated increases of effective stress within the clay. Settlements due to self-weight consolidation will be enhanced by those associated with shrinkage. The conventional environmental boundary condition, therefore, has the consequence that the pore pressures within the centrifuge model will not come into a state of equilibrium within the practical time frame of centrifuge modelling.

An environmental chamber has been designed to impose appropriate environmental boundary conditions on centrifuge models of clay embankments (Figure 1a). The chamber is sealed to prevent moisture transfer from the model to the external surroundings, and has the capability to regulate the relative humidity of the air in the chamber, and subject the embankment to model rainfall from atomising mist nozzles.

2.3 Image-based deformation measurement

An image-based system of deformation measurement system has been developed, both for use at 1-g and in the centrifuge, based on the techniques of Particle Image Velocimetry (PIV) (White et al. 2001a). This technique relies on the texture (i.e. a map of pixel intensities) of a patch of soil to identify the patch in a second image corresponding to a following time step (Figure 1b). Once the position of best-match is identified to sub-pixel accuracy, the difference between the location of the patch in two successive images is the displacement vector of the patch. Using this technique, deformations can be measured to a precision finer than $1/10^{th}$ of a pixel both in soils which have an inherent texture due to a large (i.e. visible) grain size and in fine-grained soils onto which a texture has to be applied (White et al. 2001b).

The use of image-based deformation measurement has the benefit that the measurement points need not be decided before the event occurs. In this manner, patches can be arranged to straddle a shear surface, for example, to monitor the strain mobilised along the length of the discontinuity during progressive failure. The effort involved with embedding target markers into clay models at equally spaced intervals has historically restricted the number of measurement points. Image-based deformation measurement, therefore, has the additional benefit of allowing thousands of displacement measurement points, thereby allowing the capture of shear strain localisation.

To minimise the effective pixel size over the area of interest at an economical price, a multi camera image acquisition system has been developed to capture digital images of various areas of interest in the centrifuge model at a resolution of 1760x1168 pixels. The multi-camera system is shown in Figure 1a, with the resulting camera views displayed in Figure 1b. The multiple digital cameras are controlled using a scalable scheduling engine written in Visual Basic based on the Kodak ActiveX software development kit for the DC240/280/3400/5000 family of consumer-level digital still cameras. Once captured, the images are quickly transferred to a computer in the low-g environment at the centre of the beam centrifuge before the cycle is repeated.

Image calibration provides the translation rule to convert deformations from image-space (pixels) to real space (mm) using the principles of close-range photogrammetry. The eighteen variables used in the calibration process include, but are not limited to, corrections for lens distortion, camera orientation, and refraction through the observation window. Despite the observed relative rotation with respect to the camera CCD of the camera lens under its own elevated self-weight, the process of camera calibration can correct the movement of soil patches by accounting for the apparent movement and distortion of stationary control points (White et al. 2001a). Finally, a mapping rule is applied to convert the curved embankment geometry as tested in a radial acceleration field to its equivalent in a parallel acceleration field.

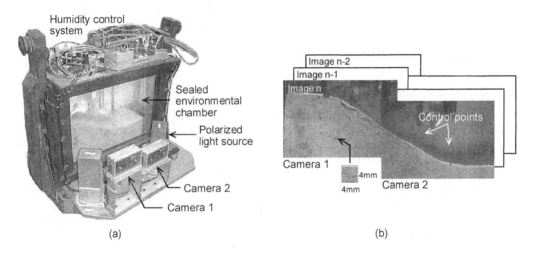

Figure 1. (a) Environmental chamber and multi-camera image acquisition system, and (b) the resulting view of the slope

3 RESULTS

The E-grade Kaolin clay used to create the model embankment was consolidated to a pressure of 500 kPa from slurry mixed at 100% moisture content. The resulting block of clay was shaped to form an embankment, underwent suction probe installation, and was inserted into the environmental chamber at a suction of 40 kPa. The clay embankment was then subjected to destabilising influences due to cycles both of increased acceleration level and of rainfall infiltration. The resulting change in embankment geometry resulting from a combination of these actions is presented in Figure 2. The original 36° slope was reduced to an angle of 30°, with a drop in the embankment surface elevation of 6.9 mm, corresponding to a field scaled displacement of 414 mm at 60g.

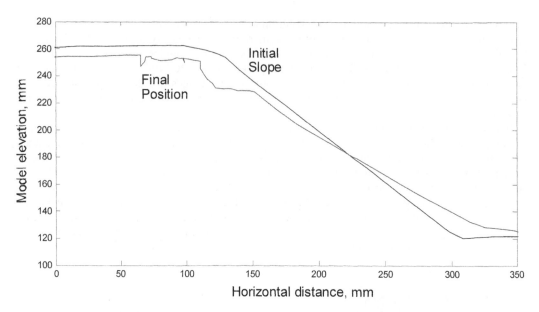

Figure 2. Embankment deformation observed in centrifuge model

3.1 *Failure due to acceleration level*

Much of the early work on centrifuge modelling of slope failures was performed by increasing the acceleration level until an undrained failure was observed. The gravity 'turn-on' technique of such a centrifuge test produced dramatic results and did not require complicated excavation or surcharging systems to bring the slope to failure. Based on the observed failure, the undrained strength of the clay could be then be back-calculated.

The 140 mm high model embankment in the current study was rapidly accelerated to 60g, causing the generation of excess pore pressures, thereby leading to some slope instability. Upon reaching an acceleration level of 60g, a failure mechanism was initiated by a crack forming at the crest of the embankment, which propagated with depth until a slip surface formed from the apex of the crack. The shear surface is very pronounced at the crest of the embankment, but gradually disappears with depth. The embankment deformations associated with this failure mechanism, as measured over 52 minutes of centrifuge testing, are shown in Figure 3. Following this slope instability, the embankment once again regained slope equilibrium and stopped deforming. This is consistent with the nature of the method in which the pore pressures were heightened. The excess pore pressures are at their maximum value at the moment the centrifuge acceleration is increased. For a relatively permeable clay embankment constituted from E-grade Kaolin, the fast dissipation of the pore water pressures leads to the embankment quickly stabilising.

The heightened pore pressures provoked by increasing the acceleration level do not model the potential increase in pore water pressures due to rainfall infiltration. As shown in Figure 3, the failure due to increasing the acceleration level will tend to be deep, and may not be driven to complete collapse. Although centrifuge models of slope instability brought about by increasing the acceleration level may behave according to the correct scaling laws, they do not necessarily reflect reality.

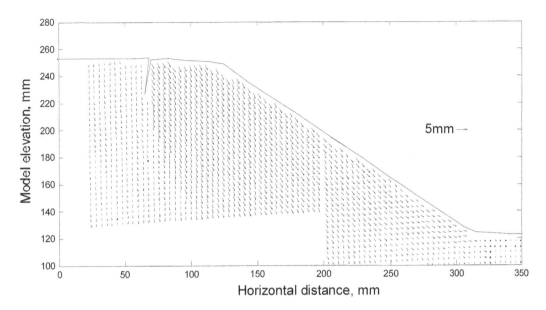

Figure 3. Observed embankment deformations associated with gravity 'turn-on' technique

3.2 *Failure due to rainfall infiltration*

Following the initial instability due to elevated g-level, the embankment was allowed to consolidate, with no additional deformations being observed on the initial failure surface. With the water table being held below the base of the embankment, and negligible moisture loss to the external environment, the pore pressures approached values of negative hydrostatic conditions. These suctions were then eliminated by the application of rainfall infiltration over the entire slope. Despite elevating the pore pressures to positive values everywhere in the embankment, only small deformations were observed. After a period of extended, continuous model rainfall, a second tension crack was formed at the crest. The deformations measured during a period of 11 minutes just prior to, and following the tension crack formation are shown in Figure 4. The deformations presented in Figure 4, therefore, do not present a complete deformation history of the embankment, but rather, a snapshot of the onset of the second instability event. The second tension crack propagated until such depth that a slip surface developed with a corresponding increase in down-slope displacements.

Shortly after the mechanism of Figure 4 formed, a third tension crack appeared at the crest, movement along the existing failure surface ceased, and the down-slope velocity was transferred to a new failure surface (Figure 5). The resulting failure surface observed over 10 minutes was also non-circular and extended slightly deeper at the toe.

4 DISCUSSION

The series of cracks and failure surfaces observed in the clay embankment together form a 'rubblisation' failure mode in which the embankment material slowly breaks up and softens. Despite the series of these rubblisation failures, the slope still retained an angle of 30° at the completion of testing. It would be unrealistic to assume, therefore, that the rubblisation process had been carried out to its final state.

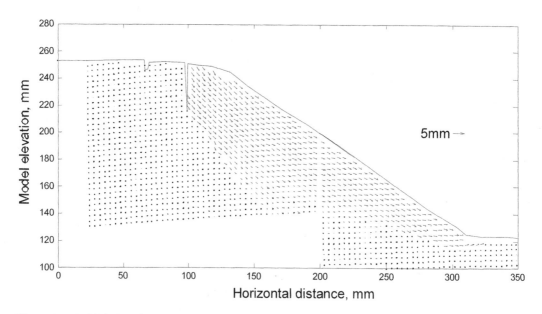

Figure 4. Rainfall induced embankment deformations

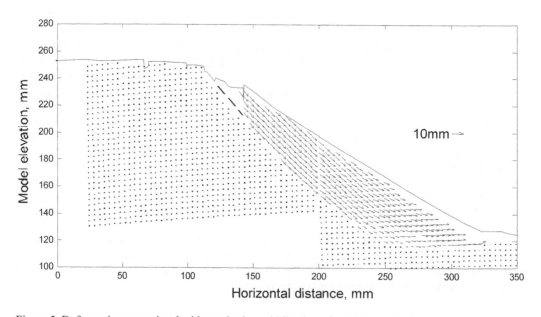

Figure 5. Deformations associated with continuing rubblisation of embankment slope

The observed rubblisation failure mode has important consequences for the design of retrofit measures for embankment stabilisation. For example, if the design of a retrofit scheme is based on soil strengths obtained from a back-analysis of a previous shallow slope failure, the rubblisation failure mode indicates that the strengths may fall to a lower value in the future. Although the calculated strengths are representative of the strength mobilised during a shallow slope failure, they may be higher than the fully softened strength of the material.

Each of the three failures presented in this paper displayed tension cracking at the crest of the embankment. The accurate modelling of the tensile strength of the clay as well as the rainfall in-filling of the tension cracks would be difficult within a numerical modelling framework. In addition, the interaction and the relevant strength parameters on each of the dormant and active slip surfaces are arguably beyond the state-of-the-art of numerical modelling. Despite these difficulties, there exists a need to design embankment retrofits. This presents a significant challenge for numerical modelling.

5 CONCLUSIONS

High quality displacement measurement technology, the ability to measure soil suctions, and strictly controlled boundary conditions allow the investigation of progressive failure in overconsolidated embankments using centrifuge tests. A rubblisation failure mode has been observed to form during rainfall infiltration in which a series of tension cracks and shear surfaces form as the embankment material softens. The complexity associated with this failure mode would present a considerable challenge to current numerical modelling codes.

REFERENCES

Alonso, E., Gens, A. & Lloret, A. 1995. Effect of rain infiltration on the stability of slopes. In *Unsaturated Soils: Proceedings of the 1st International Conference on Unsaturated Soils, Paris*: 241-249. Rotterdam: Balkema.

Guan, Y. & Fredlund, D.G. 1997. Use of the tensile strength of water for the direct measurement of high soil suction. *Canadian Geotechnical Journal* 34: 604-614.

Potts, D.M., Dounais, G.T. & Vaughan, P.R. 1990. Finite element analysis of progressive failure of Carsington Embankment. *Géotechnique* 40(1): 79-101.

Potts, D.M., Kovacevic, N. & Vaughan, P.R. 1997. Delayed collapse of cut slopes in stiff clay. *Géotechnique* 47(5): 953-982.

Ridley, A.M. 1993. The measurement of soil moisture suction. Ph.D. dissertation, Imperial College, London.

White D. J., Take W.A., Bolton M.D. & Munachen S.E. 2001a. A deformation measuring system for geotechnical testing based on digital imaging, close-range photogrammetry, and PIV image analysis. *Proc. 15th Int. Conf. Soil Mechanics and Geotechnical Engineering, Istanbul, Turkey:* 1: 539-542. Rotterdam: Balkema.

White D.J., Take W.A. & Bolton M.D. 2001b. Measuring soil deformation in geotechnical models using digital images and PIV analysis. In Chandra Desai et al. (eds.), *Proc. 10th Int. Conf. Computer Methods and Advances in Geomechanics. Tucson, Arizona*: 997-1002. Rotterdam: Balkema.

Constitutive and Centrifuge Modelling: Two Extremes, Springman (ed.)
© *2002 Swets & Zeitlinger, Lisse, ISBN 90 5809 361 1*

Localisation and critical states

H.M. Heil
Institute of Geotechnical Engineering, ETH Zürich, Switzerland

ABSTRACT: Localised deformation develops in undisturbed samples of soft, medium sensitive Kreuzlingen Clay when they are sheared in undrained triaxial extension. The two localisation modes identified were necking and shear banding, with the deformation patterns the samples displayed after the tests indicating that both localisation modes developed simultaneously. Necking is strongly linked to the specific kinematic constraints of the triaxial extension test and does not necessarily hint at an unstable path of the tested soil. Modelling of an undrained plane-strain extension test using the Modified Cam-Clay model with a mixed finite element formulation further illustrates that even the formation of shear bands is not inevitably linked to an unstable path of the material or softening. Although additionally, for pre-peak loads, the behaviour of the Kreuzlingen Clay samples could well be described by a critical state constitutive model, the abrupt softening of these samples in the post-peak stages of the triaxial extension tests then however indicates a marked degradation of the shear strength within the zones of localised shear deformation for this normally consolidated clay.

1 INTRODUCTION

It is well known that material response at failure is frequently accompanied by non-uniform or localised deformation. In his 1882 contribution, giving the graphical representation of the stress and the strain tensor by a circle and putting forward the related failure criterion that later had been called the Mohr-Coulomb criterion, Mohr stated: "the deformations which occur between reaching the limit of elastic behaviour (of the material) and fracture are usually so complicated that one must refrain from observing these deformations in this period exactly and must restrict oneself to the determination of the states of stress. Even this task is difficult because the assumed distribution (or apportionment) of the stresses in most cases doesn't match the real conditions. In my (Mohr's) opinion, the most difficult task yet is to achieve prescribed simple states of stress in the experiment." Mohr (1882) also notes: "it is possible and even probable that for rupture too, not the stresses but instead the normal and shear strains are decisive. However, for the time being, one will have to be content with deriving the representation of the rupture and yield limit from the states of stress because the deformations at this limit elude observation while the state of stress can be determined at least approximately."

Behind these statements we have the most general supposition that a failure criterion or constitutive model which describes the yielding of a material should ideally be valid on the shear planes or within the shear bands or within any other kind of zone where yielding or failure of the material actually occurs. It was in exactly this sense that Hvorslev (1937) determined the water content of small parts of overconsolidated clay samples tested in direct shear. He took these small parts of each sample from a thin zone located at the plane of shearing prescribed by the test rig of the direct shear apparatus, deeming this plane to be the most likely location for the occurrence of a slip surface or failure zone.

From an engineering point of view, the state of stress can frequently be determined with a sufficient degree of precision when peak strength of the material is the parameter sought. This is typical in direct shear tests on overconsolidated samples as performed by Hvorslev or in simple compression tests on brittle materials, cf. sketch of a cracked sample given by Mohr (1882) in Figure 1a. But due to the localisation of deformation, especially at the post-peak stages of a test, one may encounter serious difficulties even to determine the state of stress approximately. The sample of soft Kreuzlingen Clay, sheared in undrained triaxial extension (Fig. 1b), may serve as one example.

Roscoe et al. (1963) tackled a similar problem, when they carried out a special drained triaxial extension test on a sand sample, which exhibited considerable necking similar to the sample shown in Figure 1b. During their special test, Roscoe et al. measured continuously the minimum cross-sectional area at the neck that formed due to the extension of the sample, as well as the axial strains in different subregions of the sample. They found that, depending on the failure mechanism assumed, the shear strength of the sand sample in the necking region had even increased slightly to moderately after the formation of the neck, while the sample as a whole softened and the conventional mean cross-sectional sample area correction presuming uniform deformation of the whole sample would have given much lower values of shear strength.

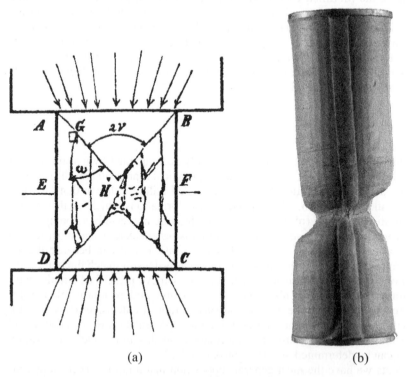

(a)   (b)

Figure 1. Localisation of deformation in a simple compression test on brittle material (Mohr 1882) and in an undrained triaxial extension test on soft Kreuzlingen Clay

Inherent in the findings of Roscoe et al. (1963) is the fundamental supposition, that a constitutive model, which in their case would be a critical state model, should ideally describe both the behaviour of the material in zones of continuously varying deformation patterns, as well as the behaviour of the same material in zones of strongly non-uniform, localised deformation. As encountered by Mohr (1882) or by Roscoe et al. (1963) and others, at least in routine tests, the possibilities of determining material parameters and of validating the model within the zones of localisation clearly are limited due to the localised deformation pattern itself, along with practical restrictions on testing technique.

2 YIELDING OF KREUZLINGEN CLAY IN UNDRAINED TRIAXIAL EXTENSION

Kreuzlingen Clay is a soft and essentially normally consolidated, post-glacial lacustrine clay, typical for Switzerland. Previous results of cone penetration tests, field vane tests, and the results of laboratory investigations including undrained triaxial compression tests and one triaxial extension test were reported by Heil et al. (1997) and Springman et al. (1999). From these data, Kreuzlingen Clay can be characterised as a medium to low plasticity clay with a clay fraction of up to 55% and a carbonate content of about 30-35%, exhibiting a sensitivity S_t between 2 and 3 in vane tests and up to about $S_t = 4$ in fall cone tests.

For the strain controlled undrained triaxial extension tests presented herein, undisturbed samples of Kreuzlingen Clay taken from different depths had been consolidated isotropically to between 2.5 to 3 times the *in situ* vertical effective overburden pressure, so that a truly isotropic state was reached prior to shearing (Houlsby & Sharma 1999). The tests were carried out using rigid filter stones along with filter paper strips speeding up consolidation and improving the reliablility of the pore pressure measurements during shear (cf. Figs 1b, 4). By using deaired water during sample insertion and by applying a backpressure of between 1 and 3 bar throughout, full saturation of the samples could be ensured.

With the truly undrained testing conditions, loading in triaxial extension then only leads to elastic-plastic shear loading of the samples developing from the initial isotropically normally consolidated state, and in contrast to drained extension tests any initial elastic unloading of the samples is prevented by the incompressibility constraint. The effective stress paths in a p'-q-plot may thus be interpreted directly in terms of plastic volumetric hardening of a critical state model or

$$\Delta v^e + \Delta v^p \equiv 0$$

where Δv^p is the plastic change in specific volume of the soil skeleton and Δv^e is its elastic counterpart.

Despite testing undisturbed, instead of more homogeneous remoulded samples, quite uniform behaviour was observed, resulting in a very similar shape of the effective stress paths given in Figure 2 (with σ'_a and σ'_r denoting the effective axial and radial stress respectively). This suggests that during undrained shear, all samples had experienced an equivalent plastic hardening process in terms of effective stress.

The stress paths in Figure 2 have been calculated using conventional mean cross-sectional sample area correction. This is seen to be adequate as long as the samples deform in a uniform way. But, as mentioned above (Roscoe et al. 1963), if necking develops in triaxial extension, then the mean cross-sectional sample area correction tends to give grossly underestimated values of the strength of the soil. With the mean cross-sectional sample area correction applied throughout, for the *post-peak* stages of the tests the stress paths given in Figure 2 thus are not reliable. In order to distinguish between non-localised and localised modes of deformation, it is helpful to take a look at the crude measurements for load P and excess pore pressure Δu, given in Figures 3a,c for two of the samples, as well as at the deformation patterns these two samples displayed after being sheared beyond peak loads, i.e. at the end of the tests (Figs 4a,b).

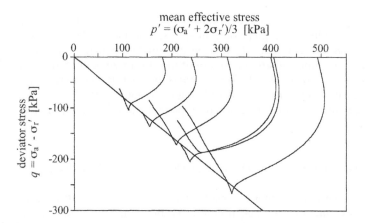

Figure 2. Effective stress paths of Kreuzlingen Clay in undrained triaxial extension

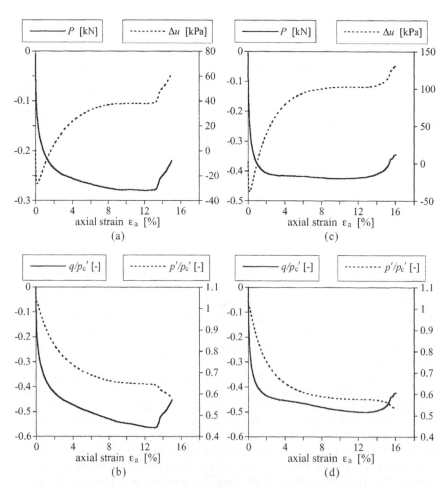

Figure 3. Load-strain and stress-strain curves in undrained triaxial extension, plots (a) and (b) correspond to the sample in Figure 4a, plots (c) and (d) correspond to the sample in Figure 4b

(a)   (b)

Figure 4. Undisturbed samples of Kreuzlingen Clay subjected to less straining after peak load: sample (a) corresponds to plots in Figures 3a,b, sample (b) corresponds to plots in Figures 3c,d

After an initial steep increase in tension load P as well as in excess pore pressure Δu, the stiffness of the samples quickly decreases, and at an axial strain of about $\varepsilon_a = 8\%$, both the load P and the excess pore pressure Δu reach a stationary value (Figs 3a,c). At this stage of the test, the calculated stress-strain response, which is plotted in Figure 3b for the sample shown in Figure 4a, indicates an increase of shear stress at a decreasing rate, while Figure 3d indicates that shear stresses remain essentially constant for the sample shown in Figure 4b. The samples can thus be regarded as having reached the critical state or at least a near critical state. The calculation of the stresses using mean cross-sectional sample area correction at this point can still be considered relatively reliable because the necking of these samples remained quite modest after even considerable additional axial straining of about another 8%, until, at the end of the tests, they finally reached the deformed state shown in Figures 4a,b. Compared to the sample of Figure 1b, these two samples had been subjected to a much lower total amount of axial strain.

Then, at an axial strain ε_a of about 13 to 14% there is a marked change in the behaviour of the two samples. The sudden decrease in tension load P (Figs 3a,c) indicates progressive softening of the samples as a whole. At the same time, a marked increase in excess pore pressure Δu can be observed (again Figs 3a,c), that is caused by relaxation and an increase of axial as well as mean total stress in the major part of the sample outside the localisation zone. This quick change in behaviour, along with the high rates at which the softening of the samples occurs, testifies to a sudden and persisting change in the deformation pattern of the samples.

The examination of both samples after the test reveals that in the sample of Figure 4a, a distinct shear plane had been formed while the necking of the sample had remained subdued, which compares with the more rapid change in behaviour indicated in Figure 3a. Conversely, for the sample in Figure 4b, necking appears to have been the dominant mode of localisation

203

while the load-strain-curve in Figure 3c shows a more smooth transition to the softening branch of the test. This hints that shear banding is the cause of the abrupt change of sample stiffness. When the samples reach the critical state with the potential for hardening being exhausted, the sensitivity of Kreuzlingen Clay, in combination with the concentrated shear deformation that sets in within the persisting shear band, then leads to a rapid degradation of the shear strength within this shear band. A similarly abrupt post-peak softening, along with the occurrence of a major persisting shear band, has been observed by Mooney et al. (1998) for dense softening sand specimens, tested in drained plane-strain compression.

It is emphasised here, that the observed behaviour of Kreuzlingen Clay during the pre-peak stages would be amenable to a plastic hardening formulation comparable to that of the Cam-Clay models. However, the marked change in behaviour exhibited at the abrupt transition to post-peak shear banding and softening then prohibits the further application of a critical state model, because the stress-strain response of Kreuzlingen Clay in the softening regime no longer conforms to the principles of critical state soil mechanics. Here again, in a sense similar to Mohr's statements cited above, the problem arises that if deformation localises, the determination of stresses and strains in a triaxial test, which would be an essential prerequisite for the analysis of the soil behaviour in the localised state, becomes difficult if not impossible.

3 IS LOCALISATION AVOIDABLE FOR AN IDEALISED CAM-CLAY SOIL ?

As pointed out by Rice (1976), imperfections of a sample play a considerable role in causing localisation of deformation, leading to a possible concentration of deformation in their vicinity and initiating a localised zone. Subsequently, this zone creates its own strain concentration, thereby traversing the material at nominal deformation conditions that are well removed from those for localisation.

For undisturbed samples, imperfections are the rule and not the exception as, unlike remoulded samples, they have not been formed from a homogenised soil. Instead, in undisturbed samples local variations of soil composition and mechanical properties are to be expected on a regular basis.

Rice (1976) further found that a rigid-plastic material with a smooth yield locus, to which normality of plastic flow applies, loses its stability against small disturbance when the plastic hardening modulus h becomes zero. This compares to the rigid-plastic straining at constant shear stress and constant volume in the critical state of the Cam-Clay model, as well as to the stable path of Kreuzlingen Clay in the hardening regime and the abrupt change of behaviour when potential for hardening of the samples in triaxial extension had been exhausted.

In order to gain some basic insight into the behaviour of the Modified Cam-Clay model in the presence of an imperfection, an undrained plane-strain extension test has been modelled in a finite element analysis using a mixed mean pressure-displacement formulation (Herrmann 1965, Zienkiewicz et al. 1995), but with no provisions made for mesh refinement or other means to capture localised deformation. Along with an initial isotropic consolidation pressure of 100kPa, a shear modulus of $G = 4000$kPa and the parameters $M = 0.9$ (slope of the critical state line in the p'-q-plot), $\lambda = 0.2$ (slope of the virgin compression line in the v-log p'-plot), $\kappa = 0.05$ (slope of the unloading-reloading lines in the v-log p'-plot) and $N = 2.8$ (specific volume at $q = 0$, $p' = 1$kPa) had been chosen (see e.g. Britto & Gunn 1987 for a more detailed description of the model and the parameters). Like in the laboratory tests, rigid end platens had been assumed, while the undrained loading of the isotropically normally consolidated sample under strain control to an axial strain $\varepsilon_a = 10\%$ was achieved by prescribed displacements on the upper boundary. A geometrical imperfection had been placed on the left hand side of the mesh.

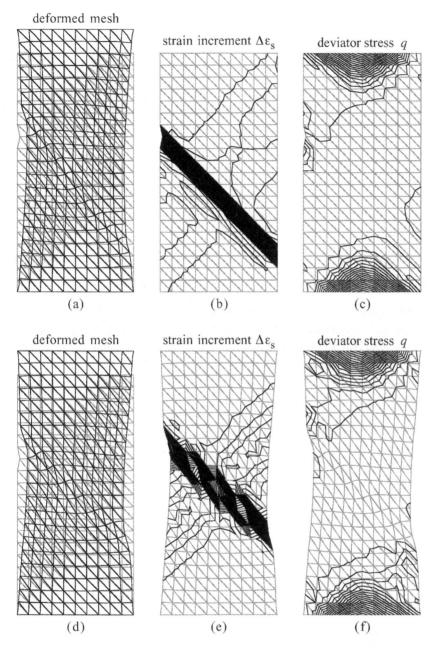

deformed mesh   strain increment $\Delta\varepsilon_s$ deviator stress q

(a)   (b)   (c)

deformed mesh   strain increment $\Delta\varepsilon_s$ deviator stress q

(d)   (e)   (f)

Figure 5. Solutions obtained with mixed mean pressure-displacement finite elements for isotropically normally consolidated Modified Cam-Clay in undrained plane strain extension, (a)-(c) without mesh updating, (d)-(f) with mesh updating

In Figures 5a-c, the results after the final loading step (i.e. at $\varepsilon_a = 10\%$) of this calculation are given, carried out using a conventional small strain approach. Figure 5a shows the deformed mesh on a true scale. Clearly the occurrence of a shear band can be recognized starting at the imperfection on the left hand side of the mesh and traversing the sample to the lower right part.

Correspondingly, the contour lines for the deviator strain increment $\Delta\varepsilon_s$ (or the second invariant of the strain increment tensor) for the last load step are concentrated in the row of elements that are located within the predicted shear band (Fig. 5b). This means that shear banding, not necking, is the dominating deformation mode when this Modified Cam-Clay model sample approaches the critical state.

Remarkably, despite the localisation of deformation, the plot of the contour lines of deviator stress q (Fig. 5c) shows a wide area in the centre of the sample where deviator stress is very uniform, translating to a surprising absence of pore pressure gradients in the vicinity of the shear band for this normally consolidated sample.

As could be expected, due to the soil behaviour having been assumed to remain unaltered by any localisation of deformation, the occurrence of a shear band in the normally consolidated model soil does not affect the load-strain response of the overall sample at all. Instead, for the small strain approach, the load-strain behaviour of the sample, given by the solid curve in Figure 6 follows a smooth path throughout while continuously approaching the analytical failure load, which has been calculated using the undrained shear strength:

$$s_u = \frac{1}{2}M \exp((\Gamma - v_0)/\lambda) \tag{1}$$

predicted by the Cam-Clay models for triaxial testing states of stress (e.g. Britto & Gunn 1987), and by multiplying with a factor of $\sqrt{4/3}$ that becomes necessary due to the plane-strain loading ($\sqrt{4/3}$ is the ratio of the shear strengths predicted by the von Mises and the Tresca criterion for a Lode-angle of $\theta = 0$, e.g. Hill 1950). In equation 1, v_0 is the specific volume of the sample at the start of the test while Γ for Modified Cam-Clay is:

$$\Gamma = N - (\lambda - \kappa)\ln 2 \tag{2}$$

(Wood 1990).

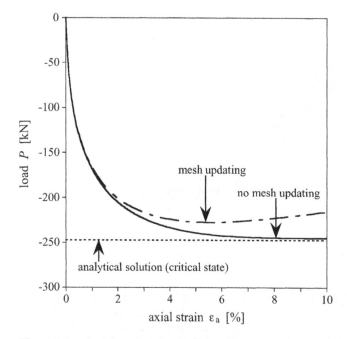

Figure 6. Load-strain-curves for the finite element solutions shown in Figure 5

In comparison with the pure small strain analysis, a different load-strain-curve is obtained by a finite element analysis using simple mesh updating (i.e. the mesh coordinates being updated by the calculated displacement increments). The corresponding dashed curve in Figure 6 clearly does not reach the analytical failure load and at later stages of the modelled plane-strain test even indicates a decrease of tension load at an increasing rate. The reason for this is the cross-sectional area reduction due to the deformation of the model sample (cf. Fig. 5d), which, in contrast to the pure small strain analysis, effectively enters the calculation.

The comparison with the load-strain curves of the triaxial laboratory samples (Figs 3a,c), on the other hand, reveals the following two differences. Firstly, although the sample modelled using Modified Cam-Clay, like the Kreuzlingen Clay samples, exhibits a decrease in tension load at the later stages of the test, there is no abrupt change in behaviour for the model sample. The updated mesh calculation had to be stopped at an axial strain of $\varepsilon_a = 10\%$ because of the already significant geometrical distortion of the finite elements. But as for the sample of non-softening sand investigated by Roscoe et al. (1963), the load-strain-curve can be expected to be continued smoothly, because within the numerically modelled sample at $\varepsilon_a = 10\%$ the localisation of deformation had already fully materialised so that no further change in the deformation mode of this sample is to be expected. Secondly, in comparison to the samples of undisturbed Kreuzlingen Clay, the relatively early decrease in tension load for the model sample in the updated mesh analysis, obvious from the corresponding load-strain-curves, is eased by the early concentration of deformation in the model sample triggered by the imperfection.

As a final result, despite non-optimal mesh alignment in the updated mesh, the contour lines for the deviator strain increment $\Delta\varepsilon_s$ of the last load step (Fig. 5e) indicate a preferred shear band inclination of 45°. This prediction is due to the assumption that associated plastic flow will take place with respect to the Modified Cam-Clay yield locus, and a similar prediction would have emerged if a drained plane-strain extension test had been modelled. However, this theoretical prediction of Modified Cam-Clay may not be verified easily in standard laboratory tests on real soil, as for any other prediction for the inclination of shear bands. When a persisting shear band is formed within a sample, then the concentrated deformation, in combination with the softening behaviour of many soils, leads to a degradation of the shear strength within this band. The weakening of the material within the persisting shear band then enforces further localisation of deformation within the same band, until the residual shear strength may be reached within this band as a lower limit. At least for non-brittle materials in standard triaxial or plane-strain tests, the inclination of the persisting shear band then changes as the sample is deformed further until the end of the test. In contrast to that, in a non-softening material, like in the Modified Cam-Clay model soil of Figure 5e, a re-orientation or re-formation of shear bands may take place even at later stages of the test.

Clearly there is mesh-dependence for these finite element calculations so that the location of the imperfection strongly affects the calculated deformation pattern, and the size of the finite elements predetermines the width of the predicted shear band. On the other hand, further finite element calculations using the same mesh show that practically identical results are obtained for the loadstrain- curves of Figure 6 if the imperfection is placed on the right hand side of the mesh instead of the left hand side, which results in a less favourable mesh alignment.

4 CONCLUSIONS

The Modified Cam-Clay model in the normally consolidated state was basicly shown not to preclude the localisation of deformation or the formation of shear bands *a priori*. The frequently expressed notion that critical state soil mechanics, at least as far as normally consolidated soils are concerned, and shear banding are mutually exclusive may just stem from the instinctive presumption that shear banding would *always* be a result of softening of the soil - a material behaviour that is not allowed for by modelling normally consolidated soil via Modified Cam-Clay.

The comparison of the load-strain-behaviour observed during the undrained triaxial extension tests on Kreuzlingen Clay samples to the pattern of combined necking and shear banding displayed by these samples at the end of the tests reveals that the occurrence of shear banding indeed appears to be one of the symptoms of softening.

For medium sensitive Kreuzlingen clay, the formation of a persisting shear band in the triaxial samples marks a distinct alteration of its stress-strain behaviour, while Modified Cam-Clay *per se* remains indifferent towards the localisation of deformation.

REFERENCES

Britto, A.M. & Gunn, M.J. 1987. *Critical state soil mechanics via finite elements*. Chichester: Ellis Horwood, Halsted Press.
Heil, H.M., Huder, J. & Amann, P. 1997. Determination of shear strength of soft lacustrine clays. *Proc. 14th ICSMFE, Hamburg* 1: 507-510.
Herrmann, L.R. 1965. Elasticiy equations for incompressible and nearly incompressible materials by a variational theorem. *Journal AIAA* 3: 1896-1900.
Hill, R. 1950. *The mathematical theory of plasticity*. London: Oxford University Press.
Houlsby, G.T. & Sharma, R.S. 1999. A conceptual model for the yielding and consolidation of clays. *Géotechnique* 49(4): 491-501.
Hvorslev, M.J. 1937. Über die Festigkeitseigenschaften gestörter bindiger Böden. *Ingeniørvidenskabelige Skrifter*, Danmarks Naturvidenskabelige Samfund, Copenhagen, A No. 45
Mohr, O. 1882. Ueber die Darstellung des Spannungszustandes und des Deformationszustandes eines Körperelements und über die Anwendung derselben in der Festigkeitslehre. *Civilingenieur* 28: 113-156.
Mooney, M.A., Finno, R.J. & Viggiani, M.G. 1998. A unique critical state for sand ? *ASCE J. Geotech. Geoenv. Eng.* 124: 1100-1108.
Rice, J.R. 1976. The localization of plastic deformation. *Theoretical and Applied Mechanics, Proc. 14th IUTAM Congr.*, Delft: 207-220.
Roscoe, K.H., Schofield, A.N. & Thurairajah, A. 1963. An evaluation of test data for selecting a yield criterion for soils. *Laboratory Shear Testing of Soils, ASTM STP* 361: 111-128.
Springman, S., Giudici-Trausch, J., Heil, H.M. & Heim, R. 1999. Strength of soft swiss lacustrine clay. *Transportation Research Record* 1675, TRB, Nat. Res. Council,Washington D.C.: 1-9.
Wood, D.M. 1990. *Soil behaviour and critical state soil mechanics*. Cambridge: Cambridge University Press.
Zienkiewicz, O.C., Huang, M. & Pastor, M. 1995. Localization problems in plasticity using finite elements with adaptive remeshing. *Int. J. Numerical and Analytical Meth. Geomechanics* 19: 127-148.

Constitutive and Centrifuge Modelling: Two Extremes, Springman (ed.)
© *2002 Swets & Zeitlinger, Lisse, ISBN 90 5809 361 1*

Influence of the structure on the mechanical behaviour of sand

N. Benahmed
Department of Civil Engineering, University of Bristol, United Kingdom

J. Canou & J.-C. Dupla
Cermes, ENPC-LCPC, France

ABSTRACT: The effect of the mode of preparation on the mechanical behaviour of the Hostun RF sand is investigated. Undrained tests have been carried out on very loose samples, which have been reconstituted on the one hand by moist tamping and on the other hand by pluviation through air. Although initial conditions (void ratio and effective confining stress) were identical, results show different responses of the soil samples for the two modes of deposition: a contractive behaviour type liquefaction for moist tamping and a dilatant behaviour for dry deposition. This difference in the behaviour is explained by a different initial structure (i.e. geometrical pattern of the soil particles), which may result from the different modes of preparation. In fact, observations with the electron microscope have allowed the existence of two different structures to be revealed, according to the depositional fabric: a grain pattern in aggregates with macro-pores for the moist tamping and a much more regular pattern without macro-pores for the dry pluviation. It then appears very important to take the "structure" parameter into account when characterising the mechanical behaviour of sands.

Combined loading of a jack-up unit

G. Vlahos
University of Western Australia, Perth, Australia

ABSTRACT: Conversion of jack-up units from drilling rigs to production platforms will demand survival in severe weather conditions for longer periods of time. In turn, this has ignited interest in investigating jack-up behaviour, particularly soil–structure interaction and load sharing among the spudcan footings. As a result, this paper presents research conducted at the University of Western Australia.

1 INTRODUCTION

Previous experimental studies of soil–structure interaction have generally been performed with single spudcan footings. This paper presents brief results from a series of 1-g experiments conducted on a scaled three-legged jack-up unit model, equipped with spudcan footings, on soft clay. The model rig is subjected to combined vertical and horizontal loads at the hull level, resulting in combined vertical V, moment M and horizontal H loads at the footings. The physical tests explore the load redistribution among the spudcan footings, hull and footing displacements at failure, and the ultimate system capacity.

The data obtained through these tests will give some indication of the amount of rotational foundation fixity that can be provided by spudcan footings on clay. This issue has been the subject of much interest in recent years, because, if a degree of foundation fixity is taken into account, member stresses at the leg-hull connection and other response values critical to the structural integrity of the jack-up are reduced (Chiba et al. 1986).

The long-term aim is to use the scaled rig described in this paper to refine existing work-hardening plasticity models that describe the combined load–displacement behaviour of spudcans on various soils. This will allow direct numerical simulation of the present model tests and evaluation of a suitable plasticity-based spudcan footing model. While this is beyond the scope of this paper, some of the experimental results will be examined within the framework of work-hardening plasticity.

2 EXPERIMENTAL APPARATUS

A scale model jack-up unit, for testing at unit gravity, has been designed and assembled at the University of Western Australia. The dimensions of the scaled model were based on information from the Friede & Goldman (1998) web site, as summarised in Prior (1999). Various sizes were

Table 1. Jack-Up model dimensions (after Prior 1999)

Model Characteristics	Prototype	1:250 Scaled Model	Built Model
Leg Length, Full and Partial	150, 50 m	600, 200 mm	600, 200 mm
Spudcan Diameter	18 m	72 mm	72 mm
Centre to Centre of Aft Leg Separation (measured from centre of legs)	51 m	204 mm	216 mm
Aft and Forward Leg Separation (measured from centreline of legs)	45 m	180 mm	187 mm
Depth of Hull	10 m	40 mm	40 mm
Breadth of Hull	78 m	312 mm	324 mm
Length of Hull	70 m	280 mm	280 mm
Second Moment of Area of Leg	7.2 m^4	1843.2 mm^4	1892 mm^4
Cross-sectional Area of Leg	1 m^2	16 mm^2	73.8 mm^2

considered, however, a 1:250 scaled version has been built due to physical constraints. Table 1 shows the chosen prototype and model jack-up dimensions. The legs have been constructed from stainless steel with a Young's modulus E of 193000 MPa and a shear modulus G of 80000 MPa.

The model jack-up test was performed on a 600 mm diameter overconsolidated kaolin specimen prepared from a 120% moisture content slurry and consolidated for 2 weeks (Vlahos et al. 2001). The pressure was applied in stages reaching a maximum of 110 kPa. Then the sample was unloaded in steps, eventually to atmospheric pressure. A 5 mm layer of water covered the sample at all times. Whereas an offshore deposit might be normally consolidated, this preparation method produced a heavily overconsolidated sample, but with a similar strength to that expected in the prototype deposit.

The sample strength was determined by using a T-bar penetrometer (Stewart & Randolph 1991), which results in a continuous soil profile over the depth of the sample. The undrained shear strength, s_u, was 10 kPa at one spudcan diameter (72 mm) depth.

3 TESTING STAGES: RESULTS

The physical test incorporated two distinct stages: installation and pushover.

3.1 *Installation*

During installation the jack-up unit is lowered at a constant rate, ensuring undrained conditions prevailed. When one diameter penetration is reached, the unit is then unloaded to the target preload level ($V/V_{preload} \approx 0.5$) and prepared for the pushover stage. Figure 1 depicts the vertical load experienced by the individual spudcan during preload penetration. The distance from the tip to the Load Reference Point (L.R.P.) of the spudcan, located at the widest cross-section of the footing, is indicated as the initial negative displacement shown in Figure 1. The undisturbed surface specifies when the L.R.P. reaches the original clay surface, prior to penetration.

3.2 *Pushover*

In the pushover stage, a monotonic horizontal load is applied at a constant rate of 0.75 mm/sec at the hull level of the jack-up model, as depicted in Figure 2. Once the initial pushover has finished, several additional slow horizontal load cycles are performed, to assess behaviour in remoulded clay, though these results are not discussed in this paper. Only then is the jack-up pulled out of the clay.

212

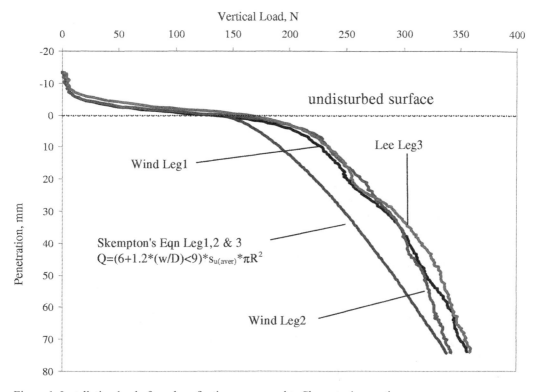

Figure 1. Installation load of spudcan footings compared to Skempton's equation

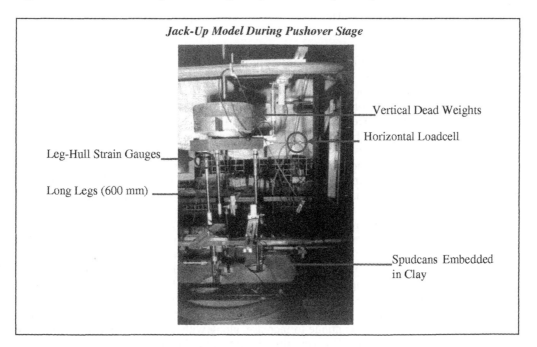

Figure 2. Pushover test of model jack-up unit

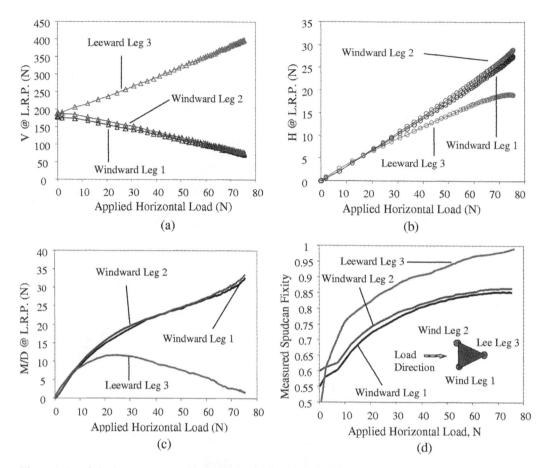

Figure 3. Load sharing among footings during monotonic pushover test

The combined load variation experienced by the footings during the pushover is shown in Figures 3a-c. As the applied horizontal load is increased, each windward footing experiences a loss of vertical load, and the leeward footing gains vertical load at twice the rate (to maintain vertical equilibrium; Fig. 3a). However, the opposite occurs with the horizontal and moment loads (Figs 3b, c) where the leeward footing gradually sheds load on to the windward spudcans.

Initially, all of the spudcan footings develop similar moment reactions under the applied loading. This is best seen in the moment paths of Figure 3c, when the applied load is less than about 10 N. If thought of within the context of plasticity theory, these loading increments would be modelled as elastic (or entirely within a theoretical yield surface in V, M and H space). However, as the load paths are different (for roughly the same M and H, the leeward footing has increasing V, whereas the windward footings have decreasing V), the behaviour is not the same once 'yield' has occurred. After about 10 N, Figure 3c indicates a quick loss of moment capacity at the leeward footing (though the vertical load (Fig. 3a) continues to increase). At the same time, the horizontal load-carrying capacity (Fig. 3b) of the leeward footing also decreases, an effect which accelerates as the test proceeds. Interestingly, while the windward footings indicate some softening in the moment response, no definite peak is detectable. Instead, the moment loads on the windward footings continue to increase, along with a steady increase in horizontal capacity throughout the pushover. This was also observed in a similar test by Prior (1999).

214

The reduction of rotational fixity provided by the footings during the pushover is shown in Figure 3d, where the degree of fixity is measured by the ratio:

$$\frac{M_{Leg-Hull}}{H_{L.R.P.} \times L} \tag{1}$$

$M_{Leg-Hull}$ is the moment at the leg–hull connection (or top of the leg), $H_{L.R.P.}$ is the horizontal reaction at the Load Reference Point of the spudcan, as defined by Martin (1994), and L is the leg length. Using this definition, a value of 1 indicates pinned footings (no rotational restraint) and a value of 0.5 represents fully fixed footings (infinite rotational restraint). Figure 3d supports the argument that fixity of the windward footings is still present even with failure of the system, while the leeward footing approaches the pinned condition.

4 CONCLUSIONS

Results from preliminary experiments on a $1/250^{th}$ scale model jack-up unit with spudcan footings on soft clay have been presented. Monotonic pushover tests allow the examination of complex soil–structure interaction processes, and specifically, the degree of rotational fixity provided by the spudcans has changed, and load redistribution between footings after initial yielding of the soil has occurred.

The main series of monotonic pushover tests to be performed with the model jack-up unit will examine factors such as different preload levels, skewed loading (not along an axis of symmetry), and different foundation types. A series of cyclic loading tests is also under preparation. Numerical simulations of the experimental work will be conducted using an existing work-hardening plasticity model of spudcan behaviour, which has been incorporated as a user element into the finite element program ABAQUS. The aims of the numerical work are (i) to investigate the suitability of such models for predicting jack-up capacity and behaviour; (ii) to enhance existing models to account for cyclic loading conditions.

REFERENCES

Chiba, S., Onuki, T. & Sao, K. 1986. Static and dynamic measurement of bottom fixity. *The Jack-up Drilling Platform Design and Operation*: 307-327. London: Collins.
Friede & Goldman. 1998. Naval Architects and Marine Engineers, Web Site www.fgh.com.
Prior, M. 1999. The testing of a full model jack-up on clay. *Honours Thesis*, University of Western Australia, Australia.
Stewart, D.P. & Randolph, M.F. 1991. A new site investigation tool for the centrifuge. In H.Y. Ko (ed.), *Proc. Int. Conf. Centrifuge '91*: 531-538. Rotterdam: Balkema.
Martin, C.M. 1994. Physical and numerical modelling of offshore foundations under combined loads, *D.Phil. Thesis*, University of Oxford, Great Britain.
Vlahos, G., Martin, C.M. & Cassidy, M.J. 2001. Experimental investigation of a jack-up model, *Proc. 11th ISOPE Conf.*, Stavanger, Norway.

Constitutive and Centrifuge Modelling: Two Extremes, Springman (ed.)
© *2002 Swets & Zeitlinger, Lisse, ISBN 90 5809 361 1*

Centrifuge modelling of caisson capacity under uplift loading

A.R. House & M.F. Randolph
Centre for Offshore Foundation Systems, Crawley, Australia

ABSTRACT: The large permanent axial foundation loads associated with deep-water hydrocarbon developments are raising concerns over the ability of suction caissons to sustain the anchoring demands in the longer-term. Recent design issues concern caisson response to long term sustained tensile loading and the influence of internal ring stiffeners on the required installation pressures. This paper presents the results from a recent centrifuge test focussed on the above-mentioned concerns for suction caisson foundations in cohesive soils.

1 INTRODUCTION

Design methods for the axial capacity of suction caissons in cohesive soils are relatively well established although there remains uncertainty in the ability of a caisson with a 'sealed' lid to rely upon reverse end bearing capacity in the longer term. Another design issue, which is not fully understood at present, is the geotechnical consequence of providing internal structural ring stiffeners.

This paper presents the results from a series of centrifuge tests modelling the axial capacity of suction caissons in normally consolidated kaolin clay. A caisson, with aspect ratio (length / diameter) of 4, was extracted under long-term sustained axial loading under a load equivalent to 75 % of the ultimate undrained capacity. Coincidentally, the capacity under long term sustained axial load was approximately equivalent to the immediate drained capacity.

2 CENTRIFUGE MODELLING OF CAISSON CAPACITY

2.1 *Apparatus and experimental methods*

Centrifuge modelling was undertaken at an acceleration of 120 *g* on models of linear scale 1:120. The UWA fixed beam geotechnical centrifuge is a model 661 Acutronic with major specifications and capabilities discussed in greater detail by Randolph et al. (1991).

In this study, 3 model caissons were fabricated from aluminium that was lightly sandblasted to produce a semi-rough surface. Each caisson had a prototype length of 14.4 m, diameter of 3.6 m, wall thickness of 60 mm and submerged weight of 230 kN. A smooth walled caisson (caisson A) was adopted for the purposes of assessing the total frictional resistance, while two stiffened caissons (caissons B and D) were modelled to investigate the influence of these on the total penetration resistance of the caissons.

Caissons were subjected to jacked installation at a model rate of 1 mm/s before a combination of either a) drained or undrained, b) immediate or soaked and c) monotonic, cyclic or sustained axial pull-out.

2.2 Sample preparation and characterisation

A normally consolidated kaolin clay sample was prepared from a slurry through centrifugal self-weight consolidation at 120 g. Miniature pore pressure transducers were positioned at various depths within the sample to monitor consolidation progress through the dissipation of excess pore pressures.

Sample strength was determined *in-flight* using a T-bar penetrometer (Stewart & Randolph 1991). The penetrometer consists of a cylindrical bar screwed perpendicularly to the end of an axially instrumented shaft. An exact plasticity solution for the limiting resistance, based on a plane-strain soil flow mechanism is used for derivation of shear strength from the measured bearing resistance. The T-bar bearing capacity factor adopted, N_T, of 10.5 represents an average between the perfectly smooth ($N_c = 9.14$) and fully rough ($N_c = 11.94$) solutions of Randolph & Houlsby (1984).

Results from penetration tests undertaken over the duration of caisson testing are presented in Figure 1 showing an average shear strength profile $s_u \approx 1.125z$ (s_u in kPa and z in metres). A slight increase in shear strength with time was recorded and was thought to be due to the cyclic consolidation and swelling of the sample as a result of stopping the centrifuge to change tools. T-bar pull-out data showed a remoulded shear strength of 0.7 times the peak undrained shear strength (equivalent to a sensitivity of 1.4). Cyclic T-bar tests indicate a sensitivity of approximately 2 while vane shear tests on the same clay have been used to determine an average sensitivity of 2.4 after 500° rotation.

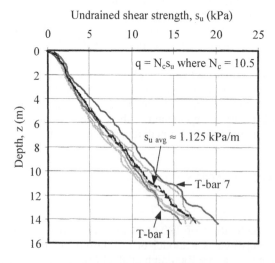

Figure 1. Undrained shear strength profiles

2.3 Caisson installation and resistance to axial pull-out

The first two tests were performed on the smooth walled caisson. Immediately following installation, the caissons were extracted monotonically with caisson IP1 having an open lid (drained capacity, $v_{pull-out} = 0.1$ mm/s), whereas IP2 was sealed (undrained capacity, $v_{pull-out} = 0.3$ mm/s).

Tests IP3 and IP4 were undertaken on stiffened caissons, and were soaked after installation for a period equivalent to a prototype consolidation time of 1 year. The caisson of test IP3 was subjected to several "packets" of cyclic tension while caisson IP4 was subjected to a sustained load of 2.5 MN (including caisson weight) for a prototype consolidation period of approximately 2.5 years. Installation and pull-out data from the 4 tests are presented in Figure 2.

Pore pressure measurements taken inside the caisson lid are shown in Figure 3. All caissons were installed by jacking, with the pressure gradient during installation representing hydrostatic

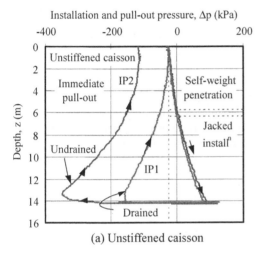

(a) Unstiffened caisson

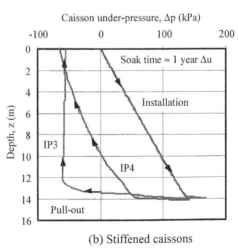

(b) Stiffened caissons

Figure 2. Installation and pull-out response

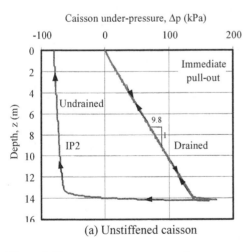

(a) Unstiffened caisson

(b) Stiffened caissons

Figure 3. Pore pressure measurements

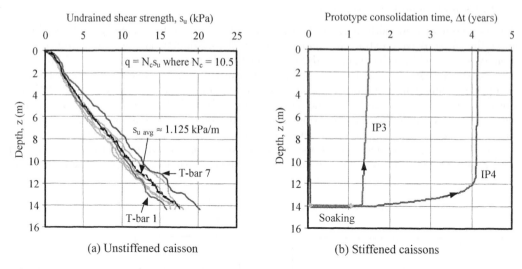

(a) Unstiffened caisson   (b) Stiffened caissons

Figure 4. Caisson depth versus prototype consolidation time

head. Figure 4 shows the caisson penetration depth plotted against prototype consolidation time. Tests IP1 and IP2 were monotonically extracted. Pull-out velocity was a function of the difference between the actual and target tensile load for the cyclic and sustained load tests IP3 and IP4.

3 DISCUSSION

Experimental installation data have been compared with a static force equilibrium solution to the caisson penetration resistance, with good agreement achieved for a coefficient of mobilised friction in the range 0.35 to 0.4.

The same friction coefficient yielded a back-analysed reverse bearing capacity factor for undrained pull-out of 25 using similar force equilibrium for pull-out. In all undrained pull-out tests, the soil plug had a submerged weight of 810 kN representing internal frictional resistance of at least 0.64 times the average undisturbed undrained shear strength over the soil plug length.

It is observed that the immediate drained pull-out of test IP2 required a force approximately 20 % greater than that required for installation. This is thought to be a rate effect combined with the rapid dissipation of localised excess pore pressures adjacent to the thin skirts. Investigations into rate effects in the same material (Randolph & House 2001) as non-dimensional velocity $V = vd / c_v$, where v is the installation velocity, d is the diameter of the caisson and is c_v, is the consolidation coefficient, show that V decreases over the range 5 to 0.5, with an increase in normalised resistance of approximately 1.7. The 20 % increase in partially drained pull-out resistance agrees well with a remoulded strength of 0.7 times the undrained strength, factored by approximately 1.7 for rate effects.

A theoretical installation model for the penetration of stiffened caissons agrees with the experimental findings by assuming complete soil flow around the internal stiffener(s) and a bearing capacity factor of approximately 6 on the first stiffener only.

4 CONCLUSIONS

Preliminary experimental data suggest that design for long term undrained (caisson lid sealed) sustained loading should not rely on full mobilisation of reverse end bearing at the caisson skirt tip. The long term tensile capacity under sustained axial loading will be generated by the external friction and soil plug weight alone, with experimental data showing pull-out progression under loads greater than 75 % of the ultimate unsoaked capacity. Further tests are required on models of different aspect ratio such that the soil plug weight represents a varying proportion of the total internal frictional resistance.

A fundamental assessment of the friction between the aluminium and clay under equivalent normal soil pressures is part of an ongoing study into suction caisson performance in cohesive soils. Further tests upon stiffened caissons in heavily over-consolidated samples are required to calibrate the theoretical installation model. Jacking and suction installation are thought to induce different mechanisms of soil displacement due to the penetrating skirts (and hence zone of effective stress generation adjacent to the skirts), thereby varying the contribution of internal and external skirt friction set-up effects. Suction installation should also be modelled to enable comparison with the set-up effects due to jacked installation presented in this test series.

REFERENCES

Randolph, M.F. & House, A.R. 2001. The complementary roles of physical and computational modelling. *International Journal of Physical Modelling in Geotechnics* 1(1): 1-8.
Randolph, M.F. & Houlsby, G.T. 1984. The limiting pressure on a circular pile loaded laterally in cohesive soil. *Géotechnique* 34(4): 613-623.
Randolph, M.F., Jewell, R.J., Stone, K.J.L. & Brown, T.A. 1991. Establishing a new centrifuge facility. In H.-Y. Ko (ed.), *Proc. Int. Conf. on Centrifuge Modelling – Centrifuge '91*: 3-9. Rotterdam: Balkema.
Stewart, D.P. & Randolph, M.F. 1991. A new site investigation tool for the centrifuge. In H.-Y. Ko (ed.), *Proc. Int. Conf. on Centrifuge Modelling – Centrifuge '91*: 531-538. Rotterdam: Balkema.

Constitutive and Centrifuge Modelling: Two Extremes, Springman (ed.)
© *2002 Swets & Zeitlinger, Lisse, ISBN 90 5809 361 1*

Discussion on problems governed by FAILURE

S.M. Springman & T. Weber
Swiss Federal Institute of Technology, Zurich, Switzerland

ABSTRACT: This paper contains the discussions to the theme lectures for problems related to failure as well as the brief introductions from the young researchers and the resulting discussion.

1 OBSERVED FAILURES OF LABORATORY SAND SAMPLES AND CONSTITUTIVE MODELLING (A. GAJO)

1.1 *Discussion*

Cino Viggiani expressed concern about the post-localisation (large strain) studies in that this analysis was based on a 'small strain' hypothesis. Alessandro Gajo agreed that at the end of the loading phase the strains were very large inside the shear bands and that this was a simplified assumption. He ventured that a more correct procedure would be to consider the change of inclination of the bands and the change of thickness but the work to date reflected an initial simplified approach, the implications of which have not been investigated yet. Viggiani approved of the suggestion and asked about the relative importance of the co-rotation of terms as compared to the stiffness of the material, which is lower when the shear bands form. He wondered if the co-rotation of terms had been neglected because they were evaluated but found to be unimportant? Gajo said he was still working on this and remarked on future plans to investigate the causes leading to band saturation, referring to the literature attributing some aspects of the response to a non-associated flow rule.

Vincenzo De Gennaro asked about how the thickness of the band was taken into account in the model in the parametric study. Gajo answered that two homogenous strain fields were assumed, with one inside the band and the other outside it. He stated that the equilibrium and the strain compatibility must be respected at the interface between the band and the material outside the band. A band thickness was assumed and then all the computation was based on a constant thickness of the shear band, with a small strain analysis.

De Gennaro also questioned the results about Hill's instability in the context of Hill's analysis and commented on past work on simulations of interface behaviour conducted with Gyan Pande in Swansea, in which instability had been checked using several methods. They found that a different constitutive matrix was developed for these two states inside and outside the shear bands, although this led to more or less, or in some cases exactly, the same solution for the Hill instability. He wondered if the speaker's analysis produced consistency of the constitutive matrix of the material inside and outside the shear band. Gajo replied that the analysis of the collapse behaviour of sand was performed assuming homogeneous behaviour, without considering the possible existence of shear bands and confirmed that only a homogeneous bifurcated solution had been analysed. He added that a homogeneous state of deformation was assumed in the case of a loose sand sample because any shear band was found from the previous analysis to be undergoing the phenomenon of band saturation in that one shear band stopped developing and another band formed.

2 CENTRIFUGE MODELLING OF FAILURES AND SAFE STRUCTURES (R. PHILLIPS)

2.1 *Discussion*

Claudio Tamagnini noted that the embankment in the CANLEX experiment was loaded very fast to create undrained conditions and to cause liquefaction inside the model in the centrifuge. He commented that consolidation in the soil mass controlled whether the problem was actually undrained or not and asked about the scaling adopted for time in the centrifuge model? Ryan Phillips replied that they had matched both the time scale for inertia and for diffusion by increasing the viscosity of the pore fluid, and making sure the pore fluid did not change the effective stress response of the sand. The time scale factor was therefore n and not n^2. Tamagnini made the point that this assumed that the permeability remained constant and asked about the influence of viscosity of the fluid on the permeability. Phillips answered that the intrinsic permeability remained constant in the analysis. Tamagnini thought that the intrinsic permeability was the same as for the prototype and that this depended on the geometry of the pores, which could not be preserved with the same scaling factor because pores would be larger. Phillips explained that the ratio of soil particles to void space was the same, which kept the void ratio and the densities constant. Tamagnini commented that even if the void ratio was the same, having pores of different dimensions could have an effect. Phillips said that there was a shorter drainage path in the centrifuge, so consolidation happened much faster and loads needed to be applied quickly, so that pore pressures would not be dissipated. Seepage velocities should be kept in the laminar range and not extend into the turbulent range.

Fook Hou Lee added that modelling the appropriate (right) permeability is not a problem and confirmed that raising the pore fluid viscosity was usually adopted to compensate for the competing influences. He stated that modelling the flow regime correctly was extremely important and that this depended upon the Reynolds number. He cited cases in which particles were suspended in fluid, which would probably fall outside the accepted limits, but felt that, in general this was not a problem, especially for sands and finer grained soils.

Lee took the opportunity to ask about the equivalence between the sand properties in the model and the prototype, in that the initial state of the sand must play a significant role in investigating a phenomenon that was related to instability. Phillips replied that the centrifuge model tests were carried out before the field event, but using the same materials. He explained that a detailed laboratory study of the stress-strain response of the oil-sand mixture had been carried out at different relative densities and that a number of those densities were targeted to give a reasonable estimate of what the stress-strain response would be in the model. He stated that the modelling was *a priori*, so it was not a direct model of the field event. The numerical analyses were rerun later for the actual field conditions.

3 THE USE OF CENTRIFUGE MODELLING TO INVESTIGATE PROGRESSIVE FAILURE OF OVERCONSOLIDATED CLAY EMBANKMENTS (W.A. TAKE)

3.1 Introduction

The speaker introduced his research into the failure of stiff overconsolidated clay embankments due to cyclical wetting and drying events as well as three experimental developments, which were necessary to support this work. He explained the three mechanisms observed successively during a series of 'rubblisation' failure modes, which incorporated dormant slip surfaces. He stated unequivocally that to be able to have any hope of modelling these embankments, appropriate boundary conditions must be provided and tools must be available for an effective stress analysis. He remarked on the challenges involved for numerical modelling of this sort of event since tension cracking, tensile stress and tensile strength, as well as the water infilling the cracks, controlled this rubblisation failure mode. He concluded by posing a question 'what is failure'?

3.2 Discussion

Cino Viggiani commented that localised failure was being observed so it was difficult to answer the question, and he would rather pose one instead about the relevance of information obtained from centrifuge tests whenever failure was localised, in that there was a scale effect. Andy Take asked whether Viggiani also had a problem with a finite element (FE) mesh to which he replied in the affirmative.

Claudio Tamagnini added that he would not trust any FE analysis in the standard Boltzmann continuum of a failure problem, which showed localisation patterns, because it would give rise to pathological mesh dependence since the model itself would have no internal length scale. He stated that it was necessary to introduce an internal length scale in the continuum model of the system and we would expect that such an internal scale would play a rôle in a centrifuge model. Clarifying this further, he said that the internal length had some relationship to the grain size, which would become important in a centrifuge test and asked how the grain size was scaled in the material under test?

Bolton was unconvinced that the grain size should be scaled in every case and asked whether the grain size would be scaled in any other soil characterisation test, for example a triaxial test. He thought that if the answer was no then why should this large lump of soil be scaled, since it contained considerably more soil grains than in a triaxial test? Tamagnini answered that he would not scale grain size in a triaxial test because he would trust the triaxial test result only within the range that the triaxial test behaved as a single element. He said that the element was not a structure, the behaviour was essentially homogeneous and the problem of localisation would not arise initially. After localisation had occured, he commented that the triaxial test results were untrustworthy.

Bolton referred back to Take's work because localisation in these failure states was triggered by having a free surface, leading to development of a tension crack. He commented that modelling a soil surface was perfectly natural and effective in a centrifuge test, whereas it was not in numerical analysis. Muir Wood remarked in terms of the contribution of peak shear strength and progressive failure to the sliding failure of the slopes that John Cookson tried to identify the characteristic length of the failure surfaces in clay in the early 70s. He concluded that the possible characteristic length was of the order of metres, which would be much greater than the size of the centrifuge model.

4 LOCALISATION AND CRITICAL STATES (H.M. HEIL)

4.1 *Introduction*

The speaker presented results of some basic, undrained, triaxial extension tests on undisturbed samples of normally consolidated Kreuzlingen Clay in which the samples had undergone a very similar hardening process in the initial stages of the tests. He proposed that the critical state model would be well suited to model the behaviour of these samples but pointed that extension tests often led to necking. Two modes of localisation were observed, which showed quite different forms of behaviour: shear banding and necking.

Shear banding in the sample led to a concentration of deformation and to a degradation of shear strength in the band due to the sensitivity of the clay, where there was a coincidence of shear banding and abrupt softening. In the post critical state behaviour, changes in the stress and strain states were remarkable, and these would not conform to the principles of critical state soil mechanics. The speaker also presented results of some preliminary FE calculations, with and without mesh updating, for a plane strain triaxial extension test on a clay sample with a small imperfection, both of which showed quite different behaviour with one mesh developing shear bands and the other necking.

4.2 *Discussion*

Malcolm Bolton was interested in the pore pressures and volumetric changes developing around the localisation in the soil test. He commented that Dave White, Andy Take and Scott Munachen monitored deformation in a triaxial test using Particle Image Velocimetry (PIV) by spraying the membrane with texture. He said that they looked at localisation events and plotted contours of volumetric strain around the localisation. He suggested that before it was assumed that the effective stress model had become very complicated, it would be a good idea to measure local effective stresses or at least to confirm that there were volumetric changes. Michael Heil thought that would be difficult because the slip surface was very thin and the pore pressure would concentrate in this band due to the localised behaviour. He was also concerned about the sensitivity of the clay, because the soil has some structure, with a carbonate content of about 35%. Cino Viggiani described a number of experiments he had carried out on clay samples, while measuring pore pressure locally, both in nominally drained and undrained conditions.

Viggiani followed up by showing a movie of a biaxial, plane strain compression test on relatively stiff, saturated clay, with local pore pressure transducers on each of the two sides. The test was run in globally undrained conditions with closed valves, but quite a strange pattern of strain localisation emerged. The typical boundary measurements were recorded as well as local measurement of excess pore pressure, which was very dependent upon the distance of the measuring points from the zones of localisation. The closer the transducer to the band, the more dilatant was the behaviour, as indicated by a local reduction of pore pressure. Experimentally it has been possible to observe significant pore pressure effects for clays, although Viggiani doubted that this would be possible for sand because the permeability would be too high and the material would be unable to sustain the necessary critical hydraulic gradient.

Jean Sulem asked about the geometry of the sample in terms of the very high slenderness ratio, which might trigger global stability in terms of buckling or necking. Heil replied that it was just convenience and did not affect the end result for an extension test because necking would occur anyway and would be more or less unpreventable.

Jan Boháč thought that it was difficult in general to test critical state in a triaxial apparatus. Heil agreed and commented on mesh dependence for extension tests. His calculations showed that the deformations in the shear band for soil modelled by Modified Cam Clay were clearly dependent on the mesh, but the failure loads calculated did not change if the imperfection was placed on the other side of the sample. A different deformation pattern was obtained but the loads were the same and the element size did not change.

Tamagnini was quite confident that the failure load would change, if the element size was varied. He noticed that a bias was induced by the mesh alignment on the shear band orientation. Heil compared the answers with those from the updated mesh analyses, where the shear plane was not perfectly aligned with the mesh. He noted that there seemed to be a trend that the shear band was at an angle of 45° to the horizontal. He took confidence from his results because relocating the imperfection, even with an unsuitable mesh, led to an answer close to the analytical solution based on the undrained shear strength from the Cam clay model with homogeneous deformation. Tamagnini commented that it is not always possible to reach the maximum load level for compression based on homogeneous deformation when localisation occurred. He said that the geometric effect induced by shear banding could produce a decrease in peak load but he was unsure about the effect on extension. Heil commented that he had chosen a stable material behaviour in his FE analysis and even if shear banding had developed, the answer would not have been very different without an imperfection, whereas the material behaviour had changed in the real sample.

5 INFLUENCE OF THE STRUCTURE ON THE MECHANICAL BEHAVIOUR OF SAND (N. BENAHMED)

5.1 Introduction

The speaker presented her ideas about the influence of the initial structure on the mechanical behaviour of sand, which she investigated by carrying out undrained triaxial tests on very loose samples of Hostun-RF sand (Fig. 1). Samples were prepared to similar void ratios by dry deposition using pluviation (Fig. 1b), and also by moist tamping (Fig. 1a). Contractive behaviour was observed for samples prepared by moist tamping, whereas dilative behaviour was noted for samples prepared by dry deposition. This difference could be explained by the micro-structure created by the different methods of preparation (Benahmed 2001a).

(a) Moist tamping deposition   (b) Dry deposition

Figure 1. Different structures due to the mode of reconstitution of Hostun-RF sand (Benahmed 2001b)

 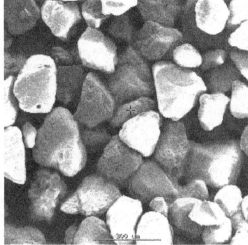

(a) Moist tamping deposition   (b) Dry deposition

Figure 2. Different structures due to the mode of reconstitution of Fontainebleau sand (Benahmed 2001b)

Scanning Electron Micrograph (SEM) pictures showed two different initial structures. Moist tamping produced an inhomogeneous structure with micro-pores, whereas a more regular structure was obtained from dry deposition. This effect was not simply a feature of Hostun-RF sand because the same results were obtained from Fontainebleau sand (Fig. 2). In terms of the steady state she saw that the initial structure had a major effect on the position of a steady state line. She stated that a structure parameter should be taken into account in the characterisation of the mechanical behaviour of sand, and asked 'how best to do it?'

5.2 *Discussion*

Cino Viggiani commented that it was very difficult to compare behaviour of dense sand (typically created from dry pluviation) with that of loose sand (derived from moist tamping), because differences in terms of relative density would be compared as well as different procedures. He thought that it would be interesting for the speaker to explain how she had been able to create such a loose specimen using dry pluviation. Nadia Benahmed responded that this was achieved using a pluviator apparatus located directly above the triaxial cell, which could give a relative density of −5% by controlling the height and the flow. She said that if the height of the pluviator was zero (controlled using a motor) and the flow rate was very high, the density would be very low.

Tom Schanz asked how the structure was retained while the sample was being moved, undisturbed, to the microscope. Benahmed replied that she fixed the matrix with water and sugar following moist tamping, and confirmed that she was sure that using the sugar solution was not detrimental to the accuracy of the resulting micrographs.

Malcolm Bolton commented that the pictures of the micro-structure were excellent but he thought that the void size distribution should be characterised as well as the particle size distribution. He noted that the void sizes were much larger than the particles and that it was obvious that the structure of the soil would be unstable because particles could fall into the voids. He proposed using positron emission tomography or computerised tomography in order to obtain the natural void size distributions, and commented that these techniques were a little expensive but were used regularly by doctors when and if we fell off our bicycles!

Philippe Nater asked which standard was used for deriving relative density and what was the lowest relative density achieved by moist tamping? He commented that he had reached –20% based on the American standard (ASTM). Benahmed replied that she had reached –20% too, also based on the ASTM. Sarah Springman said that Philippe Nater's tests had delivered this 'negative' relative density because he was dropping sand onto a spinning disk to build a model in the drum centrifuge and that the sand particles had sedimented out under water. She noted that this was another way of creating a uniform but very loose sample, and this was precisely why the determination of minimum and maximum void ratio following the ASTM test method led to a negative relative density.

David Muir Wood said he was very pleased to see the pictures of the sand micro-structures when he was examining Nadia Benahmed's thesis. He was disappointed that many people tried to test very low density soils and used moist tamping to prepare low density samples, although there would be no natural process by which a similar structure could be obtained. He felt that there was no purpose in performing tests using that mode of preparation. He also noted that much time was spent performing tests in the laboratory with purely vertical deposition, looking at very simple stress states, whereas situations where rotation of principal axes were important were common in the field. He mentioned flow slides and subsequent substructures, which could be loose for reasons other than the deposition. He also thought that the post-failure micro-structures would be unlikely to be similar to the structure obtained from moist tamping. He commented on analyses of the behaviour of partially saturated soil based on a two pore size model and thought that the structure shown here explained why these very loose structures stood up at all.

Andrzej Niemunis commented on papers by Sonja Zlatović (1995, 1997), based on similar experiments in Japan, and could not remember that such large discrepancies occurred at the steady state, although perhaps they developed at quasi steady state. He had also read that unconfined flow developed at the steady state and that all the effects of the structure disappeared at critical state. He wondered how the differences generated by the procedure of pluviation or moist tamping to the steady state behaviour could be explained. Laurent Vulliet followed up by asking whether there was a problem of wetting collapse for samples prepared by moist tamping, when the sample was saturated. Benahmed confirmed that collapse had occurred and that she had measured the density prior to saturation as well as the change in volume during saturation. Vulliet noted that the structure would be different at the end of the saturation process. Benahmed agreed that this was not the same and she had taken this variation of volume into account for all tests and that the densities were given with a correction including the first stage of compression, the saturation phase and the last stages of compression.

6   COMBINED LOADING OF A JACK-UP UNIT (G. VLAHOS)

6.1   Introduction

The speaker introduced his research on scaled model three leg jack ups on an overconsolidated clay, tested at one gravity at the University of Western Australia, to obtain the ultimate capacity for monotonic push over. Engineers from industry were using pinned conditions at the moment, which was thought to lead to an underprediction of the ultimate capacity. The focus of the experiments was to determine the extent of fixity at the footing, because previous research had indicated that some rotational fixity did exist, causing load transfer between the footings.

Vertical and horizontal loads were applied via a reaction frame, with vertical dead weights and a horizontal actuator to jack the pushover load perpendicular to the two windward footings, with the third leeward leg in front. Significant load sharing developed among the three spud can footings, which reached yield in turn and acted as a system and not just as one footing.

229

Preliminary numerical simulations of a 2-D plane frame were also carried out, using the state of the art elasto-plastic single 'cigar-shaped' surface constitutive model (model B) from Oxford University for simulating spud can footing behaviour, to predict the response obtained from the experiments. Changes of stiffness of the footing were affected by the way the model was formulated, but the pattern of predicted response was quite close to that seen from the model tests, and some load transfer was taken account of. He asked how the gradual loss of stiffness, leading to yielding of the soil due to remoulding, could be modelled most effectively, and whether there were options other than the bubble models or hypoplasticity. Another aspect that was not well modelled was the windward footing's tendency to lift-off as the vertical load reduced post-yield, and that this was important in judging the extent of fixity, which could be assumed safely for design.

6.2 *Discussion*

Bill Craig asked about the rotation of the models, since there appeared to be some indications of leg penetration. He was concerned that the lateral displacements seemed to be large as did the overall rotation and noted that these legs were very flexible in reality. He wondered if the flexibility of the legs and the leg guide superstructure interaction was being modelled directly. George Vlahos replied that he had scaled the moment of inertia of the leg of a jack up exactly, by modelling the structure and taking the flexibility/bending of the legs into account so that he could focus on the soil structure interaction. The figure indicated that the leg was straight but double curvature was observed in the leg during push over. He said that a rigid connection was assumed for the leg guide because of the constraints of the physical modelling and it had not been possible to consider the dynamic effects.

Sarah Springman mentioned centrifuge model tests carried out on a single lattice leg of a jack up structure, with a spud can instrumented with a horizontal-moment-vertical (HMV) load cell at the bottom and with another HMV load cell at the top. She commented that the fixity of the spud cans should also account for the load transfer between the lattice structure above the spud can and that this could be significant in soft clay by improving the spud can fixity. She thought that choosing a tubular leg might prove to be very conservative, since one could reach lateral pressures of up to $9s_u$ (undrained shear strength) around the tube at depth whereas her tests had indicated that only about half of this was mobilised. Vlahos agreed this was a good point and explained that the tubular legs had been chosen because of the physical problems of modelling at 1/250th scale. He commented that their research was focussing purely on the behaviour of the spud can and not on the effect of the leg submerged partially in clay. He noted that he had not seen any backfilling of the soil behind the spud can at 1g and therefore the tubular legs were not in contact with the soil. He wanted to determine the pure rotational fixity of the spud cans rather than just the load transfer through the truss and was doubtful that remoulding of the soil behind the spud (due to the surcharge of the clay) should be relied upon. Springman felt that such an interaction should be considered otherwise the outcome could be very conservative, especially if penetration of the footing was more than 1-2 spud can diameters below the seabed.

Ivo Herle asked for clarification about the numerical calculation and whether it had been carried out under plane strain conditions, to which Vlahos replied that it was a plane frame analysis. Herle was concerned because it seemed to be a 3-D problem. Vlahos agreed and commented on the simplifications adopted in the physical model by loading the jack up along the axis perpendicular to the two footings. He thought that the response of the two windward footings was very similar and assumed that this characteristic could be carried over to the numerical analysis if the moment of inertia was doubled for these windward legs. He was happy that, aside from application of an eccentric load, for which all the 3 footings would act differently, the problem would be greatly simplified with reduced computation time but without any major loss of accuracy.

Saiichi Sakajo asked about definition of the initial stress of the ground in the numerical calculation since the spud can was very shallow in a 1g test. Vlahos replied that the jack up had been preloaded up to a penetration of about one diameter of the spud can and that this was replicated in the numerical simulation to create an initial yield surface. Afterwards, he said that the footing was unloaded back to the pre-load ratio of about half of the maximum vertical load, prior to carrying out the push over test.

7 CENTRIFUGE MODELLING OF CAISSON CAPACITY UNDER UPLIFT LOADING (A.R. HOUSE)

7.1 Introduction

Economical design of deepwater foundation systems has relied traditionally upon minimising the size of the footprint. In consequence, large axial loads were imposed on the particular foundation, for example smooth walled suction caissons (prototype: 3.6 m diameter by 14.4 m length and mass of ~23 t), which needed a large aspect ratio (length to diameter ratio = 4) to withstand them. Research was carried out in the centrifuge into the capacity of (jacked-in) suction caisson foundations under long-term sustained axial loading. The influence of internal structural ring stiffeners on the penetration resistance of suction caissons was also investigated in terms of how soil would flow around the stiffeners as they were installed in very soft normally consolidated clay. If the soil flowed back onto the wall of the caisson, it would generate friction during penetration. Otherwise, a stiffer soil would form a rigid column and not make contact with the internal wall of the caisson above the stiffener, thereby requiring less suction to install the caisson in stiffer than in softer soil. It was found that additional bearing capacity was only mobilised on the lowest stiffener, the stiffeners above did not contribute to the end bearing resistance during penetration, probably due to remoulding of the soil around the lowest stiffener. The installation load correlated well with full end bearing on the lowest stiffener, with full soil flow around it and wall friction with a mobilised friction coefficient of ~0.35-0.4, which was roughly equivalent to the inverse of the sensitivity of kaolin (2.4), as determined from a vane shear test.

Undrained shear strength of the normally consolidated soil sample was characterised by a T-bar penetrometer to have zero strength at the mud line and a gradient of ~1.1 kPa/m, which was quite typical for Gulf of Mexico clay. Using a T-bar was advantageous for these softer soils in that no pore pressure or stress corrections were required, and an exact solution was available to correlate bearing resistance with undrained shear strength.

Loads mobilised due to drained pull out (with the top cap of the caisson vented) and undrained pull out were compared and found to have approximately a factor of 2 on capacity, as expected. Long-term sustained axial loading with a sealed top cap (for a prototype consolidation period of approximately 2.5 years) pulled out at 75 percent of the load of the immediate monotonic undrained pull out. This coincided with the soaked drained capacity, when the caisson was installed and left under a high groundwater table for a prototype consolidation period of one year to allow the disturbed soil to 'set up'. It was proposed that suction caissons should be designed for drained capacity only, with full soil flow around the stiffeners of caissons in very soft soils. The question was posed about how stiff should a soil be, that a rigid soil column would form without contacting the wall?

7.2 Discussion

Vincenzo De Gennaro asked how the installation was modelled from a numerical point of view? Andrew House replied that a static limit equilibrium solution was developed based on installation forces from self-weight and suction pressure (although these tests were jacked installation only because better resolution was obtained from the load cell). He added that the resistance was simply due to end bearing on the annular area at the end of the caisson combined with external and internal skirt friction and end bearing on (or flow around) the internal stiffeners.

Luc Thorel was interested in the criterion adopted to define the drained and undrained pull out velocity (v). House answered that undrained *installation* was undertaken at v = 1 mm/s (at model scale), which was normalised by the wall thickness (t) divided by the coefficient of (vertical) consolidation (c_v) to give $vt/c_v \sim 100$. He said that undrained and drained *pull out* were carried out at a factor of 3 and 10 times slower respectively and noted that drained pull out of the caisson required more force then it did to install it, which was thought to be due to rate effects. He added that tests were done on T-bar penetration resistance mobilised at variable rates and he had noticed that vt/c_v increased by a factor of about 1.4 from the drained pullout to undrained installation.

Sarah Springman wondered whether information could be obtained from the T-bar tests to see whether the hole behind the T-bar closed up. She expected that this could be related to the overconsolidation ratio and that coloured traces could be placed in the clay at the penetration site for post-test exhumation. House agreed that this would be an excellent method although he had never had a sample in a centrifuge that was stiff enough so that the soil did not flow back around the T-bar. He thought that being able to observe a half section of T-bar, penetrating against a window to a plane strain centrifuge box, would also be very helpful.

8   ON THE RELIABILITY OF CENTRIFUGE MODELLING (S. SAKAJO)

8.1   *Introduction*

The speaker showed recent experimental results for the failure mechanism for uplift bearing capacity of a belled pile in a sand in Japan, which was expected to increase with the angle of the bell (α), so that the bearing capacity increased with α. A slip plane formed under the base of the foundation. In reality, weathered cracks may develop in very weak zones and it was very difficult to represent such insitu conditions in centrifuge testing if block samples were cut from the prototype material, although often comparison of results could be very reasonable. Centrifuge model testing was found to be very useful for representing insitu conditions, although care should be taken about the sampling.

8.2   *Discussion*

David Muir Wood wondered about the scale of the modelling, which must be large by comparison with the scale of the fabric in the natural soils. He noted that a block sample would contain features with a scale separation, which, in this case, would be of the order of the pile length. Therefore he was worried about these scale effects, which must make one suspicious of the results of the model test. Saiichi Sakajo agreed but pointed out that he had checked the size effects by using different values of g (verification of models), for example at both 50g and 100g gave quite good agreement, although he felt that perhaps this was not quite sufficient.

9   SUMMARY BY ANDREW SCHOFIELD

Andrew Schofield expressed pleasure in being asked to summarise what had happened during the first day of the Workshop, in which tasks set in Cambridge so many years ago had proved to be so fruitful and had led to so much interesting work by different groups.

He recalled several visits to the Norwegian Geotechnical Institute (NGI) between 1958 and 1967. On one of the occasions he had asked Laurits Bjerrum (ed: Doctorate from IGT, ETHZ), the director, what he thought we should teach in soil mechanics to undergraduates. Bjerrum replied that we should not teach soil mechanics to undergraduates, as they cannot make judgements. Undergraduates should learn mechanics. Schofield said that he accepted that point of view and thought that the student must learn things that they can understand thoroughly. When they answer a question, they pose further and unexpected questions to the lecturer. One of his great pleasures in teaching at Cambridge had been what he had learnt from the students as he heard what they thought they knew in their answers to questions. Both their questions and their answers were important.

He recalled these events: "I went to Bjerrum in about 1967 with the page proofs of Critical State Soil Mechanics that Peter Wroth and I had just completed. I asked him if he could find time to read this book and to write a little preface for it. When I went back again he said that the conclusions we reached were interesting but he did not understand the mathematics by which we arrived at these conclusions. So he would not write a preface. The book must stand by itself."

Schofield offered his answer to the question he had posed to Bjerrum. He felt that the associated flow rule was important and was convinced that, in teaching the theory of plasticity, it was worthwhile to teach Drucker's stability criterion to students. It leads to a sound selection of materials and methods of construction. He said that if you choose your material and method of construction to give you ductility and stability in Drucker's sense, you have what David (Muir Wood) called a good material. It is in your choice. You can make decisions that are 'good' or you can make a bad choice of materials and methods. Remembering the Oklahoma Federal Centre bombing when many people were killed, if that same bomb had exploded in the centre of London, the buildings might not have collapsed in that way. Designers in London in general must take stability into account more than he believed was the case for the Federal engineers who had designed that Oklahoma building. It was inherently unstable in the sense that a perturbation, just a small vector, led to a large (out)flow of energy.

He was quite convinced that ductility and stability in Drucker's sense was something that students should learn in plastic design of structures and that they can carry over into the theory of plasticity applied to soil mechanics. He was delighted to hear from Malcolm Bolton that so much work had gone into investigating the crushing of materials and noted that rockfill was confirmed to be an aggregate, which produced ductile behaviour with plastic yielding, despite such apparently brittle unit behaviour.

He added that one choice an engineer can make about materials is how they are to be compacted. He showed Figure 3 from his Rankine lecture in 1980, noting that relative density was presented in terms of equivalent liquidity. He pointed out the zone in which Coulomb rupture occurs (e.g. AB, GE) for material on the dry side of the critical state (e.g. C, F), whereas on the wet side of critical state (e.g. BD, DE) the soil behaves in a ductile fashion. He noted that as the state of the soil moved through the zone of Coulomb behaviour with peak shear strengths due to heavy compaction, the soil would gain stiffness and strength, but made the point that "this is treacherous since all the structures which have betrayed us (Teton Dam, Baldwin Hills Dam) are found exactly in this region. They are so stiff that as they rubble-ise and break up, the rapid flow of fluid into the cracks turns the rubble-ised material into an avalanche (debris flow). We were betrayed by a sudden instability (like the failure of Oklahoma City Federal Centre); a sudden deterioration of the whole structure. Structures that are in a ductile zone have a benign nature. We have choices we can make about a site and a construction material and method."

Schofield commented on the contributions concerning the jack up spud foundation. He saw jack up spud fixity in terms of a yield locus, with vertical load and shear load (treated as generalised stresses and associated strains leading to vertical and horizontal displacements). The stability of the jack up unit depended on a ductile foundation system. The system can be plastic or

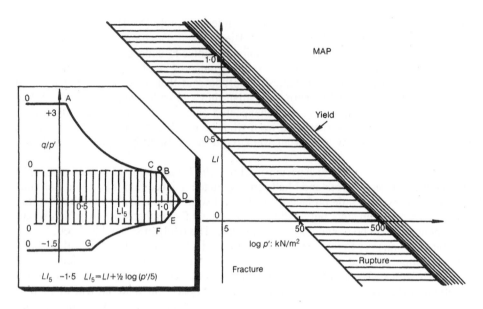

Figure 3. Remoulded soil behaviour (Schofield 1980)

it can be brittle, and crack apart under the stresses. It was not only the foundations but also the engineers who failed if they did not realise the benefits of ductility and plasticity.

His final words were a message for those teaching soil mechanics in the coming years: "Soil is plastic. Plasticos[1] is a Greek word meaning moulding, like to mould clay into a pot or a sculpture. Inherently soil should be plastic. You can get soil into a benign state if you are not too far away from critical state behaviour. When only slightly on the dry side you do not have a very high peak followed by a large fall of strength and instability. Our objective should be to teach students the benefit of designing for Drucker's stability criterion and to select materials and methods of construction in that benign region where a structure is ductile and we do not fail as engineers. The general public relies upon engineers."

10 CONCLUSIONS

Lukas Arenson had reviewed the research presented under the umbrella of problems governed by 'Failure' with Lis Bowman, Gerd Festag, Luiza Dihoru and Matthew Dietz. He declared, on their behalf, that their first reaction had been somewhat depressing since it was clear that there were significant problems with the representation of failure for both the constitutive and centrifuge modelling extremes. Finite element methods and many laboratory tests were also being challenged to establish appropriate boundary conditions and to deal with localisation (Gajo 2002). He felt, however, that there was light at the end of the tunnel in that the contributions were moving in the right direction.

He noted that failure should be defined initially, in that Chris Martin (2002) had pointed out that the way in which practising engineers and academics would describe failure was not always the same. He also thought that the difference between element failure and system failure was ex-

---

[1] plastic = πλαστικόν: characterised by moulding or giving form to clay, wax, etc; the art of shaping or moulding; pertaining to moulding or modelling; readily assuming a new shape (Shorter Oxford Dictionary Vol. II: 1601).

tremely important, in that some elements could fail (e.g. the Confederation bridge mentioned by Phillips (2002), the embankments described by Take & Bolton (2002), and the spud can jack up rig response presented by Vlahos (2002)). But, in these cases, system failure had not yet developed, although the collapse of the Oklahoma public building, described graphically by Andrew Schofield, was clearly an example of complete system failure caused by progressive failure of the elements. He thought that significant improvements were needed here and noted that occasionally local failure phenomena were also very important for some types of construction (e.g. the suction caissons introduced by House & Randolph (2002)).

Arenson commented that determining the pre-failure and the eventual failure mechanism was quite well done at present, but that modelling the failure itself was another problem, especially if progressive failure had played a major rôle. He cited shear band formation in a triaxial test (e.g. Heil 2002), and thought that despite measurement of porewater pressures at some boundaries, it was not possible to state categorically what was happening in the shear bands, particularly with a view to establishing the effective stresses mobilised there. He recalled a short movie of a biaxial shear test, shown by Cino Viggiani, in which the porewater pressure transducers on either side of the sample (one close to and one outside the shear zone) did not show the same results.

He was also concerned about issues relating to micro-structure in terms of the arrangement of the particles within the soil matrix and the effect of sample preparation on the failure mechanism (e.g. Bolton & Cheng 2002, Benahmed 2002). He thought that micro-mechanics could offer a useful insight into the failure mechanisms and noted also that David Muir Wood had insisted that sample preparation should take account of typical insitu conditions, particularly in respect of remoulded samples. In addition, Muir Wood had asked a very important question in relation to performing laboratory tests as well as numerical modelling: Do we model the real conditions?

He concluded that an improvement in testing equipment was required in both centrifuge modelling and triaxial equipment, and speculated on a most welcome future invention of a porewater 'PIV' to determine porewater pressure in a non-invasive way in the laboratory tests! But his final message was that failure had to be defined more clearly, especially for industry, otherwise practising engineers would not be interested in using the highly sophisticated models developed by the academics.

ACKNOWLEDGEMENTS

We are grateful to the Chair, Cino Viggiani, and the discussion leader, Jan Boháč, for their insightful leadership and guidance of this session, and Lukas Arenson and his team for their summary. We also thank all of the authors and contributors for their patience with our interpretation of their statements, and for their help in ensuring that this record of the discussions is as accurate as possible. Finally we wish to acknowledge the assistance of our Greek scholars, Dr Demetrios Coumoulos and Polly Caffrey (née Schofield) for their help with the basic etymology of the word plastic, as mentioned by Andrew Schofield in his summary.

REFERENCES

Benahmed, N. 2001a. Comportement mécanique d'un sable sous cisaillement monotone et cyclique: application aux phénomènes de liquéfaction et de mobilité cyclique. Thèse de doctorat de l'ENPC.

Benahmed, N. 2001b. Presentation on "Influence of the structure on the mechanical behaviour of sand". *Workshop on Constitutive and Centrifuge Modelling, Monte Verità: Two Extremes, Ticino, July 2001.*

Benahmed, N., Canou, J. & Dupla, J.-C. 2002. Influence of the structure on the mechanical behaviour of sand. In S.M. Springman (ed.) *Workshop on Constitutive and Centrifuge Modelling: Two Extremes, Monte Verità.* Lisse: Swets & Zeitlinger.

Bolton, M.D. & Cheng, Y.P. 2002. Micro-Geomechanics. In S.M. Springman (ed.) *Workshop on Constitutive and Centrifuge Modelling: Two Extremes, Monte Verità*. Lisse: Swets & Zeitlinger.

Gajo, A. 2002. Failures of laboratory sand samples and constitutive modelling. In S.M. Springman (ed.) *Workshop on Constitutive and Centrifuge Modelling: Two Extremes, Monte Verità*. Lisse: Swets & Zeitlinger.

Heil, H.M. 2002. Localisation and critical states. In S.M. Springman (ed.) *Workshop on Constitutive and Centrifuge Modelling: Two Extremes, Monte Verità*. Lisse: Swets & Zeitlinger.

House, A. & Randolph, M.F. 2002. Centrifuge modelling of caisson capacity under uplift loading. In S.M. Springman (ed.) *Workshop on Constitutive and Centrifuge Modelling: Two Extremes, Monte Verità*. Lisse: Swets & Zeitlinger.

Martin, C.M. 2002. Impact of centrifuge modelling on offshore foundation design. In S.M. Springman (ed.) *Workshop on Constitutive and Centrifuge Modelling: Two Extremes, Monte Verità*. Lisse: Swets & Zeitlinger.

Phillips, R. 2002. Centrifuge Modelling of Failures and Safe Structures. In S.M. Springman (ed.) *Workshop on Constitutive and Centrifuge Modelling: Two Extremes, Monte Verità*. Lisse: Swets & Zeitlinger.

Schofield, A. N. 1980. Cambridge Geotechnical Centrifuge Operations. Géotechnique 30(3): 227-268.

Take, W.A. & Bolton, M.D. 2002. The use of centrifuge modelling to investigate progressive failure of overconsolidated clay embankments. In S.M. Springman (ed.) *Workshop on Constitutive and Centrifuge Modelling: Two Extremes, Monte Verità*. Lisse: Swets & Zeitlinger.

Vlahos, G. 2002. Combined Loading of a Jack-up Unit. In S.M. Springman (ed.) *Workshop on Constitutive and Centrifuge Modelling: Two Extremes, Monte Verità*. Lisse: Swets & Zeitlinger.

Zlatović, S. & Ishihara, K. 1995. On the influence of nonplastic fines on residual strength. *First International Conference on Earthquake Geotechnical Engineering, Tokyo, Japan* 1: 239-244. Rotterdam: Balkema.

Zlatović, S. & Ishihara, K. 1997. Normalized behavior of very loose nonplastic soil: Effects of fabric. *Soils and Foundations, Tokyo* 37(4): 47-56.

Problems governed by deformation

Constitutive and Centrifuge Modelling: Two Extremes, Springman (ed.)
© 2002 Swets & Zeitlinger, Lisse, ISBN 90 5809 361 1

Difficulties related to numerical predictions of deformation

I. Herle
Institute of Theoretical and Applied Mechanics, Czech Academy of Sciences

ABSTRACT: Numerical simulations of boundary value problems are composed of several equally important steps. Underestimating one of those steps can result in unrealistic calculation results. A brief overview of several prediction competitions reveals an insufficient ability to predict deformations of geotechnical structures correctly. Subsequently, some factors influencing the computational results are discussed.

1 INTRODUCTION

Outputs of numerical calculations are rather impressive today. Contour plots of stresses and deformations contribute to the fast interpretation of results and create the impression of being perfect and reliable. Nevertheless, many of the end-users overlook the complicated process leading to such results. This process can be divided into several basic steps:

1. *Simplification of the reality*
   (choosing important variables, geometry, selection of substantial aspects of the problem and disregarding minor ones)

2. *Discretization*
   (space and time discretization — element size, time step, boundary conditions, construction details)

3. *Constitutive model*
   (framework for calculation of strains, selection of the appropriate material description, model for interfaces, calibration of the parameters, determination of the initial state)

4. *Mathematical and numerical aspects*
   (type of time integration, equation solver, iteration scheme, well-posedness).

All steps are equally important. If one step is not correctly implemented, the whole result may become wrong. All steps compose a chain which fails at the weakest link. The outlined steps are valid for any calculation. It is not possible to distinguish seriously between calculations of deformation and limit states. Even in problems related to deformation the material usually reaches the limit state of stress at several points! Moreover, we need deformations in order to calculate limit states since the subsoil and structure interact and we cannot distinguish between action and reaction (although it is a wish of code designers). The necessity of an overall approach was already discussed e.g. in the Rankine lecture by Roscoe: "Influence of strains in soil mechanics" (Roscoe 1970).

2 PREDICTIONS VS. MEASUREMENTS

The only way to evaluate numerical simulations is to compare them with measurements and observations. However, opportunities to compare measurements with true predictions (so-called class A predictions (Lambe 1973)) are very rare. They teach us that in most cases our predictions are rather far away from measured values.

It is useful briefly to review some competitions on predictions in geotechnical engineering, with emphasis on deformation calculations (all competitions were designed as plane strain problems):

1. *MIT trial embankment* (Wroth 1977)
   A normally consolidated soft clay layer beneath a trial embankment controlled the deformation behaviour. Laboratory experiments and field measurements for the first construction stage up to 12.2 m height were done prior to the prediction calculations. Predictions of deformations, pore pressures and maximum additional height of the embankment at subsequent rapid filling to failure were received from ten groups. There was a large scatter of the numerical results: e.g. additional height of fill at failure ranged between 2.4 m and 8.2 m (measured: 6.4 m), additional settlement at the ground surface in the centre of the embankment due to 1.8 m of fill — between 1.9 cm and 34.8 cm (measured: 1.7 cm), additional horizontal movement beneath the middle of the embankment slope due to 1.8 m of fill ranged between 0.4 cm and 21.8 cm (measured: 1.3 cm), etc. One of the best predictions, based on the Modified Cam Clay model (Wroth 1977), was very good with respect to pore pressures but still less accurate for deformations.

2. *Excavation in sand* (von Wolffersdorff 1997)
   A 5 m deep excavation in a homogeneous sand layer above the groundwater level in Germany near Karlsruhe was supported by a sheet pile wall. Struts were installed at 1.5 m depth. An additional surcharge was placed at the ground surface after the excavation and finally the struts were loosened. In situ and laboratory soil investigations were performed prior to the excavation. Results from 43 predictions concerned horizontal displacements of the wall, vertical displacements at the ground surface, earth pressures on the wall and bending moments in the wall. The comparison with measured values was very disappointing. Especially worrying was that displacements were often predicted wrong in a qualitative sense (i.e. in the opposite direction).

3. *Excavation in clay* (Lydon 2000)
   A 7 m deep excavation in soft clay and peat with a high groundwater level was located in the Netherlands near Rotterdam. The purpose of this field test was to complement the field test in Karlsruhe with another type of soil. Again, predicted values were often far away from the measured ones.

4. *Tunnel and deep excavation* (Schweiger 1998)
   Both benchmark calculations were intended to check only the numerical aspects and were not accompanied by any measurements. Geometry, constitutive model (Mohr-Coulomb), model parameters and boundary and initial conditions were specified. Still the results diverge in a considerable manner: In the case of the tunnel excavation, most calculations predicted nearly the same surface settlement only for one step excavation. Applying two steps, the maximum surface settlement ranged from 3.3 cm to 5.8 cm (10 calculations). Even worse was the case of a deep excavation supported by a diaphragm wall with two

rigid struts. Vertical displacement of the surface behind the wall ranged between 1.3 cm and 5.2 cm and the horizontal displacement of the head of wall oscillated between −1.3 and 0.5 cm (12 calculations).

5. *Excavation with a tied back diaphragm wall* (Schweiger 2001)
A geometrical specification of the 30 m deep excavation in Berlin sand supported by a diaphragm wall anchored at three levels was accompanied by a detailed specification of the simulation stages and basic laboratory experiments (index tests, oedometer and triaxial tests). Even after "filtering out" the most extreme and questionable results, the values of the maximum horizontal displacements ranged between 0.7 cm and 5.7 cm, and the surface settlements between 5.0 cm and −1.5 cm. Large scatter was obtained even for results from the same code using the same material model.

6. *Vertically loaded small model footings on sand* (Herle & Tejchman 1997, Tatsuoka et al. 1997, Tejchman & Herle 1999)
Small model strip footings ($B = 1.0$, 2.5, 5.0 and 10 cm) were placed on the surface of a dense sand and loaded vertically. Load-displacement curves were calculated prior to the experiments. Due to small dimensions of the footings, a pronounced scale effect was involved. The overall results of the prediction competition (Tatsuoka et al. 1994) have never been published. However, a personal communication to the organizer[*] confirmed that the predicted load-displacement curves (Herle & Tejchman 1997), based on the polar hypoplastic model, were the best ones. Although the scale effect and the strains at the peak of the load-displacement curve were reproduced very well, the maximum load was overpredicted by almost 40%.

The above overview of several prediction competitions shows an unsatisfactory state of the art. It seems that we are not able to predict reliably the behaviour even of the simplest geotechnical structures.

Leaving aside the question as to why there are, in reality, only a few geotechnical structures which fail (probably due to very high values of safety), I will try to analyze in more detail the steps No. 3 and 4 of Section 1, which describes the process of the creation of a numerical model.

3 CONSTITUTIVE MODEL

From the point of view of a geotechnical engineer, the question of the constitutive model and of its calibration is the crucial one. A large number of soil models exist but still no general agreement on their quality (and how the quality can be defined (Kolymbas 2000)) and on the range of their applicability. The second criterion is probably more important since every model has some limitations which are unfortunately usually hidden by the designer of the model. Moreover, the complexity of advanced models often prevents other potential users from implementing and checking them[†]. This results in applying but very simple models in most practical cases.

---

[*]Personal communication from Professor F. Tatsuoka at the IS-Nagoya, 1997.

[†]The question of checking the models is extremely important. There is a useful analogy with the open source software e.g. under the GNU license. This software is accessible to everyone who is interested and can read the source code. Consequently, a large number of people test the software and report the bugs as well as propose improvements. It is a public secret that many open source codes are today better and more reliable than commercial software which can be supported only by limited financial means for development and testing. Unfortunately, the situation in the development of constitutive models resembles more of a struggle between advertisement agencies. Users are merely convinced of superiority of each particular model even without understanding it and having a possibility to check it.

It is not the aim of this contribution to evaluate the suitability of particular soil models in describing the soil behaviour. Alongside prediction competitions on boundary value problems (BVP), prediction competitions on soil behaviour in element tests have also been performed (Gudehus et al. 1984, Saada & Bianchini 1988). Nevertheless, there are several topics closely related to constitutive models which become obvious first in calculations of BVP.

3.1 Calculation of strains

Constitutive models use strains instead of displacements in order to be scale independent. However, there is no unique definition of strain. In soil mechanics, we usually consider small strains which refer to a very small change of lengths and angles: in 1-D we write $\varepsilon = (l - l_0)/l_0$ with l_0 being the initial length and l the actual length. This approach is the simplest one, but mostly it cannot be justified. Using small strains, we refer to a reference (initial) geometrical configuration without taking into account geometrical effects. Although it may often seem that the overall deformations are small, local strains can reach high values in some cases, consider e.g. cavity expansion problem or shear zones. Another class of problems sensitive to the strain calculation refers to situations where the stiffness moduli are comparable with the stress level. We should always bear in mind that using highly non-linear constitutive models, which correspond better to the soil behaviour than the linear ones, even a small change in strains can yield a substantial change in stresses.

3.2 Strain decomposition

The vast majority of constitutive models assumes the strain decomposition into elastic and plastic parts thus implicitly assuming the existence of an elastic range. Elastic strains are recoverable (reversible) after so-called unloading. Usually linear elasticity is assumed. But it is not simple to detect an elastic range even in highly sophisticated laboratory experiments in the range of very small deformations (Tatsuoka et al. 1999).

One often argues that this decomposition is not so important for monotonic deformation processes where only one loading direction appears. However, if the strain decomposition is not important, so why do we do it? Do we want to forget the basic rule „*Keep the model as simple as possible (but not simpler)*"? Moreover, the argument mentioned above is not true. Practically all boundary value problems involve regions of simultaneous loading and other regions of unloading. Probably no BVPs exist which include only one loading direction. A classical demonstration can be a tunnel excavation or a deep excavation (Herle & Mayer 1999, Viggiani & Tamagnini 2000). Both, loading and unloading is also present in coupled problems with the generation and the subsequent dissipation of pore water pressures (Lambe 1967).

The problem of the elastic range is closely related to the non-linearity of the soil behaviour. Although, according to experimental results, the elastic strain range is almost vanishing, many models presuppose quite a large domain of (linear) elastic strains (except for very sophisticated elasto-plastic models which, on the one hand can reproduce the soil behaviour in such a small elastic range, but on the other hand are extremely complex and remain a topic of the academic research of few specialists). Even rather advanced models can suffer from the effect of an unrealistic elastic range, see Figure 1 (Herle et al. 2000).

An example of failing to reproduce the observed deformation pattern by applying linear elasticity may be the class A prediction of displacements due to the excavation for the underground car park at the palace of Westminster (Burland 1989). Although the horizontal displacements at the surface behind the wall were in a tolerable range, the predicted vertical displacements had the opposite sign to the measured values. This resulted in the wrong prediction of the direction of rotation of the Big Ben Tower. An additional calculation using a bilinear model with ten times higher initial stiffness yielded a substantial improvement (Simpson et al. 1979).

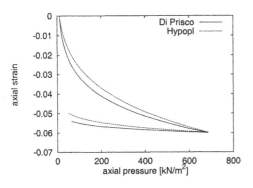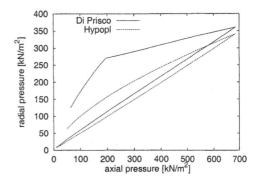

Figure 1. Elastic range of *di Prisco model* yields an unrealistic oedometer stress path during unloading (right)

3.3 *Hypoplasticity*

Let me give an example of a non-linear constitutive model without strain decomposition. There is a class of so-called hypoplastic models which abandon the strain decomposition and consequently do not use notions of yield and potential surfaces, flow rules etc. (Kolymbas 1991, Gudehus 1996). The basic idea can be explained for a 1-D case. The stress rate follows from

$$\dot{\sigma} = E_1 \dot{\varepsilon} + E_2 |\dot{\varepsilon}| \tag{1}$$

where $E_1 > E_2 > 0$ are the stiffness moduli. Defining "loading" as $\dot{\varepsilon} < 0$, we have $|\dot{\varepsilon}| = -\dot{\varepsilon}$ and therefore

$$\dot{\sigma} = (E_1 - E_2)\dot{\varepsilon} \tag{2}$$

whereas for "unloading", $|\dot{\varepsilon}| = \dot{\varepsilon} > 0$ and

$$\dot{\sigma} = (E_1 + E_2)\dot{\varepsilon}. \tag{3}$$

Obviously, the stiffness for loading is smaller than for unloading, as usually observed.

Considering two stress components, σ and τ, and the corresponding strain components, ε and γ, a hypoplastic equation can be written as

$$\begin{Bmatrix} \dot{\sigma} \\ \dot{\tau} \end{Bmatrix} = \begin{bmatrix} L_{\sigma\sigma} & L_{\sigma\tau} \\ L_{\tau\sigma} & L_{\tau\tau} \end{bmatrix} \begin{Bmatrix} \dot{\varepsilon} \\ \dot{\gamma} \end{Bmatrix} + \begin{Bmatrix} N_\sigma \\ N_\tau \end{Bmatrix} \sqrt{\dot{\varepsilon}^2 + \dot{\gamma}^2} \tag{4}$$

with L_{ij} and N_i being components of the tangent stiffness matrix. A particular representation of this equation (Herle & Nübel 1999) may be

$$\dot{\sigma} = c_2 \sigma \left(\dot{\varepsilon} + c_1^2 \frac{\sigma\dot{\varepsilon} + \tau\dot{\gamma}}{\sigma^2} \sigma - c_1 \sqrt{\dot{\varepsilon}^2 + \dot{\gamma}^2} \right) \tag{5}$$

$$\dot{\tau} = c_2 \sigma \left(\dot{\gamma} + c_1^2 \frac{\sigma\dot{\varepsilon} + \tau\dot{\gamma}}{\sigma^2} \tau - 2c_1 \frac{\tau}{\sigma} \sqrt{\dot{\varepsilon}^2 + \dot{\gamma}^2} \right) \tag{6}$$

with c_1 (=$1/\tan\varphi$, φ is a friction angle) and c_2 being the model parameters. With such a very simple model one can obtain many features of the soil behaviour, including non-linearity, shear-volumetric coupling or stress-dependent stiffness.

In the general case, the hypoplastic model includes an objective stress rate tensor $\overset{\circ}{\mathbf{T}}$ instead of $\dot{\sigma}$ and the stretching tensor $\mathbf{D}$ instead of $\dot{\varepsilon}$. The modulus $|\dot{\varepsilon}|$ is replaced by $\|\mathbf{D}\| = \sqrt{\mathbf{D}:\mathbf{D}}$. The matrices L_{ij} and N_i become tensors of the fourth and the second order, respectively, and depend on the effective stress tensor (and its invariants) and on the void ratio (von Wolffersdorff 1996).

3.4 Model calibration

3.4.1 Quality of experiments

Calibration of the model parameters is mostly based on laboratory experiments. Nowadays, the quality of test results is considered to be high (although the precision of measurements is limited) and consequently many researchers try to check constitutive equations by reproducing the experimental curves numerically as closely as possible. However, looking more in detail at basic element tests performed in different apparatuses, substantial scatter of experimental results can be observed (Bianchini et al. 1988, Muir Wood 2000). This is not a point against experiments (!), it merely suggests that the development and the validation of constitutive models should be based mainly on the qualitative aspects of the soil behaviour. The realistic solution of boundary value problems does not depend only on the reproduction of calibration curves.

3.4.2 Oversimplified models

Material parameters are always linked to a particular constitutive model. In other words, if the constitutive model does not describe particular aspects of the material behaviour properly, one cannot determine corresponding model parameters.

A classical example of this problem is the calibration of the Mohr-Coulomb model. Non-linear stress-strain curves are approximated by a straight line underpredicting thus the initial stiffness and overpredicting the tangential stiffness at higher strains. Moreover, one often does not inspect the volumetric behaviour which is uniquely linked to the stress-strain curve. Such an inspection would show that the initial (elastic) volumetric compression is grossly overpredicted (Figure 2).

Recognizing the above mentioned problem, some people try to fit the parameters of a constitutive model directly to in situ measurements, referring to the observational method (Peck 1969). Beside the fact that this approach disables a class A prediction, it usually fails to reproduce the realistic pattern of deformation. Regarding e.g. a tunnel excavation, one manages to fit the surface settlement but departs from the measured tunnel deformation and vice versa (Doležalová et al. 1998).

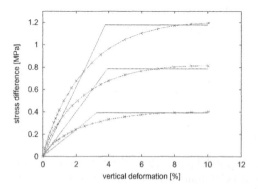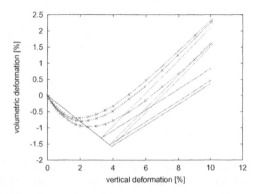

Figure 2. Questionable calibration of the Mohr-Coulomb model from the outputs of triaxial tests (Young modulus adjusted to the stress level)

The situation is even worse with respect to the subgrade reaction model. In many countries this model is standard for modelling the soil-structure interaction (design of excavation or tunnel support) in spite of its doubtful parameter, modulus of subgrade reaction. This parameter cannot be a soil constant since it is system dependent and thus it cannot be measured (except in a real 1:1 prototype in the same soil conditions).

3.4.3 *Complex models*

The calibration of complex models is a hard task since they usually have many material parameters. Nevertheless, the number of parameters is of a minor importance if a reliable calibration procedure is at the disposal. Unfortunately, this is rarely true.

Most model developers claim that one should be able to determine the model parameters from standard types of laboratory tests and they recommend usually triaxial and oedometer tests. Still, the majority of parameters are interrelated because they represent coefficients in highly non-linear functions which cannot be uniquely determined from experimental output. It is difficult to define a measure of an optimum approximation of the experimental data with such models which are also sensitive to the initial conditions in calibration tests.

A robust calibration should be based preferably on asymptotic states. Asymptotic states are e.g. critical states, proportional compression with constant ratio of the components of the stretching tensor or states with pressure-dependent minimum density due to cyclic shearing with a small amplitude. Such states are insensitive to initial conditions, eliminating thus the influence of sample preparation. They can be usually described by relatively simple equations with a low number of parameters. These parameters can be considered as true material constants because they do not depend on the actual state of soil.

Novel ways of model calibration seem to be optimization procedures including artificial neural networks. However, they are still under development and are thus not suitable for routine applications.

3.5 *Initial state*

It is astonishing how often an appropriate determination of the initial state is neglected in numerical simulations of BVP. It is standard to start calculations with a K_0 stress state, taking $K_0 = 1 - \sin \varphi$, although the geological history of the site can be reproduced at least qualitatively (Mähr 2000). The initial state can be influenced even more dramatically by construction processes, e.g. old excavations, compaction or wall installation (von Wolffersdorff 1997, Nübel et al. 1997).

The situation becomes worse with unsuitable constitutive models which usually do not distinguish between soil constants (i.e. material parameters) and state variables. An approximation of non-linear stress-strain curves with straight lines inevitably yields material "constants" which depend on the current state of soil and thus contradict the notion of a material parameter.

4 MATHEMATICAL AND NUMERICAL ASPECTS

Mathematical and numerical aspects in simulations of BVP are certainly linked to constitutive models. The general rule — the more complicated model, the more serious numerical problems — is valid.

Mathematicians have known for a long time that satisfactory results can be expected only in the case of a well-posed BVP, for which existence, uniqueness and stability of solution is guaranteed. It is not possible to perform such proofs in practical cases but one should be aware that there is sophisticated mathematics behind the contour plots and that to obtain a realistic solution is by no means obvious. According to the classification by Belytschko (1996), most of the geotechnical calculations would still fall into the category *computer games*!

On the other hand, one cannot consider numerical difficulties as mere obstacles in applications. Reality is neither unique nor fully deterministic (Prigogine & Stengers 1983). Bifurcation points in our equations often coincide with instabilities in the material response, which are far away from the limit states of stress, and a slight change in the initial conditions may result in a different material response. An instructive demonstration of this fact is obtained by analyzing the controllability of laboratory tests (Imposimato & Nova 1998). Some stress paths, which become out of control in laboratory apparatuses (e.g. behaviour after the peak of the stress-strain curve from undrained triaxial compression of water-saturated sand), cannot be controlled in numerical simulations as well.

5 CONCLUDING REMARKS

I gave some hints on several pitfalls of numerical methods in calculation of deformations. In particular, we should be aware of

1. correct calculation of strains from displacements,
2. simultaneous loading and unloading in most cases,
3. absence of linear elasticity in the soil behaviour,
4. scatter in experimental results,
5. unsatisfactory deformation patterns calculated with simple models,
6. ambiguous calibration procedures,
7. considerable role of initial state,
8. difference between material parameters (constants) and state variables,
9. indeterministic components in the soil behaviour resulting in ill-posed BVP.

There are other important factors which have not been discussed here but which can significantly affect results of numerical calculations, among them

- scatter of soil conditions in situ,
- averaging procedures in multiphase continua,
- inertial and damping effects,
- drawbacks of iterative solution of the set of equations and of time integration.

The inability to get reliable numerical predictions could seem almost hopeless. So why do we sometimes succeed in obtaining realistic results? Is it just because we look only at particular curves we want to see? In the most cases we do not perform true predictions. We rather apply the observational method in the sense of the adaptation of numerical results to measured data. If we do not have data, we try at least to adapt the numerical results from analogical cases from our experience or intuition. The prediction competitions teach us that in unknown (new) situations the reliability of our predictions is poor. Consequently, questions may arise: Should we stop with the development of constitutive models and numerical tools since they do not bring us desired results? Should we rely only on experiments (model tests, centrifuge, prototypes) and observations?

I think that the opposite is true. We should abandon the Terzaghi tradition that a soil is characterized by E, φ and c, and intensify further research. We cannot proceed only based on physical modelling since most geotechnical constructions can be considered as prototypes and model tests with granular materials bring even more questions (e.g. how to scale the sand grains, how to eliminate systematic and random errors or how to measure stresses at small scale).

ACKNOWLEDGEMENT

Valuable remarks by Prof. D. Kolymbas and the financial support by the Grant Agency of the Academy of Sciences of the Czech Republic (Grant No. B2071901) have been highly appreciated.

REFERENCES

Belytschko, T. 1996. On difficulty levels in non linear finite element analysis of solids. *IACME xpressions* (2): 6–8.

Bianchini, G., Puccini, P. & Saada, A. 1988. Test results. In Saada & Bianchini, (eds.), *Constitutive Equations for Granular Non-Cohesive Soils*: 89–97. Rotterdam, Balkema.

Burland, J. 1989. "Small is beautiful" — the stiffness of soils at small strains. *Canadian Geotechnical Journal* 26: 499–516.

Doležalová, M., Zemanová, V. & Danko, J. 1998. An approach for selecting rock mass constitutive model for surface settlement prediction, *Int. Conf. on Soil-Structure Interaction in Urban Civil Engineering, Vol. 1*: 49–65. Darmstadt.

Gudehus, G. 1996. A comprehensive constitutive equation for granular materials. *Soils and Foundations* 36(1): 1–12.

Gudehus, G., Darve, F. & Vardoulakis, I. (eds.) 1984. *Results of the International Workshop on Constitutive Relations for Soils, Grenoble*. Rotterdam, Balkema.

Herle, I., Doanh, T. & Wu, W. 2000. Comparison of hypoplastic and elastoplastic modelling of undrained triaxial tests on loose sand. In D. Kolymbas (ed.), *Constitutive Modelling of Granular Materials, Horton*: 333–351. Springer.

Herle, I. & Mayer, P.-M. 1999. Verformungsberechnung einer Unterwasserbetonbaugrube auf der Grundlage hypoplastisch ermittelter Parameter des Berliner Sandes. *Bautechnik* 76(1): 34–48.

Herle, I. & Nübel, K. 1999. Hypoplastic description of the interface behaviour, In G. Pande, S. Pietruszczak & H. Schweiger (eds.), *Int. Symp. on Numerical Models in Geomechanics, NUMOG VII, Graz*: 53–58. Rotterdam, A.A.Balkema.

Herle, I. & Tejchman, J. 1997. Effect of grain size and pressure level on bearing capacity of footings on sand. In A. Asaoka, T. Adachi & F. Oka (eds.), *IS-Nagoya'97: Deformation and Progressive Failure in Geomechanics* 781–786. Pergamon.

Imposimato, S. & Nova, R. 1998. An investigation on the uniqueness of the incremental response of elastoplastic models for virgin sand. *Mechanics of Cohesive-Frictional Materials* 3: 65–87.

Kolymbas, D. 1991. An outline of hypoplasticity. *Archive of Applied Mechanics* 61: 143–151.

Kolymbas, D. 2000. The misery of constitutive modelling. In D. Kolymbas (ed.), *Constitutive Modelling of Granular Materials, Horton*: 11–24. Springer.

Lambe, T. 1967. Stress path method. *Journal of the Soil Mechanics and Foundations Division ASCE*, 93(SM6): 309–331.

Lambe, T. 1973. Predictions in soil engineering. *Géotechnique* 23(2): 149–202.

Lydon, I. 2000. Behaviour problems. *Ground Engineering* 33(11): 31.

Mähr, M. 2000. Berechnung primärer Spannungsfelder nach der Methode der Finiten Elemente und mit dem hypoplastischen Stoffgesetz. *Diploma Thesis*. IGT, University of Innsbruck.

Muir Wood, D. 2000. The role of models in civil engineering. In D. Kolymbas (ed.), *Constitutive Modelling of Granular Materials, Horton*: 37–55. Springer.

Nübel, K., Mayer, P.-M. & Cudmani, R. 1997. Einfluß der Ausgangsspannungen im Boden auf die Berechnung von Wandverschiebungen tiefer Baugruben in Berlin. In *Ohde-Kolloquium*: 183–207. TU Dresden.

Peck, R. 1969. Advantages and limitations of the observational method in applied soil mechanics. *Géotechnique* 19(2): 171–187.

Prigogine, I. & Stengers, I. 1983. *Order Out of Chaos*. New York, Bantam Books.

Roscoe, K. 1970. The influence of strains in soil mechanics. *Géotechnique* 20(2): 129–170.

Saada, A. & Bianchini, G. (eds.) 1988. In *Constitutive Equations for Granular Non-Cohesive Soils, Cleveland*, Rotterdam, Balkema.

Schweiger, H. F. 1998. Results from two geotechnical benchmark problems. In A. Cividini (ed.), *Proc. 4th European Conf. Numerical Methods in Geotechnical Engineering*: 645–654. Springer.

Schweiger, H. F. 2001. Comparison of finite element results obtained for a geotechnical benchmark problem. In C. Desai, T. Kundu, S. Harpalani, D. Contractor & J. Kemeny (eds.), *Computer methods and advances in geomechanics — Proc. of the 10th International Conference, Tucson*. Rotterdam, A.A.Balkema.

Simpson, B., O'Riordan, N. & Croft, D. 1979. A computer model for the analysis of ground movements in London Clay. *Géotechnique* 29(2): 149–175.

Tatsuoka, F., Goto, S., Tanaka, T., Tani, K. & Kimura, Y. 1997. Particle size effects on bearing capacity of footings on granular material. In A. Asaoka, T. Adachi & F. Oka (eds.), *IS-Nagoya'97: Deformation and Progressive Failure in Geomechanics*: 133–138. Pergamon.

Tatsuoka, F., Jardine, R., Lo Presti, D., Di Benedetto, H. & Kodaka, T. 1999. Theme lecture: Characterising the pre-failure deformation properties of geomaterials. In *Proc. XIV ICSMFE, Hamburg, Vol. 4*: 2129–2164. Rotterdam, A.A.Balkema.

Tatsuoka, F., Siddiquee, M., Yoshida, T., Park, C., Kamegai, Y., Goto, S. & Kohata, Y. 1994. Testing methods and results of element tests and testing conditions of plane strain model bearing capacity tests using air-dried dense silver leighton buzzard sand. *Report prepared for class-a prediction of the bearing capacity performance of model surface footing on sand under plane strain conditions*. Institute of Industrial Science, University of Tokyo.

Tejchman, J. & Herle, I. 1999. A "class A" prediction of the bearing capacity of plane strain footings on sand. *Soils and Foundations* 39(5): 47–60.

Viggiani, G. & Tamagnini, C. 2000. Ground movements around excavations in granular soils: a few remarks on the influence of the constitutive assumptions on FE predictions. *Mechanics of Cohesive-Frictional Materials* 5: 399–423.

von Wolffersdorff, P.-A. 1996. A hypoplastic relation for granular materials with a predefined limit state surface. *Mechanics of Cohesive-Frictional Materials* 1: 251–271.

von Wolffersdorff, P.-A. 1997. *Verformungsprognosen für Stützkonstruktionen*. Veröffentlichungen des Institutes für Bodenmechanik und Felsmechanik der Universität Fridericiana in Karlsruhe. Heft 141.

Wroth, C.P. 1977. The predicted performance of soft clay under a trial embankment loading based on the Cam-clay model. In G. Gudehus (ed.), *Finite Elements in Geomechanics*: 191–208.Wiley.

Constitutive and Centrifuge Modelling: Two Extremes, Springman (ed.)
© *2002 Swets & Zeitlinger, Lisse, ISBN 90 5809 361 1*

Measurement of displacement – Trends or numbers?

J.S. Sharma
Institute for Geotechnical Engineering, Swiss Federal Institute of Technology, Zurich, Switzerland
Department of Civil Engineering, University of Saskatchewan, Saskatoon, Canada

ABSTRACT: A critical appraisal of various techniques of soil displacement measurement in a centrifuge test is undertaken. It was observed that the focus of centrifuge modelling has changed from establishing failure mechanisms to investigating pre-failure deformation behaviour of soils. In addition, a growing trend of modelling "exact" geotechnical construction processes in a centrifuge model test is also observed. This shift in focus has fuelled the need for displacement measurement techniques that are more precise. It is argued that for extremely small magnitudes of displacement, the influence of strongbox boundaries can be significant. Therefore, despite the recent advances in displacement measurements such as the particle image velocimetry (PIV) analysis, the major contribution of a centrifuge model test is in revealing qualitative behaviour of soils.

1 INTRODUCTION

The physical modelling of geotechnical processes is useful for calibration or verification of the methods of prediction and for the simulation of complex construction processes. In order to facilitate quantitative analysis of the observed behaviour, measurement of both the load and the deformation is important. Measurement of the load (or stress) in a centrifuge model test can be achieved by embedding or installing miniature instrumentation such as load cells and pore pressure transducers (PPTs). In addition, the deformation of structural components such as a model retaining wall or a model pile can be measured using high-precision strain gauges and from these deformations, bending moments and loads taken by these structural components can be inferred, provided the structural components remain in an elastic state. While the PPTs and strain gauges mounted on structural components have proved to be quite reliable and accurate, the same cannot be said about the load cells. Since the main focus of this paper is the measurement of displacement, the limitations of load cells will not be discussed further here.

Measurement of soil displacement is important for calibrating the methods of analysis and for developing simple solutions based on observed deformation mechanisms. Both purposes are equally important but the latter has more impact on geotechnical design process. For the measurement of soil displacement, probes such as Linear Variable Differential Transformer (LVDT) or potentiometers can be placed at the model boundaries to record the movement of a single point. Despite their inherently intrusive nature, these probes are able to provide accurate measurement of displacement. However, such remote measurements are of limited use from the point

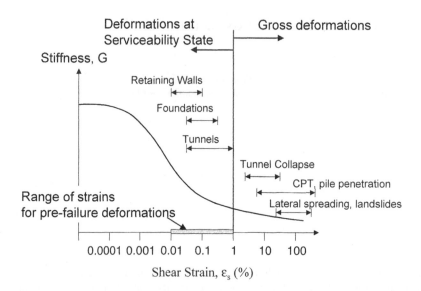

Figure 1. Typical shear strains experienced in geotechnical problems (after Mair 1993)

of view of understanding the mechanism of deformation of the soil. For this purpose, the measurement of ground movements at many different points in the model is required.

The ground movements can be categorised into two main types: (1) gross deformation and (2) deformations at serviceability state. The measurement of gross deformation is important for the understanding of the ultimate failure mechanism of a geotechnical structure. It helps in the verification of design solutions based on the upper bound collapse mechanisms. It is usually not necessary to measure gross deformations. These can be observed during the post-test investigations of a centrifuge model. Such deformation measurements represent a "before" and "after" state of the ground and give valuable insight into the mode of ground failure. In contrast to gross deformations, the deformations at Serviceability State or pre-failure deformations are required for the understanding of soil behaviour at much lower strain levels. For the measurement of serviceability state deformations, the desired accuracy should be greater by at least an order of magnitude.

Figure 1 shows the range of strain levels experienced by the ground during the various geotechnical construction processes. The pre-failure deformations are usually between 0.01 to 1% strain range. Given the reduced scale of centrifuge models (and a corresponding reduction in the magnitude of relevant ground movements), the measurement of pre-failure displacements is a task of formidable difficulty. A successful displacement measurement technique must have sufficient precision for capturing the smallest relevant displacement throughout the field of interest.

2 SOIL DISPLACEMENT MEASUREMENT TECHNIQUES

2.1 *Radiographs of Large-scale Models*

First measurement of displacements within a soil mass was carried out by Gerber (1929) using an X-ray method. Lead shots were embedded in the soil model and successive radiographs were taken to follow the movement of these shots. This system has been widely used to detect the patterns of incremental strains in large models made of sand and in shear apparatus (Roscoe et al. 1963). The technique was used in combination with geodesy principles to plot contours of

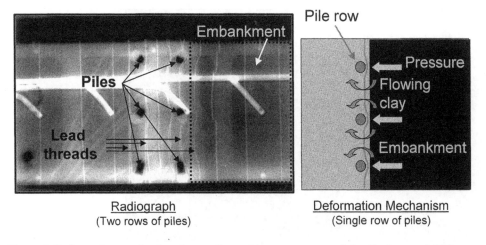

Figure 2. Deformation mechanism from radiographic measurements (after Springman 1989)

shear and volumetric strains with an accuracy of ± 0.1% by using rather large models that typically measured 2.0 m x 5.0 m (James 1965). For clays, a mixture of mineral oil and lead powder was injected at various locations to form lead threads (e.g. Phillips 1988, Springman 1989). The clay was radiographed before and after the centrifuge test to reveal the deformation mechanism (Fig. 2). One of the major advantages of this technique was that the soil displacement in any orthogonal plane could be measured. All the other subsequent techniques have not been successful in incorporating this feature.

2.2 *In-flight Conventional Photography*

It is not possible to expose radiographs in-flight in a geotechnical centrifuge. Therefore, the modellers had to adopt a new approach that involved the use of a thick, transparent perspex window to expose the frontal plane of a model, into which target markers or "spots" were placed (Fig. 3). High-resolution aerial photography film was then used to take photographs of the front of the model to record the movement of these markers at various stages of the test. These films were digitized using a film-measuring machine (Phillips 1991) to obtain the coordinates of the markers. Once the coordinates were known, the principles of geodesy were used to plot displacement vectors (Fig. 4) as well as the contours of shear and volumetric strains. Since the centrifuge models (and therefore, the exposed films) were of much smaller size compared to the large 1-g models, some loss of accuracy was inevitable. However, using a system of two high-power, small duration flashes, Sharma (1994) managed to improve the contrast between the markers and the surrounding clay and achieved accuracy comparable to that of large radiographs.

2.3 *Digital Image Processing of Video Captures*

In the early 1990s, the centrifuge modellers were concerned about the rather large amount of time and effort required for measuring soil displacements using in-flight conventional photography. Another drawback of the X-ray and the conventional photography was that it did not provide any information about the state of the soil during a centrifuge test. Around this time, with the availability of inexpensive computers with reasonable graphics capabilities, digital image processing of video images captured using closed-circuit on-board TV cameras was developed (Allersma 1991, Ethrog 1994, Allersma et al. 1994, Robson et al. 1996). This technique was

Figure 3. Installing target markers or "spots" on the front face of the model clay layer

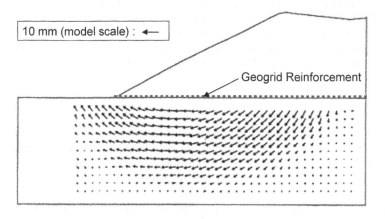

Figure 4. Displacement vectors obtained using in-flight conventional photography (after Sharma 1994).

also based on measuring coordinates of markers embedded in the exposed frontal plane of the model. Recently, Taylor et al. (1998) have modified this technique to measure 3-D displacements in soil models using three CCTV cameras capturing the images of the area of interest from three different angles.

The main advantage of this technique was that the magnitude and the direction of soil displacement could be computed in real time during a centrifuge test. In other words, it became possible to control a centrifuge test based on these measured displacements. The accuracy of measurements taken from a digital image is proportional to the object-space pixel size. Video capture in the European PAL format produces a nominal resolution of 732 x 549 pixels, although the number of photo-sensitive elements within a typical camera charge-coupled device (CCD) is often considerably less (White et al. 2001). Line jitter and the use of centrifuge slip-rings to transmit the analogue video signal can further reduce image quality. Due to these factors, the accuracy of displacement measurement achieved using digital image processing of video captures is less than that achieved using radiographs and in-flight photographs.

2.4 *Surface Scans using Laser Displacement Sensors*

Sharma & Bolton (1995) used a non-intrusive laser displacement sensor (LDS) mounted on a profilometer to scan the entire surface above and in front of a tunnel heading in sand (Fig. 5). The LDS had a measurement range from 60 mm to 140 mm at high response speeds (maximum 700 Hz or 0.7 ms). The resolution was as high as 10 µm for a relatively flat white object at relatively slow response speeds. For a surface consisting of fine sand, however, the resolution was limited to 1 grain diameter (typically 0.1 mm or 100 µm). Figure 6 shows 3-D representation of the settlement trough above a collapsed tunnel section. It is possible to scan the soil surface using two or more LDS mounted at an angle in order to obtain the 3-D displacement pattern. The major issue with laser scanning technique is the time required for one complete scan. Sharma & Bolton (1995) reported a time of approximately 12 minutes for one surface scan. Increasing the speed of scanning results in loss of accuracy. Therefore, the laser scanning technique is useful in obtaining the deformation pattern of free-draining granular soils "before" and "after" an event. It is not of much use in clays that undergo transient pore pressure changes after the occurrence of a significant event.

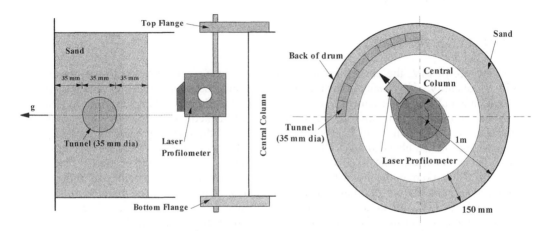

Figure 5. Surface scan using LDS - Test arrangement (after Sharma & Bolton 1995)

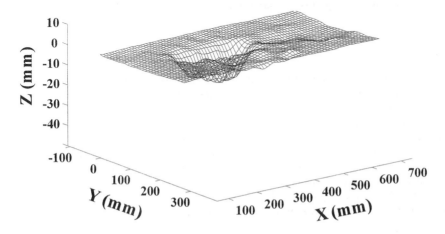

Figure 6. 3-D representation of settlement trough measured using LDS (after Sharma & Bolton 1995)

2.5 *Particle Image Velocimetry*

The latest development in soil displacement measurement comes in the form of the application of Particle Image Velocimetry or PIV (White et al. 2001). PIV is a velocity measuring technique that was originally developed for fluid mechanics applications. The flow field of a fluid can be examined by seeding the flow with marker particles and tracking the movement of small patches within a larger image. The displacement of soil is considered to be a low velocity flow process. For natural sands, texture in the form of different coloured grains and the variation in hue is already present and therefore, no seeding of markers is necessary. The PIV technique overcomes the major drawback of previous displacement measurement techniques – the reliance on discrete target markers. A suitable patch size corresponds to a couple of particle diameters. This means that for a typical area of, say 180 x 120 mm, it is possible to have over 20,000 measurement points. Embedding a comparable number of discrete markers is simply a hopeless task. Using a digital still camera that provided 1760 x 1168 pixel images, White et al. (2001) claim to have achieved an accuracy of 17 μm when viewing an area of 300 mm x 200 mm. This is a significant improvement over the previous techniques that are only able to achieve a maximum accuracy of the order of 100 μm.

3 QUALITY VS. QUANTITY

Before the use of PIV, the centrifuge modellers were always confronted with the question of quality vs. quantity – how many embedded markers should be used in order to obtain sufficient information about soil displacement? The use of a fewer markers would result in fewer amounts of data but at the same time, their movements could be tracked more accurately by zooming in on the area of interest. This will improve the resolution of the image by allocating more pixels per square inch of the physical area. On the other hand, using too many markers may influence the soil behaviour itself. The issue was resolved by using a graded marker density, i.e. using closely spaced markers in the area where large strain gradients are anticipated (e.g. in the corner of an excavation or in front of a tunnel heading) and using sparsely placed markers in the other areas. Thanks to PIV, this is no longer an issue for models that are made up of textured soils such as quartz sand. For models using non-textured soils such as white kaolin clay, White et al. (2001) have suggested the use of multi-coloured flock material to impart some texture to the clay surface. However, the technique is still far from satisfactory and one has to continue to rely on discrete markers and the issue of quality vs. quantity.

4 TRENDS VS. NUMBERS

Centrifuge model testing in the 70s and 80s mainly focussed on tests that simulated ultimate limit states, e.g. the collapse of tunnel headings in clay (Mair 1979, Taylor 1984), stability of embankments on soft clays (Davies 1981, Almeida 1984) and trenches in clays (Taylor 1984, Phillips 1987). The main focus in these tests was to establish the mechanisms of deformation in addition to stress and load measurements. The accuracy of displacement measurement was given a secondary role, mainly because of the lack of availability of sophisticated measurement techniques. Consequently, the numerical back-analyses of these tests focussed on simulating the qualitative aspects of soil behaviour. For majority of these tests, the outcome was simple design methods based on either limit equilibrium or upper-bound theorem of plasticity.

Since the early 1990s, centrifuge modelling has become increasingly more sophisticated. Nowadays, the main focus of many centrifuge tests is to establish serviceability state and to measure pre-failure deformations of soils. The development of sophisticated constitutive models that claim to be able to describe several aspects of soil behaviour has fuelled the need for good quality experimental data, especially including data on pre-failure soil deformation. Major geotechnical centrifuge centres around the world are now involved in realistic simulation of geotechnical construction processes such as pile installation (Pan et al. 1998), shield tunnelling (König 1998), sand compaction piles (Ng et al. 1998), braced deep excavation (Loh et al. 1998), etc.

Due to rapid advancements in the area of mechatronics, centrifuge modelling of complex soil-structure interaction problems is now possible. These problems include construction of a tunnel next to an adjacent tunnel, deep excavation next to existing tunnels and foundations and other similar problems encountered in an urban environment. Since the soil is usually nowhere near to failure in such tests, their interpretation relies increasingly on accurate displacement measurements. Development of precise displacement measurement techniques such as the PIV will help in revealing the mysteries of these problems. However, one has to take the outcome of these tests with a pinch of salt. Boundary effects due to crammed space inside a centrifuge strongbox are quite difficult to quantify. The presence of side friction in a plane strain strongbox is one of the main factors influencing the results of these tests. It influences both the magnitude and the direction of soil displacement. However, the influence on the direction of soil displacement can be assumed to be somewhat less than that on the magnitude. At extremely small soil displacements, the influence on the magnitude can be quite significant. If these boundary effects are not investigated carefully, these so-called sophisticated centrifuge tests may not be any more sophisticated than those tests of the 70s and the 80s, i.e. only the trends in displacement may be considered reliable and not the numbers.

5 CONCLUDING REMARKS

Traditionally, centrifuge tests have been used for studying failure mechanisms. However, the focus is now on studying the pre-failure deformation of soil models. Due to crammed space and small-scale, measurement of pre-failure soil deformation is a formidable task. For a complete understanding of the soil behaviour, measurement of both the loads and the deformations are required. Although there are still some improvements necessary for the measurement of shear loads using load cells, it can be said that we have achieved significant progress in terms of load measurement. It's the turn of the displacement measurements to catch up. It is heartening to note that there have been significant recent developments in displacement measurement techniques for centrifuge model tests. Resolutions as small as 17 μm can now be achieved. Non-intrusive measurement at several thousand points is made possible by the PIV technique.

One of the main goals of centrifuge modelling is to obtain accurate data for the verification of analytical tools and new constitutive models. One gets a false sense of confidence when a back-analysis using several simplifying assumptions in terms of boundary conditions and using rather simple constitutive models is able to reproduce the overall behaviour of a complex centrifuge test – surely a case of two or more wrongs making it right. Provided we are able to incorporate realistic boundary conditions in a back-analysis, precise displacement measurement will give us opportunities to compare the strain paths experienced by the centrifuge model with those simulated by a constitutive model. In this way, we'll be able to pinpoint the limitations and shortcomings of the constitutive model.

Centrifuge modelling is increasingly becoming more sophisticated due to technical advancements in the area of mechatronics. However, the boundary effects are much more critical for these sophisticated tests in comparison with the rather simple tests of the 70s and the 80s. One would like to think that there is no limit to what can be simulated in a centrifuge test. Whether the results will be reliable is open to question because of significant influence of the boundaries of a strongbox of limited space. The current trend of achieving "exact" simulation of geotechnical construction processes in a small-scale centrifuge test is somewhat worrying. Such tests have limited potential in fulfilling one of the main aims of centrifuge model testing – to generate reliable experimental data under controlled conditions for the calibration of methods of analysis. The time, money and effort spent on such tests may be better utilized on supplementing the results of simple but carefully planned centrifuge tests by results from sophisticated stress path testing of the model soil at various stress levels and boundary conditions. In the absence of precise soil parameters, the validation of constitutive models using the centrifuge test results is going to be a futile exercise.

A centrifuge modeller finds it exciting to play with new gadgets and equipment used in sophisticated centrifuge tests. Similarly, a numerical modeller finds it equally exciting to undertake geotechnical calculations using the state-of-the-art software, something that Herle (2001) calls "computer games" based on the classification proposed by Belytschko (1996). It is critically important that there is some understanding and collaboration between the centrifuge and the numerical modellers. These two should not just stay on opposite banks of the river that constitutes the resources of geotechnical engineering knowledge, but rather sail in the same boat and explore together.

REFERENCES

Allersma, H.G.B. 1991. Using image processing in centrifuge research. In H.Y. Ko & F.G. McLean (eds), *Proc. Centrifuge'91*: 551-558. Rotterdam: Balkema.
Allersma, H.G.B., Stuit, H.G. & Holscher, P. 1994. Using image processing in soil mechanics. *Proc. XIII ICSMFE, New Delhi*: 1341-1344. New Delhi: Oxford & IBH Publishing.
Almeida, M.S.S. 1984. Stage constructed embankment on soft clays. PhD thesis, Cambridge University Engineering Department.
Belytschko, T. 1996. On difficulty levels in non-linear finite element analysis of solids. *IACM Expressions*, 2: 6-8.
Davies, M.C.R. 1981. Centrifuge modelling of embankments on clay foundations. PhD thesis, Cambridge University Engineering Department.
Ethrog, U. 1994. Strain measurement by video image processing. *Proc. Tenth Conf. Recent Advances in Experimental Mechanics*: 411-415. Rotterdam: Balkema.
Gerber, E. 1929. Untersuchungen über die Druckverteilung im ortlich belasteten Sand. Dissertation, Technische Hochschule, Zürich, Switzerland.
Herle, I. 2002. Difficulties related to numerical predictions of deformations. In S.M. Springman (ed.), *Constitutive and Centrifuge Modelling – Two Extremes, Workshop, Ascona, Switzerland, 8-13 July 2001*, Rotterdam: Balkema.
James, R.G. 1965. Stress and strain fields in sand. PhD thesis, Cambridge University Engineering Department.
König, D. 1998. An inflight excavator to model a tunnelling process. In T. Kimura, O. Kusakabe & J. Takemura (eds), *Proc. Centrifuge'98*: 707-712. Rotterdam: Balkema.
Loh, C.K., Tan, T.S. & Lee, F.H. 1998. Three-dimensional excavation tests. In T. Kimura, O. Kusakabe & J. Takemura (eds), *Proc. Centrifuge'98*: 649-654. Rotterdam: Balkema.
Mair, R.J. 1979. Centrifugal modelling of tunnel construction in soft clay. PhD thesis, Cambridge University Engineering Department.
Mair, R.J. 1993. Developments in geotechnical engineering research: applications to tunnels and deep excavations. Unwin Memorial Lecture (1992). *Proc. ICE*, 97(2): 27-41.
Ng., Y.W., Lee, F.H. & Yong, K.Y. 1998. Development of an in-flight sand compaction pile (SCP) installer. In T. Kimura, O. Kusakabe & J. Takemura (eds), *Proc. Centrifuge'98*: 837-843. Rotterdam: Balkema.

Pan, S.S., Pu, J.L., Yin, K.T. & Liu, F.D. 1998. A new pile driver and loading set for pile group in centrifuge flight. In T. Kimura, O. Kusakabe & J. Takemura (eds), *Proc. Centrifuge'98*: 91-96. Rotterdam: Balkema.

Phillips, R. 1987. Ground deformation in the vicinity of a trench heading. PhD thesis, Cambridge University Engineering Department.

Phillips, R. 1988. Centrifuge lateral pile tests in clay: Tasks 2 and 3. A report to Exxon Product Research Corporation, USA by Lynxvale Ltd, Cambridge, UK.

Phillips, R. 1991. Film measurement machine user manual. Technical Report No. CUED/D-Soils/TR246, Cambridge University Engineering Department.

Robson, S., Chen, J., Cooper, M.A.R., Taylor, R.N. & Clarke, T.A. 1996. A photogrammetric system for measurements of ground displacements in soil models tested in a geotechnical centrifuge under gravitational forces in excess of 100g. *Photogrammetric Record*, XVI(88).

Roscoe, K.H., Arthur, J.R.F. & James, R.G. 1963. The determination of strains in soils by an X-ray method. *Civil Engineering and Public Works Review*, 58: 873-876, 1009-12.

Sharma, J.S. 1994. Behaviour of reinforced embankments on soft clay. PhD thesis, Cambridge University Engineering Department.

Sharma, J.S. & Bolton, M.D. 1995. A new technique for simulation of tunnel collapse in a drum centrifuge. Technical Report No. CUED/D-Soils/TR286, Cambridge University Engineering Department.

Springman, S.M. 1989. Lateral loading on piles due to simulated embankment construction. PhD thesis, Cambridge University Engineering Department.

Taylor, R.N. 1984. Ground movements associated with tunnels and trenches. PhD thesis, Cambridge University Engineering Department.

Taylor, R.N., Grant, R.J., Robson, S. & Kuwano, J. 1998. An image analysis system for determining plane and 3-D displacements in soil models. In T. Kimura, O. Kusakabe & J. Takemura (eds), *Proc. Centrifuge'98*, 73-78. Rotterdam: Balkema.

White, D.J., Take, W.A. & Bolton, M.D. 2001. Measuring soil deformation in geotechnical models using digital images and PIV analysis. *Proc. 10th Int. Conf. on Computer Methods and Advances in Geomechanics, Tucson, Arizona*: 997-1002. Rotterdam: Balkema.

Constitutive and Centrifuge Modelling: Two Extremes, Springman (ed.)
© *2002 Swets & Zeitlinger, Lisse, ISBN 90 5809 361 1*

Time effects in creep of sands

E.T. Bowman & K. Soga
Cambridge University Engineering Department, United Kingdom

ABSTRACT: This paper presents an investigation into creep characteristics of sand through a series of triaxial tests in compression and extension. Tests were carried out on two types of natural sand and one artificial sand. The tests highlight the link between time-dependent deformation behaviour and particle size, morphology and material strength. The results raise questions as to whether it is possible for creep response to be captured accurately by current constitutive models and whether a knowledge of the material parameters, particle shape and size distribution could be better used in understanding deformation behaviour.

1 INTRODUCTION

Sands are known to creep over time. Although, until recently, the relatively small magnitude of such creep meant that it was largely regarded as an unimportant feature in the overall geotechnical behaviour of sands. Recent research has shown, however, that creep may be an issue in the field when considering liquefaction and settlement in loose sands (Leung et al. 1996), and the time-dependent change in behaviour of freshly densified sands (Chow et al. 1998, Mitchell & Solymar 1984). The exact mechanics of such creep deformation is the cause of some debate.

The magnitude of creep in sands is thought to be a function of confining stress, stress ratio, relative density and angularity whilst log-linear and log-log relationships have been found for the time-dependent volumetric and shear strain response. Whilst much research has concentrated on the creep response of loose soils (Di Prisco et al. 2000, Tanaka & Tanimoto 1988), little research has been undertaken on creep in dense sands. Those who have undertaken such studies have suggested that there is less creep for dense sands than for loose sands (Kuwano 1999) and that dilative creep may be associated with the onset of failure (Mejia et al. 1998).

Some authors have suggested it is possible to model creep characteristics of sands accurately in a similar manner to that of clays, using isotach or elasto-viscoplastic models (Kuhn & Mitchell 1993, Murayama et al. 1984). However, the tests carried out in this series raise questions as to whether a more fundamental understanding of microscopic phenomena is necessary in order to apply constitutive models to creep in sands.

2 EXPERIMENTAL APPROACH

2.1 *Triaxial Set Up*

A hydraulically operated triaxial apparatus has been modified extensively for use with local displacement devices to measure local axial and radial strains. The apparatus uses three GDS digital pressure controllers to control cell pressure, back-pressure and ram load, and is fitted with a load cell, a pore pressure transducer and an external LVDT. End platens are frictional. Local strain devices include two submersible LDVTs (RDP D5/200W) and two proximity sensors (Kaman 2UB1 & Keyence AH-110). GDS software is used to control stress paths whilst all data is recorded using data-logging software through a 16 bit A/D converter.

2.2 *Stress Paths*

A complex stress path was chosen, to mimic that experienced by a soil element during installation of a pile. Whilst the actual stress and strain levels are low compared with those experienced in the field, it is felt that the path gives an insight into soil behaviour without adding complicating geometric and compressibility issues. It is thought that initially, the soil is heavily loaded vertically by the approach of the pile tip, as shown in Figure 1. As soil travels around the tip and up alongside the shaft however, the vertical stress reduces dramatically while radial (or horizontal) stress remains largely "locked-in", leading to a state of extension. The sense of this stress path is followed in these triaxial tests.

Included in the overall program of stress path testing are creep stages in compression and extension. The stress-controlled stress path focussed upon in this paper involves loading at 30 kPa per minute to a deviator stress of 800 kPa and mean effective stress of 600 kPa, followed by a creep stage in compression for 24 hours. The soil is then unloaded into extension, at a rate of 30 kPa per minute to a deviator stress of -70 kPa and a mean effective stress of 100 kPa, followed by a creep stage in extension for 24 hours. Cyclic creep tests are also carried out at these stages to ascertain whether applying small cycles to a pile could enhance creep-related set up. These creep stages are examined in this paper, see Figure 2.

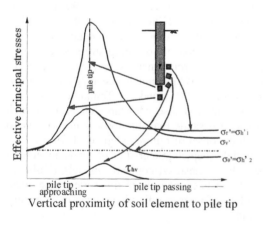

Figure 1. Hypothesised stress path of soil element: σ denotes normal stress, τ shear stress, r radial, h horizontal, θ circumferential

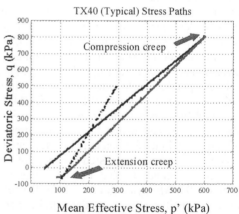

Figure 2. Triaxial stress path followed in this test series. Deviator stress is defined as ($\sigma_a - \sigma_r$) and mean effective stress as $^1/_3(\sigma_a' + 2\sigma_r')$

2.3 Materials Tested

The results of creep stages for two types of natural sand and one artificial sand are reported. The sands are a standard Leighton Buzzard 90-150 μm clean silica sand, denoted Type E, and unwashed Montpellier beach sand (silica with some shell fragments), denoted Type MP. The artificial sand is made up from glass beads of reasonably uniform size, which are imperfect spheres and include some double-particles, denoted Type GB. See Figure 3 for scanning electron microphotos of particles (the scale is given at the bottom left of each photo with the scale to each unit measuring length repeated in the figure heading). The mean particle sizes were determined using a single particle optical sensing device (Retsch Camsizer), whilst the maximum and minimum void ratios were determined using BS 1377: 1990. The Fourier descriptor results were determined using the method described in Bowman et al. (2001), with higher values of "elongation", "triangularity" or "squareness" denoting more elongate, triangular or square particles and higher values of "asymmetry" denoting more irregular particles on average. Table 1 gives e_{max}, e_{min}, D_{10}, D_{50} and Fourier descriptor analysis results for morphology of each sand.

50 mm diameter by 100 mm high soil samples were prepared by dry pluviation at a constant fall height to ensure a consistent void ratio throughout the sample. Tests were carried out at $70^{+}/-2\%$ relative density. One dry test, performed on Type E sand is shown, whilst for the four saturated tests shown, saturation was maintained by applying a back-pressure of 350 kPa. Cyclic tests are performed by applying a sinusoidal 10 kPa peak-to-peak deviatoric cycle at 6 cycles per minute about the compression deviator stress of 800 kPa and extension deviator stress of -70 kPa. The results of one cyclic creep test, performed on Type E sand are reported. These results are representative of a larger body of results, only some of these are presented for clarity.

(a)　　　　　　　　　　(b)　　　　　　　　　　(c)

Figure 3. SEM microphotographs: (a) Type E (100 μm), (b) Type MP (300μm), (c) Type GB (100 μm)

Table 1. Geometrical characteristics of the sand

Sand Type	e_{max}	e_{min}	D_{10} (μm)	D_{50} (μm)	Fourier Signature Descriptors Elongation	Triangularity	Squareness	Asymmetry
E	1.006	0.599	120	148	0.1720	0.0598	0.0331	0.0307
MP	0.934	0.716	175	237	0.1779	0.0584	0.0325	0.0239
GB	0.768	0.611	208	260	0.0344	0.0135	0.0085	0.0158

3 RESULTS

The results of a typical triaxial test are given in Figures 4a and 4b for dry pluviated Type E sand. Figure 4a shows the normalised deviator stress against axial strain, whilst Figure 4b shows volumetric strain, $\varepsilon_v = \varepsilon_a + 2\varepsilon_r$, versus shear strain, $\varepsilon_s = {}^2/_3(\varepsilon_a - \varepsilon_r)$ (compression is plotted positive). Figure 4b shows the change in strain vector direction from compressive to dilative behaviour during compression creep soon after shearing is halted and stresses are held constant. Conversely, the strain vector continues in extension creep in the direction of unloading for some time (i.e. is compressive) before becoming slightly dilative as shear creep continues. This behaviour is typical for Type E sand, with Type MP achieving a similar, if slightly more compressive overall response and Type GB a more dilative response in triaxial compression and extension.

Creep periods in compression and extension are given in Figures 5 and 6 respectively. Of note is that Type E dilates under creep compression almost immediately (albeit slowly), despite its compressive response during shearing. Type MP continues to compress for 13 hours before dilative creep begins and Type GB is already experiencing dilation during the initial loading stage. Of the non-cyclic compression creep, it can be seen that Type GB dilates and shears the most, whilst Type E dilates and shears the least (with dry and saturated sands behaving in a similar manner). Type MP initially compresses but begins to dilate after approximately 13 hours and otherwise produces intermediate shear strains. Cyclic creep produces more shear strain and more volumetric (dilative) strain than static creep for Type E sand.

The volumetric response at the end of the unloading stage is initially followed in the extension creep phase for all sands. Type GB produces the least shear strain and least volumetric strain, which is initially compressive (for 25 minutes) before becoming dilative. Type E, both for saturated and dry samples, show similar shear and volumetric strain response and begin to dilate at the same time (at 18 hours). The dry soil initially exhibits a greater rate of compressive volumetric and shear strain before levelling off and finally dilating, whilst the saturated samples develops strains more gradually before 18 hours, where dilation begins more abruptly. Type MP produces the most compression and initially the most shear strain, although after 5 hours the volumetric response levels off before dilating, whilst the shear strain rate reduces. Cyclic creep of Type E causes the soil to dilate quickly (at 60 seconds), although the net shear strain is less than for static creep at 23.8 hours.

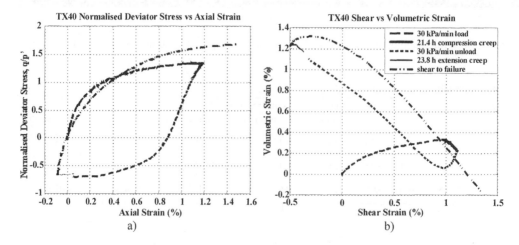

Figure 4. Typical stress path response for Type E dry pluviated sand, relative density=71%

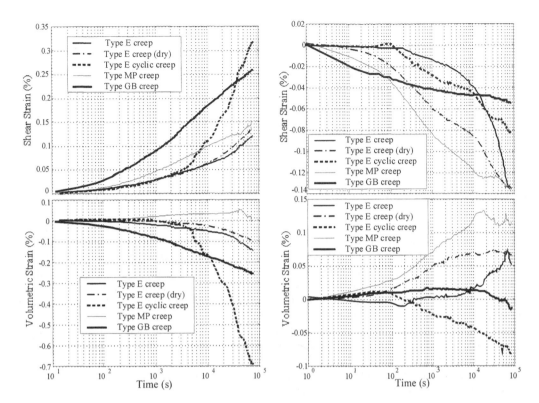

Figure 5. Shear and volumetric strains in compression against time

Figure 6. Shear and volumetric strains in extension against time

4 DISCUSSION

4.1 *Observations*

The shear strain results for compression creep and extension creep produce near-linear relations between log shear strain and log time (plots not given), although the decay of shear strain rate is faster for MP than for E, and GB decays faster in compression creep than in extension creep. Volumetric strains are rather more complex in their development and no simple linear or log law can be applied to the data. The creep in compression produces greater strain rates than in extension, and this is most likely attributable to the higher mean stress at which compression creep takes place, although the effects of initial fabric and different amounts of pre-straining due to the applied stress history cannot be ignored.

In compression creep, the two "real" sands, although from quite different sources, develop similar magnitudes of shear strain over time. Volumetrically however, Type MP spends more time compressing than Type E, which dilates almost immediately, despite the strain path being compressive in original loading. A similar trend is observed in extension creep, with MP developing greater compressive volumetric strains before dilating.

The artificial GB sand tends to dilate earlier than the real sands in both compression and extension creep. It is interesting to note however, that whilst in compression creep, GB sand shears a great deal, in extension creep it does not.

The authors consider that the difference in behaviour between the two real sands may be due to the greater compressibility of Type MP, which contains some shell fragments (weaker than silica) and a larger average particle size than E sand. The material strength and particle morphology is otherwise rather similar (Type MP sand is slightly more elongated, but less triangular, square and asymmetric than Type E). McDowell (1996) has presented a mechanism whereby larger, more irregular and elongated particles should break more easily than smaller, more round particles. That is, normal and tangential forces acting on soil grains tend to open potential cracks in more irregular particles, whereas close them in rounded particles, whilst larger particles possess more potential flaws to be propagated.

The authors suggest that, initially, local breakage of grains and asperities is the dominant mechanism whereby local shear forces on the particles are relieved, leading to global compression. Latterly, two other, related, mechanisms may become dominant.

It has been known for some time that stresses transfer through a granular material via columns of particles, sometimes called "the strong force network". These particles tend to roll over each other rather than slide during deformation (Oda et al. 1982). Oda (1997) and Kuhn (1999) have suggested that the growth of elongated voids perpendicular to the major principal applied stress is responsible for the macroscopic observation of dilation during shearing. However, Radjai et al. (1998) have also shown that a large proportion of the lightly loaded particles (the majority), which support the columns, "the weak force network", are at limiting frictional equilibrium and these tend to slide. Following these ideas, the authors suggest that the high local frictional forces on angular particles in the weak network cause sliding and rotating during creep rather than wholesale breakage. This induces an additional component of dilation as well as void growth as a result of the release of rolling angular particles in the strong-force columns.

This hypothesis is supported by the behaviour of Type GB, which has strong near-spherical particles and which produces greater shear strain than the real sands in compression creep and less in extension creep. That is, Type GB has a single dominant mechanism for dilation - the elongation of voids perpendicular to the applied major principal stress in response to local shear forces on the spherical particles. In compression, voids (due to pluviation) naturally occur in the vertical direction and their rate of growth is fast due to the spherical nature of Type GB. Types E and MP however, have an additional component of stability in compression, due to their more angular particles tending to lie horizontally (at least, initially) producing a more locked-up fabric. Therefore shear strains take longer to develop. In extension, Type GB rearranges easily during unloading, and void growth continues in a similar manner to compression creep - except in a perpendicular direction and at a slower rate. For Types E and MP however, the original arrangement of particles is no longer stable and particles rearrange themselves under the applied shear forces to become more vertical as the majority of contacts in the strong force network switch from the vertical to horizontal. Overall this may lead to a more "locked up" fabric.

It should be noted that on a microscopic level, both breakage and rotation / slippage occur simultaneously, though to an observer of macroscopic behaviour, only one, dominant, mechanism may be apparent.

It may be seen that the application of small cycles during creep for Type E sand accelerates the shear strain development in compression creep, whereas reduces the development of shear strain in extension creep, compared with static creep tests. In both compression and extension cyclic creep, volumetric dilation increases above that of the static creep tests on Type E. The main cause for this appears to be an increase in negative (i.e. dilative) radial strain. The mechanisms underlying this behaviour are currently under investigation, although it may be that the relative magnitude of the cyclic load compared with the deviator stress in compression (+/-5 kPa in 800 kPa) compared with extension (+/-5 kPa in -70 kPa) causes a different response. That is, in compression creep the cyclic load may impart fluctuations that simply allow an acceleration of angular dilative creep, whereas in extension it may be enough to produce some temporary vertical force columns, which lead to radial dilation without relieving shear forces.

5 CONCLUSIONS

Dense sands can creep a great deal and in doing so can produce complex volumetric behaviour which, to the authors' knowledge, has not yet been successfully reproduced by current constitutive models. At present it is unclear exactly what is the interrelation between material strength, particle shape, size distribution, fabric and creep mechanics, although it is clear that an understanding of creep may lead to a better understanding of overall granular deformation behaviour. This study attempts to highlight the importance of particle shape and how it may affect the creep behaviour of sands, and how small cyclic fluctuations can alter response.

A hypothesis of dilative creep is presented, whereby more angular particles may be thought of as tending, in the long term, to stable structures through grain rotation and sliding. The "delayed dilation" observed in all of the material creep tests is one which may lead to the observed time-dependent strength gain in sands known as "ageing" - or in the case of pile-driving, as "set up". This idea is currently being further explored.

REFERENCES

Bowman, E.T., Soga, K. & Drummond, T.W. 2001. Particle shape characterisation using Fourier descriptor analysis. *Géotechnique* 51(6): 545-554.

Chow, F.C., Jardine, R.J., Brucy, F. & Nauroy, J.F. 1998. Effects of time on capacity of pipe piles in dense marine sand. *Journal of Geotechnical and Geoenvironmental Engineering* 124(3): 254-264.

Di Prisco, C., Imposimato, S. & Vardoulakis, I. 2000. Mechanical modelling of drained creep triaxial tests on loose sands. *Géotechnique* 50(1): 73-82.

Kuhn, M.R. 1999. Structured deformation in granular materials. *Mechanics of Materials* 31(6): 407-429.

Kuhn, M.R. & Mitchell, J.K. 1993. New perspectives on soil creep. *Journal of Geotechnical Engineering*, 119(3): 507-524.

Kuwano, R. 1999. The stiffness and yield anisotropy of sand, PhD, Imperial Coll. Sci. & Tech., London.

Leung, C.F., Lee, F.H. & Yet, N.S. 1996. The role of particle breakage in pile creep in sand. *Canadian Geotechnical Journal* 33: 888-898.

McDowell, G,R. 1996. Clastic Soil Mechanics, PhD, Cambridge University.

Mejia, C.A., Vaid, Y.P. & Negussey, D. 1988. Time dependent behaviour of sand. In M.J. Keedwell (ed.), *Int. Conf. on Rheology and Soil Mechanics, Coventry, UK*: 312-326.

Mitchell, J.K. & Solymar, Z.V. 1984. Time-dependent strength gain in freshly deposited or densified sand. *Journal of Geotechnical Engineering* 110(11): 1559-1576.

Murayama, S., Michihro, K. & Sakagami, T. 1984. Creep characteristics of sands. *Soils and Foundations* 24(2): 1-15.

Oda, M. 1997. A micro-deformation model for dilatancy of granular materials. In C.S. Chang, A. Misra, R.Y. Liang & M. Babic (eds), *Mechanics of deformation and flow of particulate materials*: 24-37. New York: ASCE.

Oda, M., Konishi, J. & Nemat-Nasser, S. 1982. Experimental micromechanical evaluation of strength of granular materials: Effect of particle rolling. *Mechanics of Materials* 1: 267-283.

Radjai, F., Wolf, D.E., Jean, M. & Moreau, J.-J. 1998. Bimodal character of stress transmission in granular packings. *Physical Review Letters* 80(1): 61-64.

Tanaka, Y. & Tanimoto, K. 1988. Time dependent deformation of sand as measured by acoustic emission. In M.J. Keedwell (ed.), *Int. Conf. on Rheology and Soil Mechanics, Coventry, UK*: 182-193.

Constitutive and Centrifuge Modelling: Two Extremes, Springman (ed.)
© *2002 Swets & Zeitlinger, Lisse, ISBN 90 5809 361 1*

A few observations on the rheology of powder-binder systems

Luiza Dihoru
University of Bristol, United Kingdom

ABSTRACT: A rheological study has been performed in order to evaluate the variation of viscosity and the separation risk, for mixtures composed of powders and a binder system. The mixtures were prepared in a torque rheometer by mixing powders of various sizes, morphologies and size distributions, using a polymer-based binder. The effects of powder volume fraction, shear rate, and powder characteristics on viscosity, were studied by capillary rheometry. Microrheological considerations have been made in order to understand the nature of the flow process. Aspects such as the wall effect and the particle interactions in flow have been found to have a major impact on mixture viscosity and flow stability. Even if the experiments employed powder-binder systems that are used as raw materials in powder metallurgy, their results could be useful tools in understanding the creep behaviour of other types of suspensions of granular materials of various origins.

Constitutive and Centrifuge Modelling: Two Extremes, Springman (ed.)
© 2002 Swets & Zeitlinger, Lisse, ISBN 90 5809 361 1

Experimental investigations on sand under cyclic loading

G. Festag
Institute of Geotechnics, Technische Universität Darmstadt, Germany

ABSTRACT: The investigations presented here show that abrasion has a notable influence on the deformation behaviour in the small stress domain. The macroscopic behaviour of sand in tri-axial cyclic compression tests is shown. Cyclic shakedown and local failure behaviour was observed in the triaxial tests. A microscopic view shows the influence of abrasion on single grains and thus the verification of significant effects of abrasion in a microscopic and macroscopic sense is demonstrated successfully. Some questions concerning the development of hysteresis loops and constitutive modelling are raised.

1 INTRODUCTION

Cyclic loading, in the meaning of this paper, is referred to as loading with successively changing intensity of a repetitive nature, neglecting any forces of inertia. The scope is to focus on loading cycles, which are small in comparison with the ultimate failure load. Experiments show that sand under cyclic loading develops permanent plastic strain increments, and that sometimes additional plastic deformations occur suddenly. Modification of the single grains occurs in the tests and abrasion can be observed, whereby the grains become more rounded.

2 TRIAXIAL EXPERIMENTS

Triaxial tests were performed under cyclic loading to examine the influence of abrasion. A tri-axial test apparatus with a hydraulic controlled load was used for this test series. The triaxial test equipment consists of a conventional servo-controlled triaxial test apparatus and a hydraulic system, which controls the cyclic loading. A vertical load up to 10 kN, with a loading frequency of 5 Hz and a maximum vertical deformation of 1 mm per load cycle, can be obtained. The dimensions of the cylindrical test specimen were 10 cm in diameter and 20 cm in height. As a key feature of the measuring system, the load and deformation devices are plugged directly onto the test specimen. The load cell for measuring the vertical load is attached directly above the specimen and is balanced against the cell pressure, so only the deviatoric part of the load is obtained. The cell pressure is measured and controlled by a digital pressure controller. The deformation of the test specimen is measured using Hall Effect strain transducers. A Hall Effect sensor consists of a metallic or semiconductor plate through which electrical current is flowing. A permanent magnet is placed over the plate. The magnetic field has to be perpendicular to the plate and the current flow, so the charge carriers will be deflected. A voltage will thus be produced across the plate in a direction normal to the current flow. In modern transducers, the change in the voltage is linearly related to the shift of the magnet. Two axial and one radial transducer are used for measurement. The transducers are plugged onto callipers in the mid-third of the specimen so bedding and end restraint errors of the specimen will be avoided.

The transducers have an accuracy of about 10 μm and a maximum range of 10 mm. This is necessary to record the small strain in the load cycles performed.

The samples were consolidated first under an isotopic pressure of 150 kPa and then loaded with a sinusoidal deviatoric stress, cycling between 200 kPa and 350 kPa with a frequency of 1 Hz. Up to 5 million load cycles were performed. Eight tests were carried out with two different materials and several placement densities of the specimen. Only the tests on an angular grained industrial crushed gabbro sand from a quarry in the Odenwald, Germany, will be presented in this paper. The parent rock is a volcanic material. It is crushed twice in an industrial process to produce ballast, the resulting material was sieved and the medium grained sand fractions between 0.2 mm and 2 mm were used. The crushed gabbro sand is a very angular material and the grains consist of a conglomerate of different crystals, mainly feldspar, calcite and mica. Because of the heterogeneous structure and the minerals of the gabbro, it has a low grain strength (McDowell & Bolton 1998, Nakata et al. 1999).

One typical test result will be shown. Up to 4.4 million loading cycles were performed in this test. Figure 1 shows the net vertical strain in the sample against the number of loading cycles. The minimum strain ε_{min} corresponds to the strain measured at the minimum stress, whereas the maximum strain ε_{max} was measured at the maximum stress for that particular load cycle. The distance between both curves is equal to the reversible strain in a load cycle. In common elastoplastic constitutive laws, this range will be referred to as the elastic part of a load cycle. It can

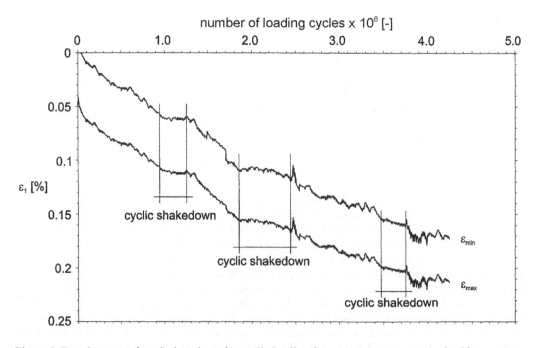

Figure 1. Development of vertical strain under cyclic loading for a triaxial test on crushed gabbro sand

be recognised that the distance between both plots of strain decreases slowly with an increasing number of loading cycles. The test result shows that the plastic strain increments become smaller with increasing number of loading cycles within a certain region. This behaviour is called cyclic shakedown and can be described by a logarithmic function in a similar way as shown e.g. by Heineke et al. (2001). Heineke shows that the development of the mean settlement with the number of loading cycles can be described by the following function:

$$s_{mean}(N) = s_{mean}(N = 1) + a \cdot \ln\left(\frac{N-1}{1000} + 1\right) \tag{1}$$

where s_{mean} = mean settlement, N = number of loading cycles and a = constant to be determined from the test data.

The cyclic shakedown state is not a steady state. After some loading cycles, in which the vertical strain rate slows significantly, increasing plastic strain increments suddenly occur once more. This form of 'local failure' has not yet been explained and also the mathematical function (1) is unable to capture this behaviour. This 'local failure' could be observed only after numerous loading cycles, which leads to long-term experiments of one or two months of duration, entailing 3 – 5 million loading cycles with 1 Hz frequency. It is remarkable that even after 4.4 million loading cycles, additional plastic strain increments still occur. A final steady state, which means a region with no additional plastic strain under constant cyclic loading, can not be observed and will probably never be reached.

The stress-strain hysteresis loops of the gabbro sand are shown in Figure 2 for the test reported in Figure 1. The number marked by the loops gives the number of loading cycles up to the loop drawn. The hysteresis loops show the typical concave-converse trend, which is characteristic for dense sand. Loose sand tends to exhibit a double convex trend. Experiments have shown that a sand that was loose initially will produce double convex hysteresis loops at the beginning but will get denser under continuing cyclic loading, developing a concave-convex trend

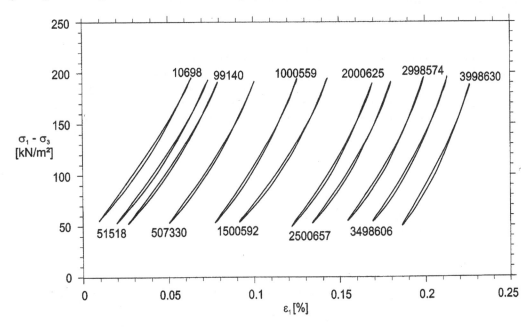

Figure 2. Selected typical stress-strain hysteresis loops for cyclic triaxial tests on crushed gabbro sand

after numerous loading cycles. The increasing gradient of the loops in Figure 2 confirms that the reversible part of the load cycles decreases slowly. The area included within each hysteresis loop gives the dissipated energy in each loop. So it can be seen that energy is dissipated during all load cycles. The area of the hysteresis loop becomes smaller in the first few hundred cycles. No significant change of this area can be observed after these first cycles. The dissipated energy is believed to lead to abrasion (Arslan et al. 2000) and to cause rearrangement of the skeleton. The allocation of the amount of dissipated energy between these effects is unknown. The area within the hysteresis loops shows no significant change in the regions of cyclic shakedown or local failure. So the initiation of a local failure region seems to be independent of the actual energy dissipation, but may be a function of the total amount of dissipated energy. This has to be investigated in the future.

3 MODIFICATION OF THE PARTICLE SHAPE

Grain crushing is usually divided into grain breakage and abrasion. Grain breakage may be described as the dissection of grains into several parts, with nearly the same dimensions. Abrasion may be described as the splintering of very small particles from the grain surface. While grain breakage only appears at high stress levels, abrasion is a phenomenon which is independent of the stress level. The idea behind this investigation was to observe the change of the texture of a grain due to abrasion.

The macroscopic behaviour of sand under repeated load was shown in the previous chapter. The gabbro sand indicated a behaviour, which can be subdivided into different regions; these are separated by discontinuous deformation behaviour. Sieve analysis before and after the triaxial tests showed no alteration of the grain size distribution, so grain breakage can be neglected. Of these two crushing mechanisms, only grain breakage can be quantified by sieve analysis because abrasion contributes minimally to changing the dimension of a grain.

3.1 Optical detection of grain shape

Abrasion is the splintering of small parts off the grain due to local contact stresses exceeding the grain strength. This is most likely to occur at the contact points in a soil matrix. An automatic, computer based procedure was developed to determine the grain shape and to check the influence of abrasion. The specification of the grain shape is mostly described by a coefficient of roundness, which is measured from a two-dimensional (2D) image of the grain. The coefficient of roundness is the relationship between the area of a grain to its perimeter and can be written as (Cox 1927):

$$K = \frac{4 \cdot \pi \cdot A}{P^2} \qquad (2)$$

K - coefficient of roundness
A - area of grain image
P - perimeter of grain image

The coefficient of roundness can have values between zero and one. When $K = 0$, the shape is a line and in the case of $K = 1$, the shape is a circle. The coefficient of roundness is the same for all figures of the same shape, regardless of size. That is to say, that a small circle is just as round as a large circle. Furthermore, the value of K for a figure of any given shape represents the percentage ratio that the area of the figure holds to the area of a circle with the same perimeter. The determination of the coefficient was done in the past via visual estimation or measurements of the grain under the microscope. A computer-based technique should be used for better repeatability. The precision is dependent upon the grain size, the resolution of the processed image and, in a statistical sense, upon the number of grains inspected and the standard error of the measured quantity. To reduce this to acceptable levels, about 600 grains were scanned for each test with a resolution of 2000 points per inch (ppi), and the area and perimeter of each grain were determined using graphic software (Ad Oculus 3.0). A grain of about 1 mm diameter is represented by more than 10,000 pixels using this resolution.

3.2 2Results of grain shape investigation

About 600 grains were scanned and analysed for the first tests, to check if the distribution of the coefficient of roundness is a normal curve of distribution. The χ^2-test showed that a normal distribution is a good approach for representing the roundness of the gabbro sand. The mean value and the standard deviation can be calculated. 600 grains are necessary in order to ensure the quality of the probabilistic analysis and achieve a representative probability of 0.95. The coefficient of roundness was determined for all test specimens after the cyclic triaxial tests and also for the original, unloaded material. Figure 3 gives an overview of the results.

Table 1 gives the main parameters for the tests shown in Figure 3. There is a clear difference between the previously loaded and the initial post-crushing particle shapes of the gabbro sand. There is a clear dependency between the number of loading cycles of the material and the coefficient of roundness in that an increase of the number of loading cycles leads to more rounded grains. Abrasion seems to be present as an ongoing effect. However in the macroscopic view, as

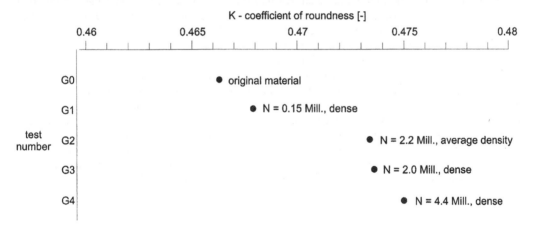

Figure 3. Development of the coefficient of roundness K (N = number of loading cycles)

Table 1. Test characteristics for the test in Figure 3

Test	Placement density	Number of loading cycles	Coefficient of roundness
G0		0	0.4663
G1	dense	0.15×10^6	0.4679
G2	average density	2.2×10^6	0.4734
G3	dense	2.0×10^6	0.4736
G4	dense	4.4×10^6	0.4750

well as in the microscopic view, an end to the abrasive influence of cyclic loading can not be seen. A dependency exists between the origin placement density of the test specimen and the roundness of the grain. The denser the specimen, the more rounding of grains occurs. This observation can be explained by considering the formation of the grain skeleton. There are more grain contacts in a dense skeleton than in a loose skeleton, so the probability of relative movement between grain contacts and therefore shear stresses developing are much higher than in a loose skeleton where movement (e.g. rolling), which does not generate high shear stresses, is possible.

4 CONSTITUTIVE MODELS

The widely used and well known elasto-plastic material model based e.g. on Drucker-Prager (Chen & Mizuno 1990) is used to represent the behaviour of granular material. If the investigation only focuses on a monotonic load path, and neglects the cyclic or dynamic loading conditions, these common models work very well and an extensive pool of parameters for different soils can be established. Hysteretic effects can more or less be introduced within a kinematic extension. If the cyclic shakedown behaviour should be replicated, even complicated kinematic models tend only to represent single phenomena.

An alternative to these is a hypoplastic model. The theory of hypoplasticity has been developed – independently of other similar proposals – by Kolymbas, Gudehus, von Wolffersdorff and their associates since the seventies (Kolymbas 1991, Kolymbas et al. 1995, von Wolffersdorff 1996). Hypoplastic models have shown that they can be used to model the deformations due to re-arrangements of the grain skeleton very well (Niemunis & Herle 1997). But applications to cyclic loading show some defects. Hysteretic behaviour can not easily be modelled and the deformations show an excessive accumulation, the so-called ratcheting effect. Niemunis & Herle (1997) proposed an extension of the hypoplastic model with a small elastic strain range called intergranular strain. This extension enables the prediction of hysteretic behaviour, but the development of the hysteresis loops with continuous loading shows some inaccuracy. E.g. the switch of the shape of the loops dependent on the density of the test specimen is not captured. The hypoplastic model, with the extension of Niemunis & Herle (1997), needs twelve material parameters. Most of them can be determined from common laboratory experiments and it can be shown that they have a physical meaning. As a conclusion it has to be said that the hypoplastic model can represent several aspects of the cyclic behaviour of granular materials, but that there are still problems in predicting the behaviour correctly, especially under cyclic loading conditions.

5 CONCLUSION

These triaxial tests have confirmed that abrasion occurs due to the deformation of sand grains under cyclic loading. A small plastic strain increment occurs in all tests, independent of the number of loading cycles. The microscopic study showed the effect of abrasion on the single grains. The gabbro sand particles tended to become rounder with increasing number of loading cycles. A dependency between the sample density and the rounding of the grains can also be observed. Further investigations have to show the influence of the extent of abrasion on certain material parameters. A mechanical model should be developed on the basis of common constitutive models, with due consideration of the principles of thermodynamics, in which the effect of abrasion is integrated. The splintering of small particles, and consequently the dissipation of energy, should be included in this model.

REFERENCES

Arslan, U., Katzenbach, R. & Festag, G. 2000. Verhalten von Sand im zyklischen Triaxialkompressionsversuch. *Proc. of "Workshop: Boden unter fast zyklischer Belastung: Erfahrungen und Forschungsergebnisse".* Schriftenreihe des Institutes für Grundbau und Bodenmechanik der Ruhr-Universität Bochum. 32: 287-300.

Chen, W.F. & Mizuno, E. 1990. *Nonlinear analysis in soil mechanics.* Amsterdam: Elsevier.

Cox, E.P. 1927. A method of assigning numerical and percentage values to the degree of roundness of sand grains. *Journal of Paleontology.* 1(3): 179-183.

McDowell, G.R. & Bolton, M.D. 1998. On the micromechanics of crushable aggregates. *Géotechnique* 48(5): 667-679.

Heineke, St., Katzenbach, R. & Arslan, U. 2001. Model scale investigations on the deformation of the subsoil under railway traffic. In E. Togrol (ed.) *Proc. of the XV. ICSMGE, Istanbul, Turkey.* Rotterdam: Balkema.

Kolymbas, D. 1991. An outline of hypoplasticity. *Archive of Applied Mechanics* 61: 143-151.

Kolymbas, D., Herle, I. & Wolffersdorff, P.-A. von 1995. Hypoplastic constitutive equation with internal variables. *Int. J. Num. Anal. Methods Geomech.* 19: 415-436.

Nakata, Y., Hyde, A.F.L., Hyodo, M. & Murata, H. 1999. A probabilistic approach to sand particle crushing in the triaxial test. *Géotechnique* 49(5): 567-583.

Niemunis, A. & Herle, I. 1997. Hypoplastic model for cohesionless soils with elastic strain range. *Mechanics of cohesive-frictional materials* 2: 279-299.

Wolffersdorff, P.-A. von. 1996. A hypoplastic relation for granular materials with a predefined limit state surface. *Mechanics of cohesive-frictional materials* 1: 251-271.

Constitutive and Centrifuge Modelling: Two Extremes, Springman (ed.)
© *2002 Swets & Zeitlinger, Lisse, ISBN 90 5809 361 1*

Deformation behaviour of a soft Swiss lacustrine clay

J.L. Trausch-Giudici
Institute for Geotechnical Engineering, Swiss Federal Institute of Technology, Zurich, Switzerland

ABSTRACT: First triaxial multi-stage tests on a Swiss lacustrine clay, the Seebodenlehm, are presented. The aim of the research is to model the deformation behaviour of the normally consolidated Seebodenlehm for plastic straining as well as for behaviour inside the current yield locus. Therefore two models are being studied: S-Clay1 for yielding and 3-SKH for elasto-plastic deformation inside the yield locus. Stress response and volumetric behaviour of specimens from 6 and 23 m depth, as well as their stiffness during shearing, are presented. The agreement of the test results with predictions based on the S-Clay1 model with rotating yield loci is discussed. The main question remaining concerns stress paths inside the yield locus: whether and how they interfere with the anisotropy of the yield surface.

1 BACKGROUND

It is common in Switzerland that regions adjacent to the major lakes, where post-glacial clays have been deposited, have later become highly populated. These soft normally consolidated clays require careful consideration during the design process for the necessary infrastructure, as well as understanding of their likely behaviour and associated properties.

Transportation projects on or in soft clay, including high embankments for a highway intersections or local link underpasses, present two types of problems related to the modelling of the behaviour of soft clays. In the first case, the onset of large plastic strains and the subsequent pattern of plastic straining is of interest. In the latter case, the question arises about how dependent is the stiffness upon the elasto-plastic strains for this normally consolidated clay, and whether a maximum elastic modulus of the soil can be taken into account for accurate modelling and economic design.

2 SOIL MODELS STUDIED

2.1 *S-Clay1*

The anisotropic elasto-plastic model with rotational hardening (S-Clay1) was developed with the aim of capturing the main features of the yielding behaviour of soft clays whilst remaining sufficiently simple (Wheeler 1997, Wheeler et al. 1999, in press). S-Clay1 is an extension of the critical state models, with anisotropy of plastic behaviour represented through an inclined ellipsoidal yield surface (as proposed by many researchers including Länsivaara (1999) and Wheeler et al. (1999, in press)) and a rotational component of hardening to model the development or erasure of fabric anisotropy during plastic straining. No attempt was made to model non-linearity of small strain stiffness or anisotropic elastic behaviour.

Compared to Modified Cam Clay, two additional parameters relating to rotational hardening (β, μ) are needed, and the state of the soil is defined by the initial values of size (p_m') and of inclination (α) of the in situ yield surface.

The suitability of S-Clay1 has been tested for four Finnish clays (plasticity index I_P in the range of 37 to 82 % and sensitivity S_t between 7 and 94) and triaxial tests have been simulated (Näätänen & Lojander 2000). The suitability of S-Clay1 for a less sensitive, siltier clay such as Seebodenlehm (I_P = 20%, S_t = 4) is of great interest.

2.2 *3-SKH*

The three-surface kinematic hardening model was formulated to simulate the behaviour of clays in overconsolidated stress states and during early stages of loading. The stress-strain response of overconsolidated soils is initially dependent on the recent stress history, and this effect gradually decreases as loading continues. The deformation of overconsolidated soil is usually considered to be elastic just after significant changes of stress path direction at very small subsequent stress or strain changes (Stallebrass & Taylor 1997).

The 3-SKH model makes use of the framework of critical state soil mechanics, and incorporates two kinematic yield surfaces within a Modified Cam Clay boundary surface model. The model requires three additional parameters to Modified Cam Clay: T and S link the size of the three surfaces and Ψ is the exponent in the hardening modulus. T and S can be inferred from multi-stage triaxial tests with different stress histories, whereas Ψ is the only parameter which has to be defined through parametric studies.

3 TRIAXIAL EQUIPMENT

Multi-stage triaxial tests were performed at IGT on Seebodenlehm using the new apparatus designed in-house (Fig. 1). The apparatus is fully computer controlled and has three stepping motors for applying axial force, cell pressure and backpressure (BP). The BP stepping motor measures the volume changes of the specimen as well, giving a high resolution of 0.02 cm^3.

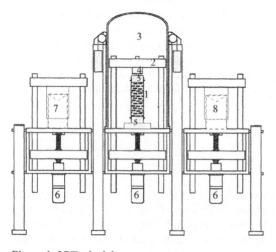

1 Sample
2 Loading frame
3 Cell
4 Load transducer
5 Top cap and bottom plate
6 Stepping motor
7 Back Pressure unit
8 Cell pressure unit

Figure 1. IGT triaxial apparatus

The axial load cell is mounted inside the cell above the top cap, and axial displacement is measured inside the cell between top cap and bottom plate. Bender and compression elements are mounted inside the top cap and bottom plate to be able to perform dynamic measurement of the maximum stiffness of the soil during loading.

4 SOIL MATERIAL

The Kreuzlinger Seebodenlehm lies at the site studied in the north of Switzerland in two layers between depths of 4 to 12 m and 22 to 24 m below ground level. Following the Swiss codes, which are based on USCS, the Kreuzlingen clay was described as either medium plasticity clay or clayey silt CL. Typically, the water content w ranged between 25 and 46% and the liquid limit between 30 - 48%, with siltier layers represented by the lower ranges of these values. Plasticity index I_p was between 14 - 32%. The liquidity index tended to decrease linearly with depth, averaging one or greater above 12 m depth, and lying between 0.5 to 0.75 in the lower clay layer (22 – 24 m) (Springman et al. 1999). Sensitivity was determined from fall cone tests to be about 4. An equation relating the undrained shear strength s_u with the in situ vertical effective stress σ'_{vo} was found from cone penetration tests to be:

$$s_u = 0.23 \, (\sigma'_{vo} + 15) \, [kPa] \tag{1}$$

The undrained shear strength of the layer varies over the depth of interest from 15 to 50 kPa (Springman et al. 1999). The Seebodenlehm is highly heterogeneous because of the alternating fine silt and clay layers, so that in a triaxial specimen of 130 mm height, the water content can vary from 22% in the silt and 30% in the clay layers, and the clay content respectively from 28% to 42%.

Undisturbed sample cores of 65 mm diameter and 800 mm length were taken at the site from a borehole drilled using triple coring techniques. These samples have been waxed and stored in the laboratory. From each of these samples, four triaxial probes of 130 mm height can be cut, the remaining material being used for classification, fall cone tests and oedometer tests.

5 RESULTS

The results of multi-stage triaxial tests on test sample 46216 from 6 m depth and on test sample 46227 from 23 m depth are shown in Figures 2 to 4. During these tests, an initial drained stress path, with an inclination η in q-p' stress space, drags the initial yield loci out to mean effective stresses which are two times the mean in situ effective stress. This stage is followed by drained unloading on the same path, after which a new inclination η will be chosen to reload the specimen to two times the previous maximum load. This test type is described in Wheeler et al. (in press). The deviatoric loading rate was chosen to be 2 kPa/h for loading and to be 5 kPa/h for unloading. Finally undrained shearing is carried out to failure.

The three samples of specimen 46216 show good agreement at critical state, also with previous tests on Kreuzlingen Seebodenlehm, where M was found to be 1.05 (Springman et al. 1999). However their volumetric behaviour is not at all similar. Sample 1 undergoes extreme volume change, whereas sample 2 behaves more stiffly. The data of sample 3 couldn't be evaluated completely because of difficulties with the software.

In testing specimen 46227, errors with the controlling software led to the following problems: sample 5 was sheared at an overconsolidation ratio OCR of 1.25, and sample 6 could not be sheared after isotropic reloading. Sample 5 gives a very high M value of 1.4: the sample was very silty, more than half the height of the sample included successive fine silt layers, which have affected the strength determined from stresses at failure. The measurement of volumetric change of both samples is not reliable, because a leakage in the pore water pipes led to water in-filtration from the cell to the pore water system. The value of v at the end of the test for sample 46227/6 is less then 1, which is by definition impossible, but is due in this test to leakage of cell water to the drainage system. However the yield pressures p' for first and second loading ex-trapolated from the v versus lnp' graph agree very well with the S-Clay1 predictions (Fig. 5) and with the results of specimen 46216.

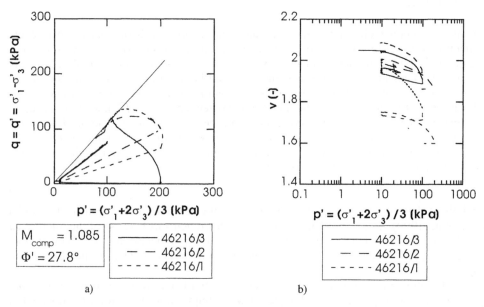

Figure 2. a) deviatoric stress q versus mean effective stress p' of samples 46216/1 to 3 b) specific volume v versus log of mean effective stress p' of samples 46216/1 to 3

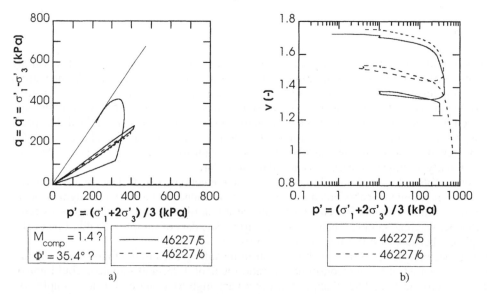

Figure 3. a) deviatoric stress q versus mean effective stress p' of samples 46227/5 and 6 b) specific vol-ume v versus log of mean effective stress p' of samples 46227/5 and 6

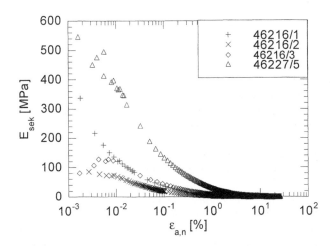

Figure 4. Secant Young's Modulus E_{sec} versus natural axial strains $\varepsilon_{a,n}$ of samples 46216/1 to 5

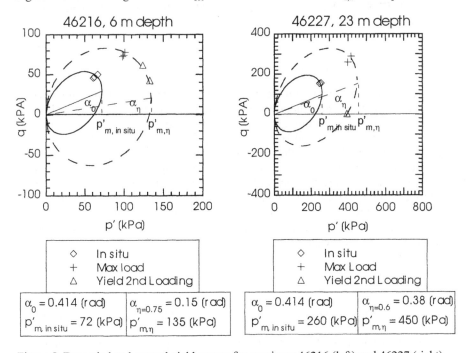

Figure 5. Expanded and rotated yield curves for specimen 46216 (left) and 46227 (right)

The dependence of the secant Young's Modulus on natural axial strain $\varepsilon_{a,n}$ (as defined by Richardson (1988) as $-\ln(1-\varepsilon_a)$, where ε_a is the axial strain) for the shearing path is shown in Figure 4. Sample 46227/5 comes from a greater depth and was lightly overconsolidated at the commencement of undrained shearing. Specimen 46216/1 was subjected to the biggest compression during the previous stress history and shows the stiffest response here. The response of sample 46216/3 agrees very well with sample 46216/1 for strains greater than 10^{-2}. Sample 46216/2 is relatively soft and shows a very small difference between small strain stiffness and secant stiffness derived during plastic straining.

6 DISCUSSION

In evaluating the results, the heterogeneity of the triaxial specimen must be taken into account. Due to the fine interspersed silt layers, the volumetric behaviour and the failure mechanism can differ from sample to sample. Sample 46216/1 was the most clayey and homogeneous, whereas samples 46216/2 and 46216/3 both had a 10 mm thick silt layer near the top or at the mid-height respectively, which influenced the stiffness measured and the failure mechanism observed. In sample 46227/5, the existence of successive very closely spaced silt layers over the half of the sample delivered a different failure mechanism. The lower clayey part of the sample bulged significantly, while the top (silty) half maintained the shape of a right cylinder. In such circumstances, the derivation of a value of M (Fig. 3a) from such a small sample is unlikely to represent conditions in the field.

The expanded and rotated yield curves are shown in Figure 5. The inclination (α_0) of the in situ yield surface has been calculated for both specimen 46216 and 46227 with a value of M of 1.08. The experimental points agree well with the sheared ellipse of S-Clay1. Loading to a high η value of 0.75 results in a major rotation of the yield curve (sample 46216: $\alpha_0 = 0.414$ to $\alpha_\eta = 0.15$). Samples 46227/5 and 6 were loaded with $\eta = 0.6$, which is close to the stress path gradient for one dimensional straining, η_{K0}. In fact the yield curve rotates only marginally (sample 46227: $\alpha_0 = 0.414$ to $\alpha_\eta = 0.38$).

7 CONCLUSION

Tests results show reasonable agreement with the predicted ellipsoidal yield surface modelled in S-Clay1. The volumetric behaviour and the stiffness are highly dependent upon the silt layer content and position in mostly clayey homogeneous specimens. More investigation of the deformation behaviour pre-yield and at yield of this soft clay is necessary. The question arises about what influence the stress paths inside the yield surface have on the rotational hardening of the yield surface of the clay. This can be investigated with multi-stage triaxial tests, which have pre-yield stress path excursions before the second loading stage. In this way, tests as described by Stallebrass (1990), could be performed inside the expanded yield locus, and an evaluation of stress history dependence of the stiffness could be made. Moreover comparison with tests on similar specimens, but without these stress path excursions, could give an appreciation of the rotational hardening due to elasto-plastic strains.

REFERENCES

Länsivaara, T. 1999. A study of the mechanical behaviour of soft clay. PhD Thesis, NTNU Trondheim, Geotechnical Institute.

Näätänen, A., Lojander, M., Wheeler, S.J. & Karstunen, M. 1999. Experimental investigation of an anisotropic hardening model for soft clays. In *Proc. 2nd Int. Symp. Pre-failure Deformation Characteristics of Geomaterials, Torino*: 541-548. Rotterdam: Balkema.

Näätänen, A. & Lojander, M. 2000. Modelling of anisotropy of Finnish clays. In *VII Finnish Symposium on Mechanics, Tampere* 2: 589-597. Tampere: Koski&Virtanen.

Richardson, D. 1988. Investigations of the threshold effects in soil deformations. Ph.D. Thesis. The City University, London.

Springman, S.M., Giudici Trausch, J., Heil, H.M. & Heim R. 1999. Strength of soft Swiss lacustrine clay, cone penetration and triaxial test data. *Transportation Research Record* 1675: 1-9. Washington D.C.

Stallebrass, S.E. & Taylor, R.N. 1997. The development and evaluation of a constitutive model for the prediction of grounds movements in overconsolidated clay. *Géotechnique* 47(2): 235-253.

Stallebrass, S.E. 1990. The effect of recent stress history on the deformation of overconsolidated soils. PhD Thesis, The City University, London.

Sivakumar, V., Doran, I.G., Graham, J. & Johnson, A. 2001. The effect of anisotropic elasticity on the yielding characteristics of overconsolidated natural clay. *Canadian Geotechnical Journal* 38: 125-137.

Wheeler, S.J. 1997. A rotational hardening elasto-plastic model for clays. In *Proc. 14th ICSMFE, Hamburg* 1: 431-434. Rotterdam: Balkema.

Wheeler, S.J., Näätänen, A., Karstunen, M. 1999. Anisotropic hardening model for normally consolidated soft clay. In *Proc. 7th Int. Symp. Numerical Models in Geomechanics (NUMOG), Graz:* 33-40. Rotterdam: Balkema.

Wheeler, S.J., Näätänen, A., Karstunen, M. & Lojander, M. In press. An anisotropic elasto-plastic model for natural soft clays. Submitted to *Géotechnique*.

Constitutive and Centrifuge Modelling: Two Extremes, Springman (ed.)
© *2002 Swets & Zeitlinger, Lisse, ISBN 90 5809 361 1*

First results of triaxial creep tests on permafrost soil samples

L. Arenson
Institute for Geotechnical Engineering, Swiss Federal Institute of Technology, Zurich, Switzerland

ABSTRACT: In order to understand the complex mechanical behaviour of creeping Alpine permafrost, such as rock glaciers, triaxial creep tests were performed on real frozen soil samples. Contrary to the general agreement that there are three well-defined creep phases, the tests showed an ongoing decrease of the strain rate with time, and depending on the stress level and time, a sudden acceleration can sometimes be observed.

1 INTRODUCTION

Perennially frozen ground, known as permafrost, has been studied for some years and many structures have been built in these environments (e.g. Andersland & Ladanyi 1994). Rock glaciers are a special geomorphological phenomenon, which also contain frozen material and creep downslope under the influence of gravity. These have been studied intensively during recent decades (e.g. Barsch 1996, Giardino et al. 1987). However, the mechanics of this ice, air, soil and unfrozen water mixture still isn't fully understood in respect of Alpine permafrost at temperatures close to the melting point, even though recent studies have identified the instabilities of melting permafrost as being an issue of national importance in Switzerland (Haeberli 1999, Haeberli et al. 1997) or at least as an engineering hazard (Giardino & Vick 1985).

Figure 1. Aerial photograph and situation of the Muragl rock glacier

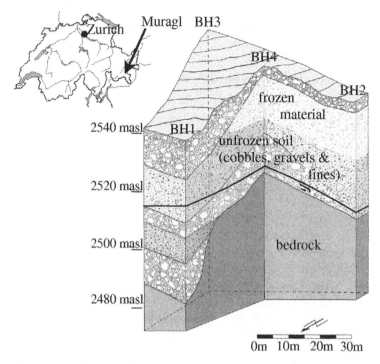

Figure 2. Stratigraphy of the Muragl rock glacier

Four boreholes were drilled in the Muragl rock glacier, Engadin, Swiss Alps in late spring, early summer 1999 (Fig. 1). Special drill bits and a triple tube drilling rig with cold air flushing were used in order to obtain "undisturbed" frozen core samples (Arenson et al. 2000). Inevitably the combination of cold temperatures, drilling friction and a coarser particle size distribution than expected contributed to a degree of disturbance. However, best efforts were made to minimise this during sampling and subsequent transport and storage. Figure 2 shows a cross section of the drill site.

The results of the first triaxial creep tests on these cored rock glacier samples are presented and discussed in this paper. A simple creep law is used in order to describe the steady state creep. However, this steady state was difficult to obtain since the tests showed an ongoing decrease in the strain rate depending on the stress level.

2 MATERIAL AND METHODS

The samples had a diameter of 74 mm and were cut to a length of about 155 mm. They were analysed after every test in order to calculate the particle, ice and air content. Since the unfrozen water content is very small and difficult to deduce, it was neglected. Table 1 gives an overview of the samples and their origin and the components. The soil can be qualified as a well graded gravel with silt (GW – GM). The temperature and confining pressure were selected as appropriate control conditions for each test. Samples 3 and 4 were tested under two different temperatures. After the third loading stage, the tests were stopped and the temperature within the cold room was changed. The next loading step was started as soon as the sample had reached the new temperature level. In the meantime, the samples could relax.

Table 1. Samples tested: Muragl rock glacier, borehole 4

Sample	Depth [m]	Temperature [°C]	Confining pressure [kPa]	Density [g/cm³]	Ice [Vol-%]	Soil [Vol-%]	Air [Vol-%]
1	5.00 – 5.15	-3.4	150	1.62	56	40	4
2	6.40 – 6.55	-3.7	200	1.31	70	24	6
3	6.65 – 6.80	-3.7 / -2.4	300	1.33	69	25	6
4	7.90 – 8.05	-3.4 / -2.1	100	1.64	56	40	4
5	15.05 – 15.20	-2.4	100	1.91	40	56	4

New triaxial apparatuses were developed and manufactured at the Institute for Geotechnical Engineering at ETH Zurich for carrying out the creep tests (Fig. 3).

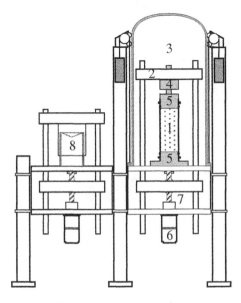

1 Sample, sealed with a rubber membrane and o-rings

2 Loading frame

3 Cell, filled with antifreeze

4 Load transducer

5 Top cap and bottom platten

6 Stepper motor

7 Spindle drive

8 Cell pressure unit

Figure 3. Triaxial test apparatus for creep tests

3 RESULTS

Figure 4 shows a typical creep test with different loading steps. The axial stresses were increased from $\sigma_{1,1}$ to $\sigma_{1,2}$ to $\sigma_{1,3}$ as soon as a "steady state" appeared to have been reached (Fig. 4a). However, constant strain rates were not observed for most of the tests. Typically, tertiary creep was reached with axial stresses higher than 600 kPa after around 100h, which is indicated by an increase in axial strain rate (Fig. 4b).

A summary of the strain rates for all stages of all the creep tests is shown in Figure 5. On a logarithmic scale, the strain rate for each sample shows almost a linear increase with the axial stress. Test No. 1 (5 m depth) shows the greatest inclination, which can be explained by a leakage that could only be established at the end of the creep test. The contact of the sample with the cell fluid (which contained antifreeze) resulted in a loss of the diameter of about 20%. This reduction, however, has not been corrected for in Figure 5.

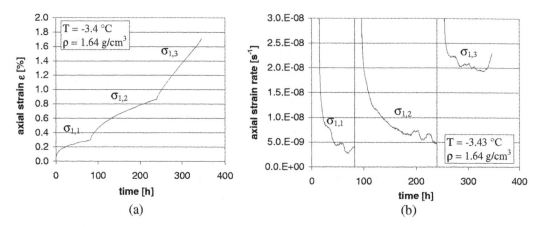

(a)   (b)

Figure 4. Typical creep curves for permafrost soils (Test: Muragl, BH4, Sample No. 4, T = -3.4°C) (a) axial strain with time, loading steps: $\sigma_{1,1}$ = 300 kPa, $\sigma_{1,2}$ = 495 kPa, $\sigma_{1,3}$ = 690 kPa (b) axial strain rate with time

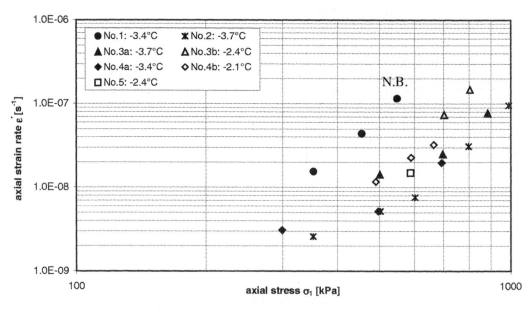

Figure 5. Axial strain rate versus axial stress in triaxial creep for the Muragl rock glacier. N.B. Sample membrane was not waterproof throughout the test

Even though different confining pressures were used (Table 1), no clear differences in the strain rates were observed. A similar trend was shown by Jones (1982), who reported nearly no effect on the yield stress as a function of the confining pressure for polycrystalline ice at low strain rates. On the other hand, an increase of the temperature may lead to a significant increase of the strain rate in terms of both gradient and intercept (Test Nos. 3 & 4). The density, which is directly linked to the ice content, also tends to control the strain rate: the denser the sample, the lower the strain rate will be.

4 DISCUSSION AND CONCLUSION

The results of the tests show that a direct comparison of the data is very difficult. This is mainly due to the heterogeneity of the samples tested. However, if the creep behaviour of rock glacier material is to be modelled successfully, the heterogeneity has to be taken into account. This means that a creep model has to consider at least the microstructure (ice content or density) and temperature. A first approach might be to use Glen's flow law (Glen 1955) for modelling the secondary strain rate $\dot{\varepsilon}$, including the influence of temperature.

$$\dot{\varepsilon} = A \cdot \sigma_1^{\ n} \cdot e^{(-Q/RT)} \qquad (1)$$

A, n: Creep parameters $= f$(microstructure)
Q: Activation energy $= f$(microstructure)
R: Universal gas constant $= 8.31 \cdot 10^{-3}$ kJ K^{-1} mole^{-1}
T: Temperature (K)

The creep parameters are summarised in Table 2 for the presented tests. However, the number of tests is not large enough to determine consistent values. In particular, an activation energy cannot be determined, since the inclination (n) for the same tests at two different temperatures is not constant. A reason for this effect might be the change in microstructure. The amount of unfrozen water can increase by 2-4% (Anderson & Morgenstern 1973) depending on the size and the distribution of the soil particles. Furthermore, no trend can be observed when comparing the creep parameters n or $A \cdot e^{(-Q/RT)}$ versus the density.

Table 2. Creep parameters: Rock glacier Muragl, borehole 4

Sample	Temperature [°C]	Density [g/cm^3]	n [-]	$A \cdot e^{(-Q/RT)}$ [kPa$^{-n} \cdot$s^{-1}]	R^2 [-]	no Tests [-]
1	-3.4	1.62	4.55	$3.93 \cdot 10^{-20}$	0.99	3
2	-3.7	1.31	3.52	$2.02 \cdot 10^{-18}$	0.90	5
3	-3.7	1.33	2.86	$2.39 \cdot 10^{-16}$	0.93	3
3	-2.4	1.33	5.14	$1.68 \cdot 10^{-22}$	1.00	2
4	-3.4	1.64	2.11	$1.55 \cdot 10^{-14}$	0.88	3
4	-2.1	1.64	3.37	$1.02 \cdot 10^{-17}$	0.99	3
5	-2.4	1.91	-	-	-	1

In order to model the entire range of creep behaviour, Fish's (1984) proposals can be used. With his model it is possible to model constant stress (creep: σ = const.) tests and constant strain rate ($\dot{\varepsilon}$ = const.) tests. Within this model, the secondary creep is reduced to a point in-between the primary and the tertiary phase. In order to use this model, the change in the entropy has to be known for the different creep phases.

$$\dot{\varepsilon} = \tilde{C} \cdot \frac{kT}{h} \cdot e^{(-Q/RT)} \cdot e^{\left(\frac{\Delta S - \Delta S^*}{k} \right)} \qquad (2)$$

k: Boltzmann's constant: $1.38 \cdot 10^{-23}$ J/K
h: Planck's constant: $6.626 \cdot 10^{-34}$ Js
ΔS: Entropy increment as a function of normalised time
ΔS^*: Entropy increment as a function of normalised stress
$\tilde{C}$: Material constant

Figure 4b also shows that acceleration occurs (after 330 h) without the strain rate levelling off beforehand, indicating that Fish's approach may prove to be most relevant for higher loading conditions.

The tests presented here are a first approach towards describing the unusual creep response of ice rich Alpine permafrost. But more tests are necessary for a better understanding of the mechanical behaviour of rock glacier material and to be able to select appropriate parameters. In particular, the effect of the change in microstructure, such as the change in unfrozen water content, on the creep parameters has to be studied. In addition, further investigations are necessary to determine if the absence of the secondary creep phase is typical for this soil or just a result of the testing procedure. Therefore, samples with similar compositions, which have to be carefully selected from the drillings, have to be analysed under only slightly different test conditions.

REFERENCES

Andersland, O. B. & Ladanyi, B. 1994. *An introduction to frozen ground engineering*. New York: Chapman & Hall.

Anderson, D.M. & Morgenstern, N.R. 1973. *Physics, chemistry, and mechanics of frozen ground: A review*. In North Am. Contrib. 2nd Int. Conference on Permafrost, Yakutsk, U.S.S.R: 257-288.

Arenson, L., Vonder Mühll, D. & Springman S. 2000. Drilling in the Muragl rock glacier. *Geophysical Research Abstracts* Vol. 2, 25th General Assembly of the European Geophysical Society, Nice.

Barsch, D. 1996. *Rockglaciers*. Berlin: Springer.

Fish, A.M. 1984. Thermodynamic model of creep at constant stress and constant strain rate. *Cold Regions Science and Technology* 45: 143-161.

Giardino, J.R. & Vick, S.G. 1985. Engineering hazards of rock glaciers. *Bulletin of the Association of Engieering Geologists* 22(2): 201-216.

Giardino, J.R., Schroder, J.F. & Vitek, J.D. 1987. *Rock Glaciers*. Boston MA: Allen and Unwin.

Glen, J.W. 1955. The creep of polycrystalline ice. Proceedings of the Royal Society of London Series A. Mathematical and Physical Science 228: 519-538.

Haeberli, W. (ed.). 1999. *Eisschwund und Naturkatastrophen im Hochgebirge*. Zürich: vdf, Hochschulverlag an der ETH.

Haeberli, W., Wegmann, M. & Vonder Mühll, D. 1997. Slope stability problems related to glacier shrinkage and permafrost degradation in the Alps. *Eclogea Geologicae Helveticae* 90: 407-414.

Jones, S.J. 1982. The confined compressive strength of polycrystalline ice. *Journal of Glaciology* 28: 171-177.

Constitutive and Centrifuge Modelling: Two Extremes, Springman (ed.)
© *2002 Swets & Zeitlinger, Lisse, ISBN 90 5809 361 1*

Discussion on problems governed by DEFORMATION

D.J. White
Cambridge University Engineering Department, United Kingdom

S.M. Springman & T. Weber
Swiss Federal Institute of Technology, Zurich, Switzerland

ABSTRACT: This paper contains the discussions to the theme lectures for problems related to deformation as well as the brief introductions to the young researchers' presentations and the resulting discussions. The paper closes with a summary of the proceedings.

1 DIFFICULTIES RELATED TO NUMERICAL PREDICTIONS OF DEFORMATION (I. HERLE)

1.1 *Discussion*

Laurent Vulliet opened the discussion on Ivo Herle's theme lecture by commenting that it had been a rather pessimistic morning given that it was so rare for people to admit the limits and the problems linked with their predictive methods. He found this refreshing to listen to but wondered why these predictions (e.g. Schweiger 2001) were made and what was the expected quality of the results? He felt that to say that the results were different did not mean that they were useless. He suggested that a probabilistic approach, for example, could be used to check if the result did not somehow represent lower and upper bound solutions to the problems and thought that maybe the conclusion could be reached that these predictions were not so bad after all! He asked the speaker to comment on the quality to be reached with any prediction. Herle suggested that it made no sense to develop a constitutive model to try to match the measured points precisely, because there would always be some scatter in the experimental results from element tests and that this was even more valid for boundary value problems. He felt that a lot could be learnt from these prediction competitions but that we could not expect to fit the measured results perfectly in situ, because it was even more difficult to do so than for more homogeneous test samples. Herle thought that it would be difficult to carry out probabilistic analysis because sometimes the resulting number would end up being positive rather than negative, even if the magnitude was exactly what was predicted (e.g. heave behind a retaining wall rather than settlement), hence this result would be far from reality. In any case, he was always very suspicious of perfect coincidence between measured and calculated values! He wondered about finding some criteria by which to judge predictions, not only based on complete agreement with the measured data, although he had not tried to do this as yet.

Sarah Springman supported Herle's statement about the importance of capturing the underlying behavioural mechanisms, recalling the paper by Burland and Hancock (1977), which discussed a numerical prediction of heave behind a retaining wall following excavation for the

House of Commons car park. She said that the reasons had been well documented and summarised that this was a function of the elastic soil model combined with the soil being 'stuck' to the wall in the numerical model, which was known not to happen to such an extent in reality. She thought that the modelling technique of the predictor should be taken into account in evaluating the predictions, so that one could eliminate results from those who had made inappropriate assumptions or had used unusual methods of analysis. She offered a more optimistic view by suggesting that a narrower band (of results) would emerge, so that a revised review of the prediction exercise would not be quite so pessimistic overall.

Cino Viggiani thought that these prediction exercises, when modelling more or less realistic boundary value problems, could be instructive if categories of geotechnical problems could be classified. He acknowledged that it was not by chance that excavation and tunnelling boundary value problems were investigated because predictions were difficult for such cases. He thought that this was when the differences between constitutive models came into the play, as opposed to other problems in which the results were less sensitive to the constitutive ingredients in the theories. Herle commented that the problems presented in the prediction competitions were focused on relatively simple cases. In reality, he noted that designers today were expected to perform predictions of settlements due to excavation within a tunnel when there was a footing load at the edge of a nearby excavation and some dynamic influence of the train passing just behind it! He said that we should admit we are not able to predict such cases, noting that we continue to struggle with simple cases. He was sceptical about adopting any form of categorisation, although it could help to organise the topic.

Jan Laue proposed that the prediction contest based on a sheet pile wall in von Wolffersdorff was not simple because a large number of stages were modelled, and the construction method was difficult to capture. His impression was that the comparison was not as easy as had been suggested. He queried Herle's statement that computational results should not be compared to field measurements in terms of trying to fit predictions to data and wondered how the calculation result would be proven otherwise. Herle corrected any misunderstanding by saying the qualitative aspects of the measured behaviour and of the calculations should be investigated more than a comparison between the precise measurements and the calculations. He recalled the presentation by Jitendra Sharma, who had preferred not to lend weight to the exact measurements in the centrifuge in some cases, but to concentrate more on the qualitative behaviour. Herle thought that this was also true for computational methods, in that mechanisms could be revealed to try to understand what was happening, but that the quantitative results of the calculations should not be believed to be 100% accurate (compared to a prototype situation), because there were so many aspects, which could not and would not be captured.

David Muir Wood questioned how the quality of a prediction made from a computational model should be judged? He commented that in the past this had been carried out on the basis of the settlement of one point on the wall or an embankment, whereas the real behaviour and the computational model both included, for example, the deformed profile of the wall. He noted that the quality must depend on what happened behind the wall, which could not actually be seen in reality. He mooted that an overall system performance should be replicated, rather then a single observation on the surface, to be confident about the prediction. He was impressed with Ivo Herle's frankness about the inability of numerical modelling to produce consistent data, in that even if they were using the same model, on exactly the same problem, and using the same parameters, they still got different results! He thought that there should be a complementary comment to make for centrifuge modelling. If exactly the same experiments were carried out in Cambridge, Delft or Zürich, using the same soil, with the same loading arrangement, he wondered what would be the spread of results obtained? Bolton answered that there was a ±15% variation in cone penetration tests (CPT) and footing tests, according to previous European collaborations. Herle was impressed with this outcome, citing shear box tests carried out in several laboratories in Germany on one material and under the same conditions, producing a deviation of the friction angle of at least 5° or more.

Andrew Schofield commented that quality assurance became necessary when a failure occurred, and that all data records would be audited and inspected, and some court action would be taken, underpinned by the question of what had been assumed and on what basis a product had been sold. He was enthusiastic about the old graphical static methods, based on transferring information from site measurements and from survey books onto a large sheet of drawing paper. He noted that links between the mechanism of failure (e.g. slip plane or slip circle) and the survey region were easy to crosscheck. He felt that this confirmed an internal consistency of the data records and that if a failure occurred, this drawing would be the basis of the subsequent audit. He supported holding such competitions for audit purposes. Herle agreed and commented that these classical methods forced everyone to look more into the mechanisms, but it was possible to get virtually any answer from the numerical methods, without worrying too much about underlying mechanisms.

Michael Heil followed up David Muir Wood's point about the difference in predictions when using identical models. He suspected that FE calculations were viewed as being the 'finished' product, always as leading to 'good' results, which was not true. He called for separation between judgement of the dependency of the results on the numerical quality of the codes as well as on the appropriateness of the selection of the constitutive model and associated parameters. He felt that the numerical aspects were important, and were often ignored by engineers.

2 MEASUREMENT OF DISPLACEMENT – TRENDS OR NUMBERS? (J.S. SHARMA)

2.1 *Discussion*

The discussion moved on to reflect the second theme lecture, which had been presented by Jitendra Sharma. Heil drew a parallel between the increasing complexity and sophistication in centrifuge modelling equipment and a similar recent development in cone penetration testing. He was convinced that time would tell what was a good development and what was not. Sharma replied that he was not worried about the sophisticated tests as such, they could be very good indeed, and that he was excited by the new developments but was instead worried about how to make sense of the emerging data, i.e. the back analysis.

Andrzej Niemunis took issue with the original title of the presentation, 'measurement of deformation', stating that using the word deformation was a misleading use of vocabulary. He suggested that *rate* of deformation should be the basic variable, so that the integral of this would be the deformation. He suggested that deformation should not be treated as a state variable in any constitutive model. Herle acknowledged that this had also been the title of his lecture. He agreed that deformation or strain could not be state variables because they did not describe a unique state, but on the other hand, the notion of deformation was a very useful one. The calculation would be started at some reference state, which would not be stress free. Since the constitutive model should not be linear elasticity, stress free states would not be necessary!

Sharma recalled that for a complete description of a physical system, knowledge of both load as well as deformation were required. He felt that it was not necessary to be fussy about the definition of deformation because deformations were what were actually seen in reality and this was what would be recorded in a physical test. Niemunis replied that even if *displacements* were recorded, it was still incorrect to speak about deformation. Muir Wood responded that the English usage of the words deformation and displacement was sloppy and suggested using the word displacement in future to remove most of the ambiguity.

Jacques Garnier commented on the question of scatter from modelling in centrifuge tests. He said that Malcolm Bolton had started to discuss this issue with reference to two different sets of results, the first example resulting from a European co-operation. He described the first programme on shallow foundations and the second one on CPT testing and he confirmed that all aspects of model preparation and test execution were well controlled in both cases, and that the scatter was, as noted previously by Malcolm Bolton, about ±15%. On the other hand, he mentioned a more recent, and less successful, international co-operation led by Japanese colleagues and the International Society (ISSMGE) Technical Committee 2, in which simulations of tests on shallow footings had been attempted. He said that the scatter was much larger than ±15% because details such as the sand density, which mainly governed the behaviour, were not so well controlled. He wondered whether this observation could be linked to the German shear box tests mentioned by Ivo Herle and suggested that the 5° difference (in ϕ') was due to differing densities of the samples.

Garnier was also concerned about the classification of the problems and suggested adopting different ways of doing this. He felt that there were reasons why excavation processes, slopes and ground response to tunnelling were more difficult to predict using numerical modelling than foundation behaviour. He mentioned that the latter case usually only involved a loading process with increasing stresses, whereas in the other classes of problems, such as excavation, slopes and tunnels, some soil would be unloaded, which would produce a different response than for a straightforward loading process. He suggested that it was more difficult to predict unloading events since the response was more dependent on the stress history of the soil. He moved on to moot a second classification as either prediction under working load conditions, far from failure, or whether prediction of the ultimate load or failure mechanisms was required. He thought that FE analysis would be more interesting for calculations at working load than for predicting the ultimate load, since most of the strains would concentrate in small bands, which would challenge FE models without an adaptive meshing capability, where the mesh could be refined throughout the process.

Andy Take countered Jitendra Sharma's scepticism of the ability of image-based deformation measurement methods to provide quantitative data from centrifuge tests. He noted that the precision of image based measurement techniques was defined in image space (pixels) and not in object space (e.g. mm), so no limit to the precision of Particle Image Velocimetry (PIV) could be quoted in object space. He confirmed that this precision was therefore a function of the pixel size, so that keeping the pixel size small (therefore increasing precision) would lead to a sacrifice in the field of view, although much smaller displacements could be detected. He commented on methods for providing clay soils with texture, in that the PIV patch size should be related to the texture density regardless of whether the soil was clay or sand. He pointed out that if a patch were smaller than a soil grain, problems would develop in trying to track that movement since a non-continuous displacement field would emerge, and added that in imparting texture using a powdered material, it should be spread out to such a density that there would be sufficient texture for the PIV method to operate. He contradicted Sharma on the issue of numbers versus trends, warning that intrusive measurement techniques, in contrast to non-intrusive image-based methods, would change what was actually recorded, mentioning earth pressure measurement in this respect. In relation to pore pressure transducer (PPT) measurements, he acknowledged that the transducer was also an inclusion, but felt that it would change the value measured to a lesser extent than for earth pressure measurement. He thought that problems with image based deformation measurement arising from friction at the soil-window interface would be of a similar magnitude to those for PPTs (i.e. manageable) and concluded by saying that the truth was a lot closer than the speaker had given credit for. Sharma agreed that rigid inclusions would distort the stress field around these models. He also noted that the deformation measurements recorded by image processing in a plane were influenced by the stress distribution on the window, and thought that establishing the normal stress distribution at the wall was almost more important than worrying about the frictional characteristics of the interface.

David White made a more general comment on Sharma's conclusion that it was tempting to get carried away with too much new technology, without understanding the implications of new modelling and measurement techniques. He felt that the improved precision derived from PIV was a prime example of a useful technique that could be exploited erroneously. Although PIV allows movements within digital images to be measured to a very high precision in image space (pixels), he acknowledged that this improved precision would reveal errors which were previously insignificant. He cited the various sources of image distortion as possible sources of error if the conversion of displacement data from image space to object space was not carried out correctly. However, he felt that the close range photogrammetry implemented by Neil Taylor (and colleagues) at City University (Taylor et al. 1998) provided the solution, but required additional experimental methods and understanding, which were even more unfamiliar to geotechnical modellers than PIV.

Ryan Phillips followed on from the comments of David Muir Wood in distinguishing between quality and accuracy. He suggested that Jitendra Sharma had been talking about accuracy of measurement and referred to Andrew Schofield's remarks on quality and auditing of data. He wondered if 'the wrong thing' was being accurately measured and pleaded for regular comparison e.g. between centrifuge models and field tests. He cited the example of the piles and the stabilising strap added to a platform in the Bass Straits (Australia) to improve the response to storm loading and said that deformations had been very reasonably predicted for an onshore lateral pile load test, based on centrifuge modelling. He had noted Saiichi Sakajo's (Sakajo 2002) predictions of deformation to failure for uplift piles, which had also seemed reasonable. He concluded that comparisons between the output of numerical and physical models, which had a different set of assumptions, indicated the quality of the modelling methodology and not how accurately measurements could be made.

Sharma thought that the quality, in terms of obtaining excellent predictions or agreement, would come when the numerical modelling represented and quantified all the effects experienced in the centrifuge, mentioning those who adopted plane strain or axially symmetric geometry (perhaps inappropriately) for representing extremely complex physical tests. He doubted that the quality would be improved if it were not possible to quantify these effects although he thought that improving the accuracy of physical measurements would lead to higher quality results, unless important boundary effects had been ignored completely.

3 SUMMARY

Sarah Stallebrass set the scene for the discussion session on deformations. She proposed that constitutive and centrifuge modelling included a hierarchy of testing, normalisation, framework, mathematical theories, prediction and evaluation, which had been developed in a sequence of characterisation, prediction and evaluation. She thought that the forthcoming presentations focused mainly on the testing, normalisation, and framework 'end of the market' rather than on the mathematical modelling or evaluation and that this represented the current interest in more complex types of materials (e.g. natural soils, behaviour of sand associated with deformation of particles and rearrangement of particles).

4 TIME EFFECTS IN CREEP OF SANDS (E.T. BOWMAN)

4.1 *Introduction*

The speaker introduced her triaxial tests, which aimed to investigate creep in sands in the context of pile set-up, in that greater shaft capacity developed over time (e.g. up to 5 years). She measured creep of 3 different sand-sized materials (angular quartz sand, shelly beach sand and uniform glass balls), each sheared in triaxial compression and extension. Some materials dilated immediately whereas others underwent compression initially before dilating.

She questioned why this happened and whether this response was a function of particle shape or material characteristics, since compressive behaviour only developed for more angular particles. She suggested that long-term set-up was related to particle shape and postulated that this would arise from the combined effects of breakage, locked in stresses, particle rotation and dilation. She was critical of continuum constitutive models based on curve-fitting to laboratory test data, if the micro-mechanical aspects of particulate soil were completely ignored.

She took issue with the centrifuge modellers by criticising model preparation methods and the fabric of pluviated centrifuge model samples, having looked at an impregnated cross section following one dimensional loading and unloading. She showed a typical fabric following pluviation in which the majority of particles are lying with their long axes in a horizontal plane. She demonstrated the effects on fabric of loading, showing particle rotation during loading, with clear differences apparent between samples unloaded immediately and after various time intervals. She concluded that time effects were important in the way particles arranged themselves under load, and challenged the constitutive and centrifuge modellers to take more account of the micro-mechanical aspects of soil behaviour.

4.2 Discussion

Cino Viggiani responded to the issue of using a continuum mechanics model for a material that was not a continuum by referring to the introduction of a book written by Truesdell & Noll (1965) called "The non-linear field theories of mechanics". He noted that it was possible to develop a continuum mechanics theory for metals, even though this material was not a continuum, but was doubtful that it was necessary to look at the crystals of a metal.

Michael Heil was unconvinced by the conclusion that all samples would dilate based on the results presented, and singled out the triaxial extension data. He asked how the samples were prepared and how they behaved when sheared, before permitting creep to occur? Bowman referred to the paper (Bowman & Soga 2002) and confirmed that both the angular quartz sand and the shelly beach sand compressed during the shearing phase and then they dilated during the creep phase after some delay, whereas the uniform glass balls dilated during both the shear and creep stages. She added that the soil samples compressed, even when being loaded in extension, but that they all dilated towards the end of the creep stage.

Jan Laue commented on evidence of creep under shallow foundations on sand, which was quantified to a certain extent by the European CPT and footings programme. He explained that this was the reason why the centrifuge was accelerated, then stopped for a certain amount of time and accelerated again prior to any CPT or footing tests being carried out. He confirmed that the resistance derived from CPT tests, which were carried out after the foundation tests, was more or less the same after a certain number of cycles. He felt this was partly responsible for the 15% scatter. Bowman expanded on this point, citing the drop in CPT resistance obtained in comparison with initial conditions to that measured after soil densification by blasting or vibratory compaction, but noted that there was an increase after 6 months or so. She concluded by saying that the CPT was clearly useful for estimating density, but that microstructure should be considered as well.

5 A FEW OBSERVATIONS ON THE RHEOLOGY OF POWDER-BINDER SYSTEMS (L. DIHORU)

5.1 Introduction

The speaker showed some experimental results for characterising the flow behaviour of suspensions formed from granular materials, focusing on viscosity as a key parameter in describing the behaviour at macroscopic level. She explained that the micro-rheological aspects, such as particle interactions, separation phenomena, wall effects and redistribution of particles in flow had to be investigated in detail, because they could all have a major impact on viscosity.

She had studied various powders with a range of mean diameters, size distribution and shapes, which were embedded in a binder system. She pointed out that the binder viscosity did not change at the testing temperature and that there were no chemical interactions between the powders and the binder. She had neglected Brownian forces because the particles were larger than 2 microns, therefore only viscous forces came into play. She had chosen fairly 'irregular' powders, but also powders with particles very close to a spherical shape, to investigate how the particle characteristics affected the evolution of viscosity at various shear rate regimes.

She described the preparation of these powders in a torque rheometer, which could be used to measure the mixing torque of a suspension as a function of the volume fraction of particles added to that suspension. Above a critical volume fraction, the mixing torque was shown not to stabilise and the viscosity of mixture was seen to increase very steeply, so all the suspensions subsequently tested in the capillary rheometer had a volume fraction lower than this critical volume fraction.

She explained the principles behind the capillary rheometry methods, in which the materials were forced to flow through a very narrow tube. She noted that the response would have been less complex had the material behaved according to Newtonian principles, mentioning the advantages of the uniqueness of the known function in shear stress and shear rate, and that the distribution of velocities of the cross section of the tube would be parabolic. However, she said that most materials, and particularly her suspensions, did not behave in this (Newtonian) way. She thought it was tempting to assume that shear thinning behaviour and decrease of viscosity occurred at micro-rheological level, with particles assumed to align along the flow lines, accompanied by some breakage of agglomerates and release of binder retained in these agglomerates.

She listed other factors that would cause a departure from ideal conditions, such as wall effects due to a layer of lower viscosity near the wall of the capillary tube, since the particles tended to migrate towards the centrelines of flow. In consequence, she said that the true shear rate was actually higher than the apparent shear rate measured with this device due to the changes in viscosity, necessitating corrections to calculate the true shear rate. The wall effects led to a non-zero velocity at the wall, so that a factor needed to be added when integrating the velocity gradient of the cross section of the tube to account for the extra shear stress at the wall. She also mentioned the pressure drop in the barrel of the viscometer as well as a pressure drop in the tube itself, both of which have to be accounted for. She showed details of corrections made for deriving the true shear stress and the barrel pressure drop (using tubes of different lengths), that allowed the true shear rate to be calculated, by plotting the true shear stress against the apparent shear rate.

The results confirmed that shear thinning behaviour was taking place as the shear rate increased (Fig. 1). This figure also shows the influence of solids loading on viscosity, together with phenomena of particle-binder separation and particle agglomeration, which occur for high concentration of solid particles (Dihoru 2000). Larger solids loading led to a very steep reduction in viscosity, so that both tests involving concentrations beyond the critical threshold determined by means of torque rheometry had to be aborted. Dihoru considered that a particle in flow had both a rotational and translational motion, and postulated that the torque associated

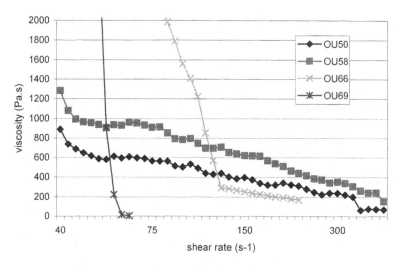

Figure 1. Evolution of viscosity in a variable shear-rate regime for feedstocks made of bi-component mixtures of powders (O - (40% solids by mass) irregular particles, U - (60% solids by mass) spherical particles, OU50- mixture with 50% vol. solid particles, constant temperature of 160 °C). (Dihoru 2000)

with a bigger particle was greater, therefore the surrounding layer of binder around the particle was subjected to larger shear stresses. She thought that this increased torque would increase the risk of separation between the two phases, whereas more irregular particle shapes would provide more friction, which would lower the risk of separation.

The evolution of viscosity was also shown for suspensions made of mono-component as well as bi-component powders of different size distributions. Enlarging the size distribution in a suspension increased the risk of size segregation in flow at the same shear rate, and indicated that the viscosity would be less stable.

The speaker emphasised that the interplay of the factors influencing the viscosity of a suspension made of particulate materials and a binder was extremely complex, and asked how the audience would be tempted to model viscosity once so many parameters had come into play. She posed a secondary question about the validity of using statistical methods and pattern recognition techniques for predicting the evolution of viscosity.

5.2 *Discussion*

Sarah Stallebrass called for thoughts on modelling viscosity and wondered whether 'creep' was happening or not. She also asked whether the tests were carried out under constant deformation rate or constant torque. Dihoru replied that it depended on the shear rates, and thought that creep would be an acceptable description of response at the lower shear rate regimes. She added that there were several types of tests, either a constant shear rate could be adopted with the velocity of the ram/plunger held constant, or the velocity could be varied for working at variable shear rate regimes.

Michael Davies presented a case for centrifuge modelling in commenting on work he had carried out on a large-scale silty soil model in a 5 m by 5 m by 3 m test chamber to investigate solifluction-gelifluction processes. These took place during freezing and thawing cycles over a period of a couple of years. Originally he said that he had supported the general belief that gelifluction was a viscous process and that viscosity-based models were appropriate for this type of problem. However, he said that centrifuge model tests were carried out for a range of gravities for comparison with the large-scale 1g test, and that subsequent numerical modelling had predicted the deformation rates to within 10-15% of the field data only when an elasto-plastic constitutive relationship, rather than a viscous constitutive model, had been used. Centrifuge modelling had shown that an intuitive type of approach was appropriate and had been used successfully for looking at processes to explain what was happening in nature. Stallebrass summarised this contribution as suggesting that viscous processes were essentially elasto-plastic and that these viscous processes should be modelled with an elasto-plastic continuum model.

Malcolm Bolton expressed puzzlement about the need for modelling the dissipation of energy within what seemed to be a slurry, because the fluid paths around the particles and the rotations and the velocities of the particles were different, and wondered if the problem was not due to material characterisation. He asked why a simple annular viscometer hadn't been used so that the slurry had been characterised in a standard way. Dihoru explained that the materials were to be adopted for further development of injection moulding techniques, which were mainly used for manufacturing very small and intricately shaped parts, and hence the interest was in how the material flowed in restricted zones at sub-millimetre scale. She said that the quality of the powder injection moulding process was enhanced when the solids loading in the mixture was as high as possible. Dihoru replied, to a point of clarification from the Chair, that the materials were more like pastes (with a consistency of a toothpaste) than slurries, to which Stallebrass suggested that there would be more particle contact in this case.

Vincenzo De Gennaro pointed out that this problem was very similar to the rheology of debris flows, for which a mud or slurry would flow very rapidly, while bringing big blocks with it. He commented on the paradigm of apparent stability and extremely instability and thought that some approaches in terms of visco-plastic behaviour might be applicable to the speaker's problem.

6 EXPERIMENTAL INVESTIGATIONS ON SAND UNDER CYCLIC LOADING (G. FESTAG)

6.1 *Introduction*

The speaker introduced his topic, which was relevant for the cyclic loading of road or railway foundations. He had tested a weak angular gabbro sand in dry conditions in triaxial compression for up to 4.4 million load cycles, with an isotropic stress of 150 kPa and an additional deviatoric stress of 150 kPa applied at 1 Hz frequency. He noted that there had not been uniform behaviour of settlement or strain, which could have been described by continuum mechanics, but that cyclic shakedown had been evident on several occasions during the loading history. During some periods, the strain had remained constant, whereas at other times, additional plastic strains had developed from one cycle to the next.

A single (yellow painted) grain was observed under the microscope after a 4,500,000-cycle test. There were two locations at which damage had occurred, including a new, smaller particle close to the original grain. Having washed the grains before the test, he had been sure that there were no such small particles in the sample at the beginning of the test and so it had to have been produced while loading. Having observed this particle breakage, the speaker investigated whether this particle abrasion led to a change in mean particle shape. He defined the coefficient of roundness as the ration of the projected area of a grain to the perimeter and showed that (based on the average of 600 grains), they had become rounder by the end of the cyclic loading. He noted that most constitutive models assumed that the material would maintain a constant friction angle, which was not dependent on the loading history, whereas this investigation has shown that the grains became more round during maintained cyclic loading and therefore that the mobilised friction angle would be smaller.

To prompt discussion, the speaker asked whether there was a way to capture this inconsistent behaviour within classical constitutive models in the framework of continuum mechanics and to replicate the microscopic observations of the effect of the change of grain shape due to loading, or was it necessary to find another framework?

6.2 *Discussion*

Malcolm Bolton commented that this problem was of prime importance to railway authorities, which require their railway lines to be maintained at elevation within a tight tolerance. He reviewed the mechanism by which pores in the ballast sub-base were blocked by small flakes of broken gravel washed from beneath the sleepers, which supported the railway tracks, leading to a loss of drainage. When drainage fails, the characterisation of the problem becomes completely different, as the saturated ballast liquefies under the cyclic loading of the wheels of the railway trains passing over the top. The continuum was altered because small particles could be washed around by internal erosion and would collect in one place.

Bolton felt that it was not always necessary to make a model for prediction. What was wanted was a creative response. He had suggested that road and rail authorities might consider making granular beds out of well graded rather than uniform sizes in future, but that this had led to reactions of complete horror! He wondered whether this was because the authorities had simply got used to adopting a particular style of only using large stones, or whether there may have been a particular reason for this (Ed. frost heave in some more wintry countries). He thought that it would be better to offer practising engineers a possible way forward by which research could save some of the hundreds of millions of dollars wasted every year on replacing granular aggregates under rail beds and roads, instead of simply trying to model the situation with constitutive relationships.

Sarah Stallebrass differentiated between particle abrasion, in which broken asperities formed new, small particles at relatively low stresses, compared to fractal breakage (into more or less equal smaller size fractions). She wondered whether these two modes of breakage should be accommodated in a constitutive model. Festag replied that his conclusions derived from abrasion of small asperities and not from breaking of grains into equal parts. He emphasised that the common assumption that a constant settlement was reached under low level maintained cyclic loading was not true.

Tom Schanz questioned how the speaker was able to measurement the roundness of 3000 particles, which Festag clarified as being a computer based technique using a high-resolution scanner with a resolution of 2000 dots per inch. He said that a grain of about 1 mm diameter filled about 800 or 900 pixels, and that software was used to find the area and perimeter of each scanned particle in terms of pixels so that the roundness coefficient could be calculated.

Lis Bowman queried the change in particle roundness observed during loading, having discussed additional measurements made by the speaker on quartz sand, which showed that this material abraded less than the gabbro sand. She wondered whether this might have been a function of the material type, in light of her own results from crushing tests, which found that different sands crushed in different ways: quartz sand produced more angular particles whereas cemented sands produced more rounded particles. She had characterised this using Fourier descriptor analysis techniques, which gave additional information beyond roundness, and asked whether the speaker had considered other descriptors. Festag responded that he had made tests on other materials (e.g. rounded quartz sand from the Rhine) and hadn't found such extensive abrasion of small particles, whereby the quartz sand broke into parts of more or less equal area. He acknowledged that the response to be expected would depend on the crystals and texture of the material.

7 EFFECT OF INITIAL FABRIC ANISOTROPY ON CYCLIC SHEAR CHARACTERISTICS OF DENSE SAND (K. SATO)

7.1 *Introduction*

The speaker referred to a paper published by Sato et al. (1999). He presented these investigations into the effect of initial fabric anisotropy on cyclic shear characteristics of dense saturated Toyoura standard sand, in which the angle between the major principal stress direction to the horizontal bedding planes in the sample was varied in undrained torsional shear tests using a hollow cylinder apparatus. He stated that the inclination of the principal stresses would change during an earthquake, depending on the location of the soil element relative to the structure.

He described how the specimen was prepared by air pluviation to achieve a relative density of 80% with dimensions of height as 200 mm and inner and outer radii of 30 mm and 50 mm respectively. He noted that this produced an initial fabric anisotropy. Several loading schemes were applied in the undrained cyclic shear tests, with a cyclic stress ratio $\tau/p'_c = 0.2$ to 0.35, where τ and p'_c were the shear stress applied to the top of the cylinder and the isotropic consolidation pressure respectively. The angle α defined the inclination of the major principal stress to the vertical axis at the beginning of the sequence of cyclic loading and this ranged between 15° and 75°. The effective stress paths for $\alpha = 15°$ and $\alpha = 60°$ for $\tau/p_c = 0.25$ showed that the rate of decrease of effective stress (and increase in pore pressure u) varied in that more than 20 2-way stress cycles were required before liquefaction occurred for $\alpha = 15°$, whereas this only required 2 2-way cycles for the test with the major principal stress more closely aligned with the bedding planes ($\alpha = 60°$). The magnitude of the decrease in effective stresses over the first cycle was greatly influenced by the initial fabric anisotropy.

The effect of initial fabric anisotropy on cyclic strength was quantified at each value of τ/p'_c in terms of the number of cycles required to mobilise a specified amount of maximum shear strain (e.g. $\gamma_{max} = 5\%$). The number of cycles (N) tended to decrease with increasing α, with $\alpha = 60°$ providing the minimum value ($\tau/p'_c = 0.25$, N = 4, $u/p'_c \sim 0.95$). He concluded that the cyclic shear strength of sand tended to decrease with increasing rotation of the major principal stress direction from the vertical for the first cycle of loading and that the cyclic shear behaviour was determined by the major principal stress direction at the beginning of cyclic loading and the initial fabric anisotropy.

7.2 *Discussion*

Sarah Stallebrass asked for an explanation about the implication of the results in practice for example for the design of a pile. Kenichi Sato replied that it was relevant for design for earthquake loading. Fook Hou Lee asked for clarification about whether a normal stress and a shear stress had been applied to the ends of the sample to create an inclined major principal stress. Sato confirmed that this was true, and that the degree of inclination of the principal stress was controlled by torsion, axial stress and the internal and external pressures. Lee sought further clarification about the difference between the strength and stiffness determined for the various specimens, wondering whether this could be explained by the inclination of the major principal stress to the bedding planes or due to the rotation of the principal stresses. It was later confirmed that the major principal stress direction was held constant for each set of cycles at a specific value of τ/p'_c.

Andrzej Niemunis thought that structure would be conserved more if undrained tests were performed because the volume would be constant, so that the orientation of contact planes would be more likely to be preserved. He wondered if this observation could be extended to drained tests for small cycles on dry sand and asked whether the settlements would be the same and would preclude pore pressure build up at the same time. David Muir Wood commented that the structure was controlled by initial deposition and thought that it didn't seem to matter if the conditions were undrained or drained, in terms of the disturbance caused to that structure by applying a rotation of the major principal stress. He felt that by applying shear stresses to planes that had not experienced shear stresses in the past, some collapse would occur. He confirmed that larger deformations would be expected as the axis was rotated under both drained as well as undrained conditions, and this was seen in both simulations and in experimental work.

8 DEFORMATION BEHAVIOUR OF A SOFT SWISS LACUSTRINE CLAY (J.L. TRAUSCH-GIUDICI)

8.1 Introduction

The speaker introduced two practical problems, both of which were related to construction in or on deep layers of normally consolidated soft lacustrine silty clay (Seebodenlehm). One site was located in Kreuzlingen in the north east of the Swiss plateau, for construction of a railway underpass for a link road for which displacements behind the wall due to the excavation could have been critical. The other site concerned a complicated highway intersection in Birmensdorf, near Zürich, which was to be built on high embankments on a Seebodenlehm layer. The clayey layer extended from a few metres in depth to more than 30 m. Problems expected here were the differential settlements, the magnitude of the final settlements as well as lengthy drainage times to dissipate the excess pore pressure.

She commented that the models used in practice to describe and quantify soil behaviour were generally unsatisfactory in capturing the overall response, so that the aim was to upgrade the capability to be able to model the deformation behaviour of this normally consolidated lacustrine clay in states of both pre-yielding and yielding. She hoped that this would lead to more economic design.

She showed the classification details of the Swiss silty clays and compared their characteristic properties with a sensitive and more plastic Finnish clay from Otaniemi. She also demonstrated the inhomogeneity of the triaxial specimens of the Swiss clay, which had fine horizontal silt layers alternating with the clay, and variable water (22-30%) and clay (28-42%) content within these layers. She concluded that the position and thickness of the concentrated silt layers had influenced the stiffness measured, the failure mechanism observed, as well as the critical state strength of these triaxial specimens.

She presented results of 3 recent multistage tests on 3 specimens from Kreuzlingen clay, which had been designed to explore the erasure of fabric anisotropy during plastic straining. These included excursions along two different drained stress paths, each time expanding the yield locus, followed by undrained shearing to failure. She found good agreement with the critical state parameter in compression (M = 1.08) from previous tests, but the volumetric behaviour was not at all similar. The sheared ellipse adopted by Wheeler and co-workers in Glasgow for the SCLAY1 anisotropic elasto-plastic constitutive model (Wheeler 1987) for soft clay was fitted to the points in q-p' space[1] where yielding began. Rotational hardening for the yielding behaviour, which depended explicitly upon stress path direction as well as both plastic volumetric and shear strains, was a key feature of this model and showed reasonable agreement with the predicted ellipsoidal yield surfaces of SCLAY1.

She was concerned about the method used to extract constitutive parameters from her laboratory tests in order to conduct numerical analysis. Her concerns started with the representativity of these 3 specimens, which were extruded from a sample of only 80 cm in height, noting that the sample structure contained several silt layers. She had wanted to be able to describe the response of the layered material using a single model with one set of constitutive parameters, but

---

[1] $q = \sigma_1 - \sigma_3$; $p' = (\sigma'_1 + 2\sigma'_3)/3$

knew that this could not cover the range of volumetric behaviour shown by the samples, let alone a 20 m deep layer of Seebodenlehm.

She also thought that since the soil was not behaving as an isotropic elastic material inside the yield surface, it would be very difficult to defend a bilinear fit to a graph of stress versus strain to obtain the yield point. She posed a question about how the pre-yield stress path could influence the rotation of these yield loci if it was assumed that elasto-plastic non-linear straining occurred from the beginning of the new loading path.

8.2 *Discussion*

Claudio Tamagnini questioned the size of the sample in terms of a representative elementary volume (REV), in that the material displayed a microstructure, and probably a mesostructure, with visible microscopic features such as bedding planes and stratification. He returned to the need for a suitable length scale in any application of continuum field theory to an inhomogeneous material. He commented that the length scale should be sufficiently small with respect to the size of the problem and sufficiently large to contain all the mesoscopic and microscopic features in the REV, and suggested running tests on much larger samples and specimens. Trausch Giudici replied that she had been limited by the size of the platens of the triaxial apparatus (diameter of 5 cm) and the undisturbed samples (diameter of 6 cm), which precluded an increase in specimen size for this test series. She also thought that it was difficult to choose an appropriate length scale because the structure was random in that it was possible to find 80 cm of almost pure clay as well as samples with closely spaced silt layers. She favoured testing in situ, where possible, to add to the knowledge obtained from the laboratory tests. Muir Wood agreed with Tamagnini, pointing out that the test sample size should be big in comparison with the variability in the natural material, or that a lot of tests should be carried out, so that any variability could be smeared out. He accepted that sometimes it was not possible to do either of these two things, in which case a statistical optimisation was required to fit a model to the results obtained, and acknowledged that this was the paradox of laboratory testing.

Tamagnini also proposed that a generalised plasticity (bounding surface) model by Dafalias or a generalised plasticity model by Pastor could be adopted instead of a classical plasticity model, even if the latter considered anisotropic hardening. He pointed out that using classical plasticity would lead to inadequate response inside the bounding surface if the material was not allowed to develop irreversible strains. He noted that an equation would be introduced within a generalised model, which would link the evolution of an anisotropic variable to the amount of plastic strain and the amount of plastic strain within the bounding surface would then modify the nature of the internal variable.

Malcolm Bolton wondered about the relevance of the mass permeability of the layered silty clay, which would have concerned Peter Rowe greatly, in that he discussed this type of problem extensively in his Rankine lecture (Rowe 1972). He commented that the particular application mentioned by the speaker was an excavation, in which case the more permeable material would lose effective stress more quickly, which would not be advantageous. He suggested that the way forward would be to perform insitu permeability tests, for example with the self-boring permeameter, which has been developed by Cambridge Insitu and Kenichi Soga, and then to choose the most permeable profile for the characterisation. He didn't think that it was the details of the plastic behaviour viewed as an undrained problem that would be the critical problem, but the drainage. Trausch Giudici replied that quite a number of insitu permeability tests had been carried out on site. (N.B. the triaxial stress path tests had been carried out following a *drained* path during the stress path excursions until the final *undrained* shearing to failure).

9 FIRST RESULTS OF TRIAXIAL CREEP TESTS ON PERMAFROST SOIL SAMPLES (L. ARENSON)

9.1 *Introduction*

The speaker introduced his research into mountain (alpine) permafrost at elevations of ~2500 m above sea level. He explained that rock glaciers, typically of the order of several hundred metres long and to 300 m wide, were geomorphological phenomena of permafrost, which crept downslope at velocities of up to ~0.5 m per year. He showed where boreholes had been drilled in the Muragl rock glacier, together with a cross section showing an active layer of up to 5 m, which could be susceptible to instability in a warming climatic scenario. Permafrost was found below this to a depth of approximately 20 m, with unfrozen sediments underneath and a groundwater table lying just above bedrock. Inclinometer measurements confirmed that a distinct shear horizon had developed at the lower boundary of the permafrost, so that around 80 – 90% of the creep deformation took place at this depth.

He displayed pictures of frozen soil samples, and noted that some unfrozen water would have been present at temperatures of just under 0°C. Results were presented from triaxial creep tests on such samples under several deviatoric loading conditions. He identified the critical point from plots of strain or strain rate versus time where the strain rate increased and hence tertiary creep had commenced, which had directly followed the primary creep phase, defined by a decrease in strain rate. He noted that secondary creep, represented by a constant strain rate, was reduced to one point and compared trends in the development of the creep strain at fixed times of 20 h, 40 h, 60 h or 80 h. He showed that, despite the heterogeneous material, the axial strain rate related linearly to the axial stress at semi-log scale. He challenged the audience to explain where the secondary creep had disappeared to and wondered whether it could have been due to a feature of the apparatus or the microstructure.

He also commented on two traditional types of creep model, including the simple power law introduced within glaciology by Glen (1955), and the associated debates about the power factor *n* as well as the relevance of a model based on an exponential component of the stresses. He noted that the former would be considered to be a rather simple model, whereas the latter was a more advanced model, and he wondered about the degree of sophistication necessary and appropriate for such a heterogeneous soil.

9.2 *Discussion*

Sarah Stallebrass noted that the soil was close to melting and asked if this had affected the secondary creep? Lukas Arenson thought that this might be the case and noted that high stress concentrations would be likely to lead to pressure melting, preventing a steady state from developing in the material, so that there were ongoing changes during the entire creep test.

Ivo Herle wondered whether it wasn't too restrictive to consider a simple 1D model for future modelling of the in situ case. Arenson pointed out that the first goal would be to model the laboratory tests: modelling the entire 3D in-situ case effectively would not be possible yet. Herle still thought that the speaker should concentrate on adopting a model, which could capture the salient effects in terms of axial and volumetric deformation, and this usually wasn't possible with a 1D model. Arenson was not convinced that it would be easy to bring in more complexity without knowing, for example, the effective stresses. He noted, however, that total stresses could be calculated and thought that it was worthwhile starting with a simple 1D model instead of going straight to a 'Rolls Royce version'.

Saiichi Sakajo asked whether testing at higher axial stresses led to a greater tendency for failure to occur due to creep-induced rupture, which Arenson acknowledged to be the case. Sakajo was interested in the mechanical conditions in which creep rupture happened as well as, for example, the ice content and temperature. Arenson commented that failure due to creep-induced rupture was only a function of stress and time. He added that the ice content was about 40 % and temperature was around -2°C, which might affect the failure.

Luc Thorel mentioned that rock salt was another geomaterial that 'crept', so that a damaged sample would lead to rupture during creep, whereas stress levels lower than a damage criterion would only lead to secondary creep. He added that measuring the volumetric deformation was very helpful in spotting advancement towards creep rupture. Arenson was grateful for the suggestion but drew attention to the ability of frozen soils to undergo self-healing, which could influence the potential for rupture quite dramatically.

10 CONCLUSIONS

10.1 *Introduction: an audit process*

David White had reviewed the research presented under the umbrella of problems governed by 'Deformation' with Philippe Nater, Thomas Weber and George Vlahos. He started by acknowledging that Ivo Herle (2002) had described a most useful framework for assessing the validity and the limitations of the output from computational models, backed up by recent results from the European COST C7 project (e.g. Schweiger 2001). These highlighted the large variation in output from computational models of well-prescribed boundary value problems due to operator-selected parameters such as mesh-size and boundary distance. The dangers of embarking on computational modelling without following such a framework were highlighted, and the ease of producing output, which was "heavy on colourful contour plots and light on robustness and validity" when using modern user-friendly modelling software, was noted.

With Herle's framework as a central feature (Fig. 2), White extended a similar assessment process to centrifuge modelling, with an equivalent set of considerations. The selection of an appropriate constitutive model for the 'virtual' computational soil is replaced by consideration of whether the 'actual' soil used in the centrifuge model would replicate the prototype behaviour; scaling laws should be examined, and the preparation methods reviewed. Mathematical and numerical aspects of computational modelling would be replaced by concerns over data extraction methods and physical model geometry. White noted that issues such as boundary distances and geometry simplifications in computational modelling have direct parallels in centrifuge modelling; box boundary conditions must be chosen, and the soil model stratigraphy and construction sequences must be idealised.

Based on the process for assessing the validity and limitations of both centrifuge and computational modelling, he proposed that a framework for making an engineering decision would emerge, which Andrew Schofield had already described as being part of an audit process. He felt that this process could be undertaken for one of two reasons: to improve a prediction for design purposes or to improve the performance of the construction technology. He noted that the centrifuge had been outvoted in the debate as being a sufficient means for design, meaning that both sides in the analysis were necessary, although Martin (2002) had argued that one or the other could be sufficient. In terms of improving performance, he complimented Lee (2002) on presenting a clear example of how performance had been optimised through centrifuge modelling. The installation of sand compaction piles was more suited to physical modelling than computational modelling and this achieved the goal of improving performance, rather than simply improving prediction. In this situation, improving the performance of a novel construction technology was the goal rather than establishing a prediction method.

305

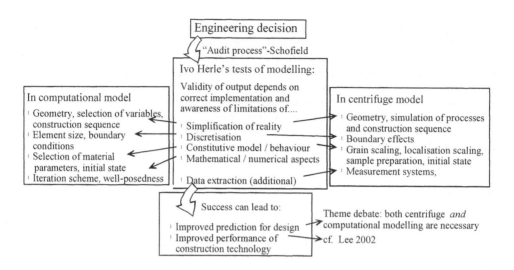

Figure 2. Ivo Herle's tests of computational modelling extended to include centrifuge modelling

10.2 *Numbers, as well as trends*

White identified the corresponding engineering decision in the subsequent presentations by the Young Researchers, including which aspect of the audit process provided the greatest challenge (Table 1). He focused initially on Jitendra Sharma's theme lecture, which reviewed recent progress in displacement measurement techniques in centrifuge modelling, asking whether numbers or merely trends could be deduced. Sharma (2002) cited boundary effects (i.e. interface friction on the viewing window) as the obstacle to improved precision and thought it was fair to state that looking for trends was the best possible goal, when referring to Ivo Herle's framework of auditing and measuring limitations and validity. White suggested that this was overly conservative given recent advances in image-based displacement methods.

Following Herle's audit of limitations (Fig. 2), White concluded that PIV alone, under the heading of 'data extraction', may be insufficient to extract reliable Serviceability Limit State (SLS) deformation, because of image distortion and boundary effects. He noted that previous techniques of image-based deformation measurement, such as spot-chasing (e.g. Sharma 1994), had operated at sufficiently low precision for these errors to be neglected. The order-of-magnitude increase in precision offered by PIV increased the relevance of sources of error which were previously insignificant. However, he added that Matthew Dietz (2000, 2002) had demonstrated the constitutive simplicity with which boundary effects could be incorporated into back analysis or numerical modelling of a centrifuge test and that close range photogrammetry allowed image distortion to be corrected for (e.g. White et al. 2001, Taylor et al. 1998). He pointed out that these advances permitted centrifuge modelling to reveal numbers as well as trends, in examining both serviceability and ultimate limit states.

Table 1. Young Researchers' papers / presentations during the 'Deformations' session

Young Researcher	Research technique?....	....relates to the engineering problem of....
Elisabeth Bowman	Constant stress TX creep	Set-up of displacement piles in sand
Gerd Festag	Cyclic TX creep	Railway ballast settlement under cyclic loading
Jolanda Trausch-Giudici	Multi-stage TX tests	Excavation in an anisotropic clay with silt layers
Lukas Arenson	Constant stress TX creep	Creep of alpine permafrost

10.3 Review of Young Researchers' presentations

White noted that a common theme ran through the next series of presentations (Table 1) in that they were all concerned with the selection of appropriate material parameters, which was one of Ivo Herle's audit tests under the heading of the constitutive model (Fig. 2). He began by introducing pile set-up, noting the complexity of this boundary value problem, which had been investigated by Lis Bowman (Bowman & Soga 2002), who had mimicked dilatant creep in soil around a displacement pile. He was convinced that pile installation could not be captured currently in a computational model because of the very high stress and strain ranges and rigid body rotations inflicted on the soil elements close to a pile during installation. Laboratory testing can not fulfil a rôle as a source of material parameters to describe a full constitutive model.

However he complimented the alternative approach of Bowman, who used the laboratory test itself to simulate the stress path during pile installation. The data from the subsequent creep stage was used to provide constitutive parameters relevant to the regime of soil behaviour exhibited during set-up. This approach had bypassed the computationally impossible installation sequence. He thought that some progress had been made in terms of determining trends to identify mechanisms and micro-mechanisms of set-up and creep under constant load, even though the simulation of pile installation was not exact. White noted that the phenomenon of pile set-up was ripe for exploitation if the underlying mechanisms (or micro-mechanisms) could be identified and accelerated. Such engineering progress would fall under the heading of improving performance rather than improving prediction.

Continuing the review of earlier presentations, White discussed the triaxial cyclic creep tests linked to the engineering problem associated with railway ballast shakedown, carried out by Gerd Festag. These had produced complex hysteretic data of stress-strain paths to which a 12-parameter hypoplastic constitutive model had been fitted. He noted that particle abrasion was identified as the underlying mechanism, using an image analysis technique to measure increasing particle roundness with cumulative axial strain.

White echoed Malcolm Bolton's suggestion that the engineering or construction process could be addressed more effectively by exploiting a qualitative understanding of the micro-mechanism than by introducing extra parameters to improve the fit of the constitutive model. This point could be illustrated as follows, with reference to Festag's study of track ballast degradation. If, after completing a full test programme, all twelve parameters of the hypoplastic model could be quoted with confidence and precision, it might be possible to model the engineering problem numerically, for example, a railbed under the action of cyclic axle loading. A good fit to some field data of a settling embankment might be achievable so that an engineering decision could be made. He postulated the following case and asked some penetrating questions: the operator wishes to reduce or eliminate cyclic settlement. Adjusting some of the twelve input parameters may provide the desired response, but how does this translate to the real world? What is the physical significance of these changes? In other words, how should the operator change his construction materials or his construction process? Perhaps it would then be time to return to the triaxial apparatus, and search for a material, which displays those twelve adjusted parameters; where would he begin his search?

An alternative approach would be to take advantage of the (micro-) mechanistic observations. Festag (2002) linked cyclic compaction to the abrasion of asperities, leading to progressive rounding of the particles. This insight into the mechanism of cyclic creep could generate immediate suggestions for improved performance, e.g. to use more rounded particles, or harder particles. Alternative engineering decisions would immediately present themselves, and quicker progress may be possible. This could be achieved without developing a full constitutive model for the material.

The next design and construction problem concerned an excavation and an embankment on soft clay, which had required Jolanda Trausch Giudici (2002) to deduce appropriate constitutive parameters from triaxial stress path tests. Here White noted that there were limitations in fitting conventional constitutive models to the behaviour of this clay, because it exhibited a very inhomogeneous fabric caused by horizontal silt layers at various spacings. He noted the suggestion made by Malcolm Bolton that the engineering decision to be addressed was dominated by consolidation, so that only compressibility and permeability were important, which would simplify the engineering process considerably. Bolton proposed that the process of material parameter selection should concentrate only on parameters that would influence the boundary value problem under investigation.

White thought that Lukas Arenson (2002) was engaged in the most difficult research project of those presented, in that this was the first attempt to explore geotechnical aspects of alpine permafrost using field monitoring and laboratory testing. In addition, the consequences of climate change on permafrost stability lent this project great ecological and economic significance. He noted that added complexity arose since there was an extra phase in the material (soil, water, air and ice) and one less property to measure in the triaxial apparatus because pore pressure, and hence effective stress, could not be determined. He approved of the pragmatic approach adopted by following a simple loading path and using a basic creep model to rationalise the behaviour. He accepted that there were enormous difficulties in clarifying the behaviour of this material and yet this melting permafrost presented a huge engineering risk, particularly in times of climate change. Arenson (2002) identified a significant increase in creep rate associated with a change in temperature of only $1.3°C$. This result was thrown into sharp relief by recent climate change data, which indicated a mean global surface temperature increase of $0.4°C$ during the period 1975-2000 (IPCC 2001).

White closed his discussion by noting the common theme of each presentation in this session. Each engineering problem was on the boundaries of our current understanding of soil constitutive behaviour. None of the research students could expect to find a constitutive model to fit all aspects of the behaviour of their material. He recalled that the projects ranged from pile set- up in sands to cyclic shakedown of railway embankments, landslides in melting permafrost and anisotropic soft clay, none of which would be found in a text book or design guide.

White countered Ivo Herle's view that, even for a problem governed by deformation, some soil was at ultimate limit state, and postulated that most of the soil within each of the four engineering problems was operating within a small range of its full constitutive behaviour. He supported the pragmatic approach by which the researchers were attempting to solve each engineering problem without recourse to a complete constitutive model. Instead, the testing was tailored to address the engineering decision and to evaluate the relevant properties only. He thought this was rather a daunting simplification for the Young Researchers to admit, given the audience of expert constitutive modellers! Nonetheless, whilst all the speakers were open to criticism for the narrowness of their considerations, each was focusing their research on the engineering decision in hand. He concluded that their awareness of the validity and limitations of this approach was greatly assisted by assessment within the framework proposed by Herle (2002), and provided a valuable and enjoyable day of discussion.

ACKNOWLEDGEMENTS

We are grateful to the Chair, Tom Schanz, and the discussion leader, Sarah Stallebrass, for their strong leadership and inspired guidance of this session, and David White's team - Philippe Nater, Thomas Weber and George Vlahos - for their assistance with the preparation of the concluding summary. We also thank all of the authors and contributors for their patience with our interpretation of their statements, and for their help in ensuring that this record of the discussions is as accurate as possible.

REFERENCES

Arenson, L. 2002. First results of triaxial creep tests on permafrost soil samples. In S.M. Springman (ed.) *Workshop on Constitutive and Centrifuge Modelling: Two Extremes, Monte Verità*. Lisse: Swets & Zeitlinger.

Bowman, E.T. & Soga, K. 2002. Time effects in creep of sands. In S.M. Springman (ed.) *Workshop on Constitutive and Centrifuge Modelling: Two Extremes, Monte Verità*. Lisse: Swets & Zeitlinger.

Burland, J.B. & Hancock, R.J.R. 1977. Underground car park at the House of Commons, London: geotechnical aspects. *Structural Engineer* 55(2): 87-100.

Dietz, M.S. 2000. Developing an holistic understanding of interface friction using sand within the direct shear apparatus. PhD thesis. University of Bristol. UK.

Dietz, M.S. 2002. Experimental observations of interface friction. In S.M. Springman (ed.) *Workshop on Constitutive and Centrifuge Modelling: Two Extremes, Monte Verità*. Lisse: Swets & Zeitlinger.

Dihoru, L. 2000. Research on the Behaviour of Stainless Steel Powders in the Injection Moulding Process, PhD thesis. University of Cluj. Romania.

Festag, G. 2002. Experimental investigations on sand under cyclic loading. In S.M. Springman (ed.) *Workshop on Constitutive and Centrifuge Modelling: Two Extremes, Monte Verità*. Lisse: Swets & Zeitlinger.

Glen, B.J. 1955. The creep of polycrystalline ice. *Proceedings of the Royal Society of London, Series A. Mathematical and Physical Science* 228: 519-538.

Herle, I. 2002. Difficulties related to numerical predictions of deformation. In S.M. Springman (ed.) *Workshop on Constitutive and Centrifuge Modelling: Two Extremes, Monte Verità*. Lisse: Swets & Zeitlinger.

IPCC, 2001. Third Assessment Report of Working Group 1 of the Intergovernmental Panel on Climate Change. www.ipcc.ch.

Lee, F.H. 2002. The philosophy of modelling versus testing. In S.M. Springman (ed.) *Workshop on Constitutive and Centrifuge Modelling: Two Extremes, Monte Verità*. Lisse: Swets & Zeitlinger.

Martin, C.M. 2002. Impact of centrifuge modelling on offshore foundation design. In S.M. Springman (ed.) *Workshop on Constitutive and Centrifuge Modelling: Two Extremes, Monte Verità*. Lisse: Swets & Zeitlinger.

Martinenghi, L.S. 1992. A study of pile behavior in different soil types including installation effects. Thèse no. 1074. EPFL.

Rowe, P.W. 1972. The relevance of soil fabric to site investigation practice. *Géotechnique* 22(2): 195-300.

Sakajo, S. 2001. Discussion on "Centrifuge testing in Japan". *Workshop on Constitutive and Centrifuge Modelling: Two Extremes, Monte Verità, Ticino, July 2001*.

Sato, K., Yoshida, N. & Yasuhara, K. 1999. Effect of initial fabric anisotropy on cyclic shear characteristics of dense sand. In P.S. Sêco e Pinto (ed.) *II Int. Conf. Earthquake Geotechnical Engineering, Lisbon, Portugal*: 53-58. Rotterdam: Balkema.

Schweiger, H.F. 2001. Comparison of finite element results obtained for a geotechnical benchmark problem. *X Int. Conf. Computer Methods and Advances in Geomechanics*: 697-702. Rotterdam: Balkema.

Sharma, J.S. 2002. Measurement of displacement – Trends or numbers? In S.M. Springman (ed.) *Workshop on Constitutive and Centrifuge Modelling: Two Extremes, Monte Verità*. Lisse: Swets & Zeitlinger.

Sharma, J.S. 1994. Behaviour of reinforced embankments on soft clay. PhD thesis. Cambridge University. United Kingdom.

Taylor, R.N., Grant, R.J., Robson, S. & Kuwano, J. 1998. An image analysis system for determining plane and 3-D displacements in soil models. In Kimura et al. (eds) *Proceedings of Centrifuge '98*: 73-78. Rotterdam: Balkema.

Trausch-Guidici, J.L. 2002. Deformation behaviour of a soft Swiss lacustrine clay. In S.M. Springman (ed.) *Workshop on Constitutive and Centrifuge Modelling: Two Extremes, Monte Verità*. Lisse: Swets & Zeitlinger.

Truesdell, C. & Noll, W. 1965. *The non-linear field theories of mechanics*. Berlin: Springer Verlag.

Wheeler, S.J. 1997. A rotational hardening elasto-plastic model for clays. *XIV Int. Conf. Soil Mechanics ad Foundation Engineering* 1: 431-434. Rotterdam: Balkema.

White, D.J., Take, W.A., Bolton, M.D. & Munachen, S.E. 2001. A deformation measuring system for geotechnical testing based on digital imaging, close-range photogrammetry, and PIV image analysis. *XV Int. Conf. Soil Mechanics and Geotechnical Engineering*: 539-542. Rotterdam: Balkema.

White, D.J. & Bolton, M.D. 2002. Observing friction fatigue on a jacked pile. In S.M. Springman (ed.) *Workshop on Constitutive and Centrifuge Modelling: Two Extremes, Monte Verità*. Lisse: Swets & Zeitlinger.

Problems governed by interfaces

Constitutive and Centrifuge Modelling: Two Extremes, Springman (ed.)
© *2002 Swets & Zeitlinger, Lisse, ISBN 90 5809 361 1*

Constitutive modelling of soil and rock interacting with interfaces

J. Sulem & V. De Gennaro
Cermes, Ecole Nationale des Ponts et Chaussées, LCPC, France

ABSTRACT: In this paper, the general framework of an elasto-plastic constitutive model for interface behaviour is presented. Important features such as hardening/softening behaviour, phase transformation state (compaction/dilatancy) and critical state are discussed. Emphasis is given to the necessity to take the material microstructure for interface modelling into account. This is possible by resorting to higher order continuum theories such as Cosserat and second-gradient model. The discussion is enlarged to some contact problems as encountered in rock mechanics and to rate and state friction laws along faults for geophysical applications.

1 INTRODUCTION

Interfaces play a major role in the response of engineering structures where materials of different stiffness are in contact. This is the case for soil-structure interaction such as for piles, anchors, reinforcements, or rock-tool interaction as for drilling engineering. Interface behaviour also plays a major role in the mechanics of jointed and faulted rock.

In this paper we review some new advances of interface modelling. Section 2 is devoted to the presentation of the elasto-plastic framework of constitutive modelling of soil-structure interfaces. Based on some results of a new experimental device, important features of constitutive modelling such as hardening/softening, phase transformation state (compaction/dilatancy) and critical state are discussed. Most frequently, the interface involves only a thin layer of material with a defined thickness adjacent to the contact surface. The two materials in contact thus interact at a scale where microstructure becomes important.

Incorporation of microstructure features of a geomaterial within the frame of continuum models can be achieved by resorting to the so-called generalised continuum theories. Cosserat and second-gradient models are discussed in sections 3 and 4. The main feature of these models is that they contain material parameters with a dimension of length in relation to the grain size of the geomaterial in the formulation of the constitutive relations. Such models allow the thickness of the shear zone along the interface to be accounted for and its relation to the roughness of the contact surface. These higher-order models introduce also additional boundary conditions which, as discussed later, cannot be considered independently of the behaviour of the material. Friction laws for fault mechanics, as recently developed in geophysics, are discussed in the last section.

2 ELASTO-PLASTIC CONSTITUTIVE MODELLING OF SOIL-STRUCTURE INTERFACE

2.1 *Observational background*

Starting from the pioneering work of Potyondy (1961), we can mention many experimental studies dealing with the mechanical behaviour of soil-structure interfaces (inter alia: Brumund & Leonards 1973, Yoshimi & Kishida 1981, Desai et al. 1985, Uesugi & Kishida 1986, Hoteit 1990, Tabucanon & Airey 1992, Tejchman & Wu 1995, Zong-Ze et al. 1995, Fakharian & Evgin 1996, Lerat et al. 1997, Zervos et al. 2000, Corfdir et al. 2001).

The most used, and probably the simplest, apparatus to perform interface tests is the modified direct shear box. It is obtained from the Casagrande direct shear box by replacing the bottom half with the material simulating the structure. Typical results of interface tests at constant normal stress performed on Fontainebleau sand are presented in Figure 1. Such an apparatus has some obvious shortcomings. Probably the major one is related to the limited values of tangential displacements attainable during the test, which depend on the dimensions of the box. During the tests, tilting of the upper part of the box containing the soil sample is also frequently mentioned and indicated as a dominant source of non-homogeneity of the experiment (Wernick 1978, Tabucanon & Airey 1992). It is worth noting, however, that an interface test is inherently non-homogeneous, the interface being the result of a localisation process.

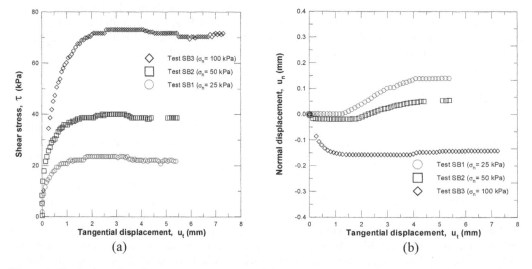

Figure 1. Typical results of tests at constant normal stress performed in a modified direct shear box on Fontainebleau sand with a rough interface (sample density $I_D = 0.46$, $e_o = 0.753$)

These and other reasons have induced various researchers to conceive more sophisticated laboratory equipment, among others: the simple shear apparatus, the ring torsion apparatus, the pull out apparatus, the cyclic three-dimensional simple shear interface apparatus and the ring simple shear apparatus. It should be mentioned that, nowadays, a large part of the experimental investigations have been devoted to the study of the interface behaviour between dry granular media and structures. Indeed, there is a need for studies involving interfacing with soil of different compositions, as well as the analysis of the effect of the pore water pressure on the overall behaviour of the soil-structure interfaces.

However, two main aspects related to the interface testing, and independent of the apparatuses, seem to be well established at the moment. Firstly, the fact that the observable (and

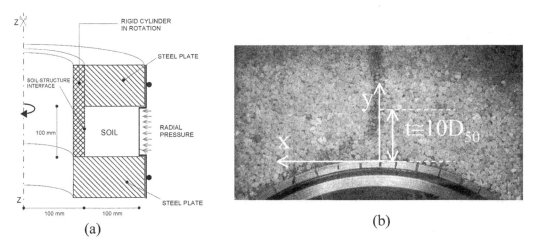

Figure 2. (a) Schematic view of the ACSA apparatus, (b) Visualisation of the interface layer

measurable) kinematic state variables during tests are the relative displacements. Secondly, the impossibility of getting an explicit response from the interface layer, the latter being impossible to isolate from the rest of the soil sample. The first aspect has direct and obvious consequences in the way of approaching the theoretical formulation of the problem (Boulon & Jarzebowski 1991, Desai & Fishman 1991). The second one, partially linked to the first one, is directly related to the problem of the visualisation of the interface layer and the evaluation of its thickness.

In order to discuss briefly the two points above mentioned, we present in Figure 2 a typical result of an interface test showing a visualisation of the interface layer forming at the contact with the structure.

Tests have been performed by Lerat (1996) on Hostun gravel, using the ring simple shear apparatus (ACSA – Appareil de Cisaillement Simple Annulaire). In this device a rigid cylinder in rotation simulates the structure (Fig. 2a). As is observable, the interface appears as a narrow remoulded zone parallel to the surface of the rotating cylinder, forced to develop between the latter and the remaining part of the surrounding soil. When available, the results of analyses permitting an estimation of the thickness "t" of the interface seem to relate the latter to the average diameter (D_{50}) of the grains. Such a thickness is obviously strictly associated with the roughness of the surface, in the same way of the shear stress mobilised during the test. The roughness is usually referred to the average grain dimension by means of the normalised roughness coefficient $R_n = R_{max}/D_{50}$, R_{max} being the maximum gauge depth of the surface (Yoshimi & Kishida 1981). For a rough surface, available experimental results suggest values ranging between 5 and 10 D_{50}.

The definition of the surface roughness of the structural element is clearly fundamental for a correct interpretation of the phenomena coupled to the soil-structure interaction. As discussed by Tatsuoka (1985), two extreme situations can be encountered when soil is in contact with a structural element. The first one corresponds to a smooth contact between soil and structure. In this case, the relative displacement in the tangential direction between the two media is possible even if the soil mass has not deformed. For this condition, the interface acts as a "velocity discontinuity" but the value of mobilised shear stress, as well as the volumetric behaviour of the layer, are independent of the mechanical characteristics of the soil. Large relative displacements are then possible although soil is not at the limit state (failure). Evidently the opposite situation holds if the surface is rough. The deformation of the soil in the vicinity of the

315

structural element is then the main phenomenon leading to the relative displacement of the two materials in contact.

If we consider the case of a two-dimensional problem (plane strain or axisymmetry), the kinematic state variables are the normal relative displacement u_n (y-direction, Fig. 2b) and the tangential relative displacement u_t (x-direction, Fig. 2b), and the corresponding stress variables are the normal stress σ_n and shear stress τ. Based on the results shown in Figure 2, it is straightforward to infer that the concept of deformation for an interface test is rather confused. If one supposes a mechanism of deformation in simple shear the distortion $\gamma = u_t/t$ and the normal deformation $\varepsilon_n = u_n/t$ depend on the thickness of the interface, which in turn is a function of the average size of the grains. Consequently an unquestionable scale effect is automatically embodied in this type of experiment.

Taking account of this possible circumstance clearly delimits the conventional approach in the context of the continuum soil mechanics. This will be discussed in the next section.

2.2 Constitutive model

In order to describe the behaviour of the granular soil-structure interface within the framework of the theory of incremental elasto-plasticity, some preliminary assumptions are needed. As already mentioned, the experimental results obtained using the standard interface testing devices are expressed as a function of the relative displacements and are influenced by a scale effect correlated to the thickness of the interface layer or, equivalently, to the average size of the grains. In the first gradient approach of standard continuum mechanics presented in this section, no internal length appears in the constitutive equations. This parameter is however required to achieve a consistent view of constitutive modelling. We will discuss the role of the microstructure in the next section and propose a way of accounting for it in the formulation of the interface problem. Let us now describe the general approach to the problem in elasto-plasticity.

2.2.1 Elasto-plastic-based formulation

For the time being, we will consider a two-dimensional problem, the kinematic state variables of the interface are the normal relative displacement, u_n, and the tangential relative displacement, u_t. The normal stress, σ_n, and the shear stress, τ, parallel to the direction of the interface, are the associated stress variables. Stresses and displacements are taken as positive in compression, and considered homogeneous within the layer of the interface. The incremental constitutive relationship between displacements and stresses at the interface is written:

$$d\underline{\Sigma} = \mathbf{K}^{ep} d\underline{U} \tag{1}$$

where $\mathbf{K}^{ep}$ is the elasto-plastic stiffness matrix, and $d\underline{\Sigma}$ and $d\underline{U}$ are the increments of stresses and displacements, respectively. This matrix is divided into two distinct parts: an elastic part $\mathbf{K}^e$ and a plastic part $\mathbf{K}^p$, so that:

$$d\underline{\Sigma} = (\mathbf{K}^e + \mathbf{K}^p)d\underline{U} \tag{2}$$

$\mathbf{K}^e$ is related to the reversible components of the displacement rates, $\mathbf{K}^p$ accounts for all the irreversible "permanent" displacements of the interface. Admitting the usual decomposition of the displacement increments into its elastic (e superscript) and plastic (p superscript) components, the displacement increment vector $d\underline{U}$ can be written as:

$$d\underline{U} = d\underline{U}^e + d\underline{U}^p \tag{3}$$

The elastic displacement increment vector $d\underline{U}^e$ of the interface under an applied stress increment $d\underline{\Sigma}$ is obtained, assuming linear elasticity and the $\mathbf{K}^e$ constant, as follows:

$$\underline{dU}^e = \mathbf{K}^{e-1}\underline{d\Sigma} \tag{4}$$

In order to obtain the plastic displacement increment vector $\underline{dU}^p$ it is necessary to define a plastic potential Q and a flow rule. The plastic potential is a function of the stress components σ_n and τ. It also depends on a number of internal variables V_k which allow some specific features of the plastic volumetric behaviour of the interface to be taken into account. Consequently:

$$Q = Q(\sigma_n, \tau, V_k) \quad \text{and} \quad V_k = V_k(u_n^p, u_t^p) \tag{5}$$

Note that the hardening parameter is here the accumulated plastic displacements and not the accumulated plastic strains as in classical flow theory of plasticity. The direction and the magnitude of the plastic displacement increments of the interface are given by the flow rule, this latter can be written as:

$$\begin{bmatrix} du_n^p \\ du_t^p \end{bmatrix} = d\lambda \begin{bmatrix} \dfrac{\partial Q(\sigma_n, \tau, V_k)}{\partial \sigma_n} \\ \dfrac{\partial Q(\sigma_n, \tau, V_k)}{\partial \tau} \end{bmatrix} \tag{6}$$

where $d\lambda$ is the plastic multiplier (scalar positive), u_n^p and u_t^p are the normal and tangential plastic displacements, respectively. Plastic displacements are generated only if the state of stress of the interface verify the failure criterion given by the yield surface F and remains on the yield surface during the loading process. This surface is a function of the stress components σ_n and τ and depends, in the same way on the plastic potential Q, on various internal variables which allow its evolution during the loading history. It is therefore given by the following equation to be taken into account:

$$F = F(\sigma_n, \tau, V_k) = 0 \quad \text{and} \quad V_k = V_k(u_n^p, u_t^p) \tag{7}$$

During yielding, the condition of consistency ensures that the stress state induces plastic flow, this condition reads as follow:

$$dF = \frac{\partial F(\sigma_n, \tau, V_k)}{\partial \sigma_n}d\sigma_n + \frac{\partial F(\sigma_n, \tau, V_k)}{\partial \tau}d\tau + \frac{\partial F(\sigma_n, \tau, V_k)}{\partial V_k}dV_k = 0 \tag{8}$$

It is worth noting that, assuming distinct expressions for the yield surface and the potential function, the flow rule (6) is non-associated. Following the standard procedure of elasto-plasticity, equation (2) can be written as follows:

$$\underline{d\Sigma} = \mathbf{K}^{ep}\underline{dU} \quad ; \quad \mathbf{K}^{ep} = \mathbf{K}^e - \frac{\mathbf{K}^e\left(\dfrac{\partial Q}{\partial \Sigma}\right)\left(\dfrac{\partial F}{\partial \Sigma}\right)^T \mathbf{K}^e}{H + \left(\dfrac{\partial F}{\partial \Sigma}\right)^T \mathbf{K}^e\left(\dfrac{\partial Q}{\partial \Sigma}\right)} \tag{9}$$

where H is the hardening modulus. In the following, this general formulation, which is based on experimental evidence, will be adapted to the specific problem of the soil-structure interface.

2.2.2 *Model presentation and evaluation*
A number of elasto-plastic models have been presented for the analysis of interface behaviour (inter alia: Boulon & Nova 1990, Gens et al. 1990, Desai & Fishman 1991, Desai & Ma 1992,

Sharma & Desai 1992, Day & Potts 1994, Shahrour & Rezaie 1997, De Gennaro 1999). Their theoretical structure is, in general, consistent with the one summarised previously. As discussed by Boulon & Nova (1990), the mechanical behaviour of interfaces is qualitatively very similar to the mechanical behaviour of the same material tested in the triaxial apparatus. Moving from this consideration, some pertinent hypotheses allow for appropriate constitutive modelling.

In granular materials, two main factors have to be considered in view of a description of their mechanical behaviour, namely: the influence of mean effective pressure and the influence of density. It is well recognised that increasing levels of mean effective pressure will induce higher values of maximum shear resistance. At same time, high density states induce dilatant mechanical responses, while compaction and reduced values of shear resistance are typical of loose samples. The same holds for the interface, if the latter is adjacent to a granular material. Consequently loose interfaces will compact if sheared until failure is reached, showing continuous hardening. On the other hand, dense interfaces undergo low initial compaction during initial loading and significant dilatancy at yield, exhibiting sometimes softening behaviour after the point of peak shear resistance. For both density states, no volume changes are detectable at large tangential displacements (critical state), except in the case of the existence of wearing phenomena (e.g. grain crushing), inducing often a second phase of compaction (Plytas 1985, Hoteit 1990, Lerat 1996).

We will try to combine all these elements into a consistent formulation. The proposed model includes deviatoric hardening behaviour, and integrates the phase transformation state (concept of compaction-dilatancy) and the critical state at large tangential displacements (parallel to the direction of shear). The soil is assumed to be dry, so that analysis can be performed in terms of total as well as effective stresses. For the sake of brevity, softening is not introduced in the present formulation.

The elastic behaviour of the interface is given by the following linear relationship:

$$d\underline{\Sigma} = \mathbf{K}^e d\underline{U}^e \quad ; \quad \mathbf{K}^e = \begin{bmatrix} K_n & 0 \\ 0 & K_t \end{bmatrix} \tag{10}$$

where K_n and K_t are the normal and tangential stiffness of the interface, respectively.

At failure it is assumed that normal and shear stresses fulfill the relationship dictated by the Mohr-Coulomb failure criterion. Neglecting cohesion, the failure condition is then given by:

$$\tau_f = \tan\delta_f \sigma_n = \mu_f \sigma_n \tag{11}$$

where δ_f is the friction angle of the interface at failure and $\mu_f = \tan\delta_f$ is the coefficient of friction. The hardening response of the interface corresponds to increasing values of the coefficient of friction μ, which grows towards μ_f, corresponding at large displacements to the coefficient of friction at constant volume (critical state). In the plane σ_n-τ such an evolution of the stress state during hardening, in agreement with the frictional failure criterion (11), corresponds to a counter-clockwise rotation of the locus $\tau = \mu\sigma_n$, starting from the initial position coinciding with the axis $\tau = 0$, until the failure line ($\mu = \mu_f$) defined by equation (11), as shown in Fig. 3(a). In order to account for this yield mechanism the Mohr-Coulomb failure criterion can be generalised as follows (Poorooshasb & Pietruszczak 1985, Sadrnejad & Pande 1989, Pietruszczak & Niu 1993, Bencheikh 1991):

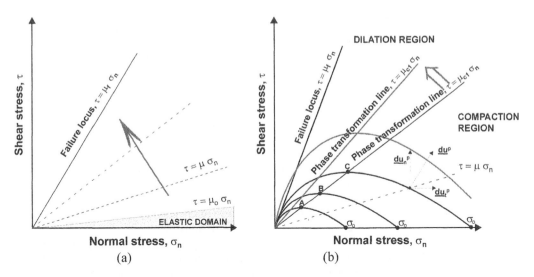

Figure 3. (a) Deviatoric hardening mechanism of the yield surface (b) Plastic potential function

$$F = \tau - \mu(\xi^p)\,\sigma_n = 0 \tag{12}$$

$\mu(\xi^p)$ is the hardening function, it describes the evolution rule of the mobilised friction coefficient during loading. It is assumed that:

$$\mu(\xi^p) = \mu_o + (\mu_f - \mu_o)\dfrac{\xi^p}{A\left(\dfrac{\sigma_{ni}}{p_o}\right)t + \xi^p} \tag{13}$$

In equation (13), μ_f is the coefficient of friction at failure, $\mu_o = \tan\delta_o$ is the friction coefficient delimiting the initial elastic region (δ_o is the initial friction angle). The parameter t is the thickness of the interface layer, A is a parameter of the model governing the shape of the hardening function, σ_{ni} is the initial normal stress and p_o is a reference pressure. The introduction of t into equation (13) allows an internal length parameter to be considered for the interface. The ratio σ_{ni}/p_o, is introduced in order to take the effect of the normal stress σ_n on the shape of the curves of mobilised shear stress into account. The variable ξ^p is the hardening parameter; it is given by the following relationship:

$$\xi^p = \frac{1}{2}\sqrt{(u_n^p)^2 + (u_t^p)^2} \tag{14}$$

where u_n^p and u_t^p are the plastic normal and tangential relative displacements accumulated during loading.

In order to represent the typical volumetric behaviour observed during interface tests, the following potential, Q, is assumed:

$$Q = \tau + \mu_c\sigma_n \ln\left(\frac{\sigma_n}{\sigma_o}\right) \tag{15}$$

The parameter μ_c is the slope of the phase transformation line $\tau = \mu_c\sigma_n$ and σ_o is the intersection of the plastic potential with the normal stress axis. This quantity depends upon the current state of stress acting on the interface. Typical plots of the function Q in the σ_n-τ plane are presented in Fig. 3b. This function describes a series of continuous surfaces, which expand progressively during deformation. The dilatancy of the interface, D, is given by the following relation, obtained by differentiation of equation (15):

$$D = \frac{du_n^p}{du_t^p} = \mu_c - \mu \tag{16}$$

At failure, $\mu = \mu_f$, and D tends asymptotically to the constant value (μ_c - μ_f). Then the residual zero rate of volumetric deformation of the interface at large tangential displacements, in agreement with experimental observations (i.e. critical state), cannot be reproduced if we consider equation (16). The condition of zero dilatancy at the interface at critical state can be obtained if the coefficient μ_c, the stress ratio at phase transformation state, increases after phase transformation towards the final value μ_f (i.e. towards the stress ratio at critical state). This corresponds to an evolution of the size of both compaction and dilation regions in Fig. 3b. Such a mechanism has a direct physical interpretation. Yield of dense interface layers due to shearing causes plastic dilation, resulting in an increase of the voids in the sample (shrinkage of the dilation region). On the other hand, shearing on loose interface layers causes an opposite effect, leading to an overall compaction of the interface and a reduction of the void ratio (shrinkage of the compaction region). As suggested by the experimental results, in order to introduce such a mechanism in the model formulation, it has been chosen to consider the void ratio as the internal state variable of the potential Q, enabling the volumetric plastic behaviour of the sandy interface to be represented.

The evolution rule of the void ratio would be:

$$e(\xi^p) = e_{cr} + (e_{co} - e_{cr}) \exp\left[-\frac{B}{t}(\xi^p - \xi^o)\right] \tag{17}$$

when $\mu \leq \mu_{co}$ et $e_{co} > e_{cr}$ (i.e. loose interface), or:

$$e(\xi^p) = e_{co} + (e_{cr} - e_{co}) \tanh\left[\frac{B}{t}(\xi^p - \xi^o)\right] \tag{18}$$

when $\mu \geq \mu_{co}$ and $e_{co} < e_{cr}$ (i.e. dense interface).

In equations (17) and (18), t is the thickness of the interface, μ_{co} defines the extension of the initial contracting region of the interface (i.e. the slope of the phase transformation line), B is a constitutive parameter controlling the shape of the evolution rules (17) or (18), ξ^o is the cumulative plastic displacement at phase transformation (when $\mu = \mu_{co}$), e_{co} and e_{cr} are the void ratio of the interface at phase transformation and the critical void ratio (asymptotic value of e when $u_t \rightarrow \infty$), respectively.

Consequently, in the dilatant regime ($\mu \geq \mu_{co}$), the following expression is proposed for the parameter μ_c:

$$\mu_c(e) = \mu_{co} + (\mu - \mu_{co})\mathcal{D}(e) \tag{19}$$

In a loose interface ($e_{co} > e_{cr}$), where only compaction is expected, the function $\mathcal{D}(e)$ can be assumed equal to unity when $\mu \geq \mu_{co}$. In a dense interface the function $\mathcal{D}(e)$ becomes:

$$\mathcal{D}(e) = 1 - \text{sech}\left[C\left(\frac{\sigma_{ni}}{p_o}\right)\text{Arc}\tanh\left(\frac{e}{e_{cr}}\right)\right] \tag{20}$$

where C is a constitutive parameter of the model, μ is the coefficient of friction mobilised during shearing, σ_{ni} is the initial normal stress and p_o is a reference pressure. The ratio σ_{ni}/p_o is introduced in order to take the observed reduction in dilatancy at higher normal stresses into account.

In the presented version of the model, nine parameters are required; these are: K_n, K_t, μ_o, μ_f, μ_{co}, A, B, C, e_{cr}. For their determination, one can use results of interface tests at constant normal stress or constant volume. An example of the model performance is given in Figure 4 (De Gennaro & Lerat 2001). Three test results corresponding to initial external radial pressure of 100 kPa (test CS1), 200 kPa (test CS2) and 400 kPa (test CS3) are simulated. The sand samples tested have an initial density index I_D of about 0.49 ($e_o = 0.743$). The numerical computations have been performed assuming the set of parameters given in Table 1, determined from the results of test CS1 at $\sigma_{ni} = 100$ kPa.

Table 1. Parameters of Fontainebleau sand obtained from test CS1 ($\sigma_{ni} = 100$ kPa, $e_o = 0.743$)

K_n (kPa m^{-1})	K_t (kPa m^{-1})	μ_o	μ_f	μ_{co}	A	B (m)	C	e_{cr}	t (mm)
3.1×10^5	1.7×10^5	0	0.45	0.37	0.0003	0.04	1.2	0.84	2

It can be seen that the computed responses match the observed experimental results quite well. With regard to test CS3, a divergence between the predicted value of the shear stress at failure and the one obtained in the test can be observed. It is believed that the restrained contractancy due to grain crushing (local reduction of σ_n) for the higher level of normal (radial) stress acting on the interface could be at the origin of this difference. This aspect is not captured by the proposed approach. It could be considered by introducing grain breakage into the constitutive model (e.g. Daouadji et al. 2001).

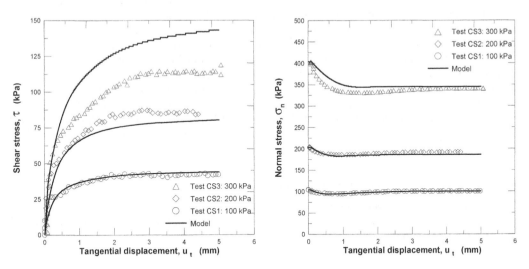

Figure 4. Comparison of model predictions and experimental results: interface tests at constant volume carried out on a simple ring shear box (De Gennaro & Lerat 2001)

3 THE ROLE OF MICROSTRUCTURE IN INTERFACE PROBLEMS

3.1 *Strain localisation and microstructure*

The various drawbacks and shortcomings of the classical continuum theory in connection with strain localisation inside a material or along an interface have been discussed extensively in many recent papers (see Vardoulakis & Sulem 1995 for a literature review). The origin of this undesirable situation can be traced back to the fact that conventional constitutive models do not contain material parameters with a dimension of length, so that the extent of the shear zone (i.e. the extent of the plastically softening region) is undetermined. We can say that localisation of deformation leads to a change in scale of the problem so that phenomena occurring at the scale of the grain cannot be ignored anymore in the modelling of the macroscopic behaviour of the material. Then it appears necessary to resort to continuum models with microstructure to describe localisation phenomena correctly. These generalised continua usually contain additional kinematical degrees of freedom (Cosserat continuum) and/or higher deformation gradients (higher grade continuum). Cosserat continua and higher-grade continua belong to a general class of constitutive models which account for the material <u>micro-structure</u>.

The description of statics and kinematics of continuous media with microstructure has been studied in a systematic way by Germain (1973a, b) through the application of the virtual work principle. In a classical description, a continuum is a continuous distribution of particles, each of them being represented geometrically by a point X and characterised kinematically by a velocity v. In a theory which takes into account the microstructure of the material, each particle is viewed as a continuum C(X) of small extent around the point X. Consequently the deformation of the volume C(X) of the particle is called the *micro-deformation*. For example, a Cosserat continuum is a 'micro-polar' medium obtained by assuming that the particle C(X) moves as a rigid body, characterised by a velocity vector v and a particle rotation vector ω^c. The corresponding kinematical quantities, velocity and rotation gradient (curvature), ∇v and $\nabla \omega^c$, are associated through the principle of virtual work with a non-symmetric stress tensor and a couple stress tensor, respectively. Similarly, in a second gradient continuum through the principle of virtual work, a symmetric second-order stress tensor and a third-order (double) stress tensor are defined which are dual in energy to ∇v and $\nabla \nabla v$ respectively.

Rotation gradients and higher velocity gradients introduce a material length scale into the problem. These higher order continuum theories introduce additional boundary conditions, which can account for the roughness of the contact surface. Moreover, Cosserat and gradient theories improve the computational stability and allow for robust post-localisation computations in the strain-softening regime.

3.2 *The boundary layer effect for a Cosserat continuum*

In a two-dimensional Cosserat continuum, each material point has two translational degrees of freedom (u_1, u_2) and one rotational degree of freedom ω^c. The index c is used to distinguish the Cosserat rotation from the rotation:

$$\omega = \frac{1}{2}\left(u_{2,1} - u_{1,2}\right) \quad ; \quad (.)_{,i} = \frac{\partial(.)}{\partial x_i} \quad i = 1,2 \tag{21}$$

For the formulation of the constitutive relationships, we need deformation measures, which are invariant with respect to rigid body motions, which are the conventional strain tensor:

$$\varepsilon_{ij} = \frac{1}{2}\left(u_{i,j} + u_{j,i}\right) \tag{22}$$

the relative rotation:

$$\omega^{rel} = \omega - \omega^c \tag{23}$$

and the gradient of the Cosserat rotation which is called the curvature of the deformation:

$$\kappa_i = \partial \omega^c / \partial x_i \tag{24}$$

The stress-strain relationships of a 2D-linear isotropic Cosserat medium are:

$$\sigma_{11} = (K+G)\varepsilon_{11} + (K-G)\varepsilon_{22}$$
$$\sigma_{22} = (K-G)\varepsilon_{11} + (K+G)\varepsilon_{22}$$
$$\sigma_{(12)} = \sigma_{(21)} = 2G\varepsilon_{12} \tag{25}$$
$$\sigma_{[12]} = -\sigma_{[21]} = -2G^c(\omega - \omega^c)$$
$$m_1 = M\kappa_1 \; ; m_2 = M\kappa_2$$

In these constitutive equations, K is the 2D-compression modulus and G is the macroscopic shear modulus that links the (symmetric) macroscopic shear strain ε_{12} to the symmetric part of the shear stress $\sigma_{(12)}$. The Cosserat shear modulus G^c links the relative deformation to the anti-symmetric part of the shear stress $\sigma_{[12]}$. Stress couples, which are conjugate in energy to the corresponding curvatures, are linked to them through a bending modulus M, which has the dimension of force. Thus in 2D Cosserat elasticity, the problem is governed by four material constants. As such, we may select the following: (a) the shear modulus G, (b) the material length for bending:

$$\ell = \sqrt{M/G} \tag{26}$$

(c) the Poisson's ratio, $-1 \le v \le 1/2$, with:

$$K/G = 1/(1-2v) \tag{27}$$

and (d) the coupling number:

$$\alpha = 1/\sqrt{1+G/G^c} \tag{28}$$

In the absence of body force and body moment fields, force and moment equilibrium lead to:

$$\sigma_{11,1} + \sigma_{12,2} = 0$$
$$\sigma_{21,1} + \sigma_{22,2} = 0 \tag{29}$$
$$m_{1,1} + m_{2,2} + \sigma_{21} - \sigma_{12} = 0$$

For the boundary conditions, in addition to kinematical or statical conditions of imposed displacements or tractions on the boundary, Cosserat rotation or couple-forces must also be prescribed to define a boundary value problem completely.

In order to illustrate the effect of Cosserat theory, we consider here the example of small-strain simple shear of a long layer of thickness H consisting of linear-elastic Cosserat material (Fig. 5). The thickness H of the Cosserat-elastic strip is assumed to be at least one order of magnitude larger than the elastic bending length ℓ. In the case of simple shear of a long layer, all mechanical properties are assumed to be independent of the x_1-coordinate, which runs parallel to the considered strip of material, i.e. $\partial / \partial x_1 \equiv 0$, which results directly in $\varepsilon_{11} = \kappa_1 = 0$, and in $m_1 = 0$. Assuming also that $\sigma_{22} = 0$ at one boundary and $u_2 = 0$ at the other boundary, then the

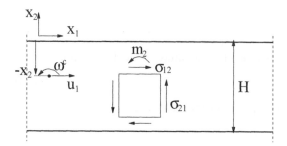

Figure 5. Simple shear of a strip consisting of linear elastic Cosserat material

fields σ_{11}, σ_{22}, and ε_{22} vanish identically everywhere in the considered domain. In that case the remaining equilibrium conditions become:

$$\frac{d\sigma_{12}}{dx_2} = 0 \text{ or } \sigma_{12} = \tau = \text{const} \tag{30}$$

$$\frac{dm_2}{dx_2} + \sigma_{21} - \sigma_{12} = 0 \tag{31}$$

At the 'interface boundary', $x_2 = 0$, we assume that both displacement and Cosserat rotation are prescribed. At the 'remote boundary', $x_2 = H, (H \gg \ell)$ the displacement is set equal to zero and the Cosserat rotation is constrained to be equal to the rigid body rotation, so that conditions for classical continuum are enforced at this boundary, i.e.:

$$
\begin{array}{llll}
\text{for } x_2 = 0: & u_1 = u_{10}; & \omega^c = \omega_0^c \\
\text{for } x_2 = H \gg \ell: & u_1 = 0; & \omega^c = \omega = \omega_H; & \omega_H = -\tau/(2G)
\end{array}
\tag{32}
$$

We observe that in simple shear, both macroscopic rigid-body rotation and shear strain are expressed by the macroscopic displacement gradient:

$$\omega = -\varepsilon_{12} = -\frac{1}{2}\frac{du_1}{dx_2} \tag{33}$$

The problem is simplified if one introduces the quantity $\Omega^c = \omega^c - \omega_H$, which measures the amount of Cosserat rotation in excess of the macro-continuum rotation, which is observed at the remote boundary.

With the above notation, the equilibrium conditions (30) and (31) result in the following equations for the unknowns of the problem:

$$\omega = \omega_H + \alpha^2 \Omega^c$$

$$\ell^2 \frac{d^2\Omega^c}{dx_2^2} - 4\alpha^2 \Omega^c = 0 \tag{34}$$

The solution of the above boundary-value problem may be expressed in terms of the dimensionless re-scaled coordinate of a point in the layer, measured from the interface boundary $x = -2\alpha x_2 / \ell$, the scaling factor $\eta = (\alpha\ell)/(2H) \ll 1$, the dimensionless displacement imposed at the boundary $u_0^* = u_{10}/2H$. With this notation, the dimensionless displacement $u^* = u_1/2H$ is given by:

$$u^* \approx \frac{u_0^*}{1-\eta}\left\{1-\left(\eta/\alpha^2\right)x - \eta e^{-x}\right\} + \frac{\eta\omega_0^c}{1-\eta}\left\{1-\left(\eta/\alpha^2\right)x - e^{-x}\right\} \qquad (35)$$

$$\omega^c \approx \omega_H + \left(u_0^* + \omega_0^c\right)\frac{e^{-x}}{1-\eta} \qquad (36)$$

where the common rotation ω_H at the remote boundary is expressed in terms of the interface boundary data and the scaling factor $\omega_H \approx -\left(u_0^* + \eta\omega_0^c\right)$.

From the above solutions, one can clearly see that the Cosserat effect is localised close to the interface boundary. Accordingly a boundary layer is formed, where the displacement profile is non-linear and micro-rotations differ significantly from macro-rotations. In this boundary layer, asymmetric shear stresses are equilibrated by the gradient of the coupling stress.

The Cosserat solution is plotted in Figure 6. The classical continuum solution is retrieved, if the boundary data are restricted such that $\omega_0^c = -u_0^*$ (curve (a)). In this case the condition $\omega^c = \omega$ is enforced also at the interface boundary. This solution is compared to the non-degenerate one with $\omega_c^0 = -10u_0^*$ (curve (b)). We observe that the Cosserat effect is confined to a boundary layer of about 3ℓ thickness.

For an elastoplastic Cosserat continuum, the influence of the additional Cosserat boundary condition, in terms of prescribed Cosserat rotation or prescribed coupling force at the interface, on the thickness of the boundary layer has been studied numerically by several authors (e.g. Unterreiner 1994, Tejchman & Wu 1995, Bauer & Huang 1999, Huang 2000). Considering a plate which moves horizontally on the top of a granular layer, one can distinguish in the relative displacement between the moving plate and the grains the contribution of grains sliding and grains rolling. A proportional relation is generally postulated between Cosserat rotation and shear displacement at the boundary of the granular layer. The proportional factor is related to the wall roughness. It is found that the thickness of the shear zone increases with increasing grain size, and increasing roughness of the bounding plate, which is in accordance with experimental observations.

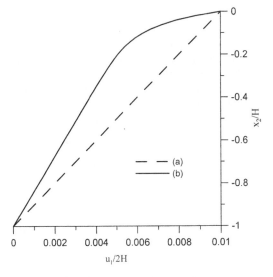

Figure 6. Displacement profile for: (a) linear displacement field for $\omega_c^0 = -u_0^* = -0.01$ (b) non-linear displacement field for $\omega_c^0 = -10u_0^* = -0.1$ $\left(G^c/G = 2.0, \ell/H = 0.1\right)$

4 ROCK-TOOL INTERACTION

4.1 *Constitutive boundary conditions*

Failure criteria for rocks usually involve only stresses and are thus suited primarily for homogeneous states of stresses. Since highly inhomogeneous stresses must be represented in rock mechanics, it is possible that stress-gradients have some effects on the failure mechanism (e.g. Sulem et al. 2001). As mentioned by Mindlin (1963), the apparent strength of rock-type materials is affected by strain gradient. It is observed that brittle failure and the onset of static yielding in the presence of stress concentration occur at higher loads than might be expected on the basis of stress concentration factors calculated from the theory of elasticity. In general, increasing strain gradients appear to make some materials stronger, and to a degree that depends upon grain size.

Rock drilling is an example where micromechanics is important to interface modelling because drilling tool and rock interact at a scale where the microstructure becomes important. The rock strength, which represents the resistance to drilling and thickness of the degrading layer and which influences the drilling advancement, are governed by a 'material length' of the rock and the effective 'roughness' and hardness of the tool. As discussed in the previous section, both material length and tool roughness are factors that cannot be accounted for within classical continuum mechanics and thus higher grade continua must be considered. A second-grade flow theory of plasticity for cohesive softening rock was considered by Vardoulakis et al. (1992) to investigate, analytically and numerically, the problem of interfacial localisation under shear. The analysis of these authors confirmed that localisation is taking place in a thin boundary layer, which is independent of the layer thickness. The authors also discuss the role of the additional boundary condition for gradient plasticity in terms of prescribed first normal derivative of the velocity or prescribed double-stress rate on the surface. These prescribed quantities are measures of interface roughness. As pointed out in the aforementioned paper by Vardoulakis et al. (1992) these boundary data must fulfil some restrictions that involve the softening rate of the material. This observation strengthens Aifantis' (1978) conjecture of the constitutive character of boundary constraints in materials with microstructure.

4.2 *Contact problems and stress singularity*

In order to analyse further the meaning of the additional boundary conditions of higher grade constitutive models, we discuss the problem of rock indentation in the following section. If the surface of a solid is deformed by pressure from a rigid indenter, then the appropriate boundary condition corresponds to the prescription of the normal component of the surface displacement under the indenter. However for the indentation of a semi-elastic medium $(-\infty < x < +\infty, -\infty < y \leq 0)$ with a rectangular indenter, for example, the corresponding normal stress on the free surface is of the form (Johnson 1985):

$$\sigma_{yy}(x,0) = p(x) \tag{37}$$

with

$$p(x) = \begin{cases} -\dfrac{P}{\pi}\dfrac{1}{\sqrt{a^2 - x^2}} & \text{if } |x| \leq a \\ 0 & \text{if } |x| > a \end{cases} \tag{38}$$

where P is the total force applied. This normal stress distribution is singular at the corners of the indenter (x = ±a) which is in contradiction to the small strain elasticity assumption.

The punching process with a rigid indenter can be represented in gradient constitutive models, by separating the effect of the deformation of the surface as the result of a prescribed (regular) normal stress distribution on the free surface. The effect of the opening of the material at the corners of the indenter can be modelled by considering appropriate higher order boundary conditions. In the following, the effect of prescribing a double force at the surface of the semi-infinite elastic medium will be investigated.

The 2nd-gradient elasticity model considered here is based on an original idea of Casal (1961) who introduced in the global strain-energy of a one-dimensional tension bar, both a 'volumetric energy' term which includes the contribution of strain gradient, and a 'surface energy' term. Accordingly, in Casal's model, two material constants with dimension of length ℓ and ℓ' are introduced to characterise the internal and surface capillarity with the following condition for positive elastic strain energy density:

$$|\ell'| < \ell \tag{39}$$

This means in particular that if surface energy terms are included, the volume strain-gradient must be also included. It is worth noticing that in Griffith's theory of cracks, only surface energy is considered, which is of course inadmissible in the sense of inequality (39). The one-dimensional Casal's model was generalised into a three dimensional anisotropic gradient-dependent elasticity (Vardoulakis & Sulem 1995) where the following expression for the strain energy function is considered:

$$w = \frac{1}{2}\left(\lambda\varepsilon_{ii}\varepsilon_{jj} + G\varepsilon_{ij}\varepsilon_{ji} + G\ell^2\partial_k\varepsilon_{ij}\partial_k\varepsilon_{ji} + G\ell_k\partial_k\left(\varepsilon_{ij}\varepsilon_{ji}\right)\right) \tag{40}$$

where λ and G are Lamé's constants, ℓ and ℓ' are the characteristic lengths of the material introduced above, $\varepsilon_{ij} = \frac{1}{2}\left(\partial_i u_j + \partial_j u_i\right)$ are the strains, u_i the displacements and:

$$\ell_k = \ell' n_k, \quad n_k n_k = 1 \tag{41}$$

is a director. Notice that such a model is, by essence, anisotropic as it includes the effect of surface tension.

With the above expression for the strain energy function (equation 40) and following Mindlin's (1964) formalism for materials with microstructure, the following constitutive equations are obtained for the total, Cauchy and double stress tensors (respectively σ, τ and μ):

$$\sigma_{ij} = \lambda\delta_{ij}\varepsilon_{kk} + 2G\varepsilon_{ij} - \ell^2\nabla^2\left(\lambda\delta_{ij}\varepsilon_{kk} + 2G\varepsilon_{ij}\right)$$

$$\tau_{ij} = \lambda\delta_{ij}\varepsilon_{kk} + 2G\varepsilon_{ij} + \ell_k\partial_k\left(\lambda\delta_{ij}\varepsilon_{kk} + 2G\varepsilon_{ij}\right) \tag{42}$$

$$\mu_{kij} = \ell_k\left(\lambda\delta_{ij}\varepsilon_{ll} + 2G\varepsilon_{ij}\right) + \ell^2\partial_k\left(\lambda\delta_{ij}\varepsilon_{ll} + 2G\varepsilon_{ij}\right)$$

Equilibrium equations and boundary conditions can be derived from the principle of virtual work leading to the following set of differential equations (see Mindlin & Eshel 1968):

$$\partial_i\sigma_{ij} + f_i = 0$$

$$\partial_k\mu_{kij} + \left(\sigma_{ij} - \tau_{ij}\right) = 0 \tag{43}$$

Boundary conditions for the stresses and double stresses are expressed in terms of imposed tractions **T** and double tractions **M**. Let **n** be the outwards normal to a smooth boundary surface:

$$n_j\sigma_{jk} - D_j\left(n_i\mu_{ijk}\right) + \left(D_l n_l\right)n_i n_j\mu_{ijk} = T_k \tag{44}$$

327

$$n_i n_j \mu_{ijk} = M_k \tag{45}$$

where **n** is the outwards normal to a smooth boundary surface and D and D_i are the normal and tangential differential operators:

$$D(\cdot) = n_i \partial_i (\cdot) \tag{46}$$

$$D_i (\cdot) = \partial_i (\cdot) - n_i D(\cdot) \tag{47}$$

The displacement-equation equilibrium of the present gradient dependent elasticity theory with surface energy in the absence of body forces for plane strain problem is (Exadaktylos 1998):

$$\overline{D}^2 \left[(\lambda + 2G) \nabla \nabla \cdot \mathbf{u} + G \nabla^2 \mathbf{u} \right] = 0, \quad \overline{D} \equiv 1 - \ell^2 \nabla^2 \tag{48}$$

and the boundary conditions (44, 45) become:

$$\text{for } y = 0 \text{ and } -\infty < x < \infty \quad \begin{cases} \sigma_{yy} - \dfrac{\partial}{\partial x} \mu_{yxy} = \sigma_0 (x) \\ \sigma_{xy} - \dfrac{\partial}{\partial x} \mu_{yxx} = \tau_0 (x) \\ \mu_{yyy} (x,0) = M_1 (x) \\ \mu_{yyx} (x,0) = M_2 (x) \end{cases} \tag{49}$$

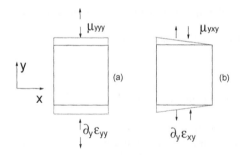

Figure 7. Double-stress components and strain gradients

We consider first a Dirac double force μ_{yyy} acting on the boundary of the half-space (Fig. 7a):

$$\sigma_0 (x) = 0 \; ; \; \tau_0 (x) = 0 \; ; \; M_1 (x) = M \delta_0 (x) \; ; \; M_2 (x) = 0 \tag{50}$$

The quantity $2M/G$ has the dimension of a square-length:

$$\ell''^2 = \frac{2M}{G} \; ; \; \frac{\ell''^2}{\ell} = \overline{\ell} \tag{51}$$

We define the following normalised displacement in x and y direction respectively:

$$u^* = u(x,0) / \overline{\ell} \; ; \; v^* = v(x,0) / \overline{\ell} \tag{52}$$

For the considered loading condition, regular solutions are obtained for u^* and v^* which are anti-symmetric and symmetric functions of x respectively, as shown on Figure 8.

328

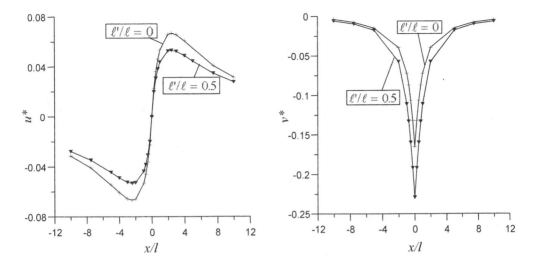

Figure 8. Dirac double force μ_{yyy} acting on the boundary of the half-space

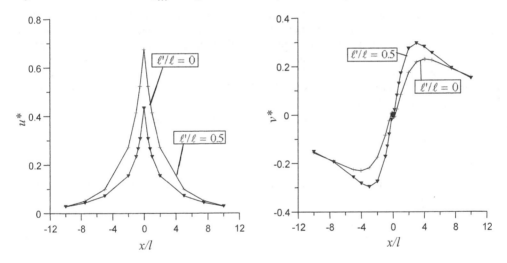

Figure 9. Dirac double force μ_{yxy} acting on the boundary of the half-space

We consider now a Dirac double force μ_{yxy} acting on the boundary of the half-space (Fig. 7b):

$$\sigma_0(x) = 0 \; ; \; \tau_0(x) = 0 \; ; \; M_1(x) = 0 \; ; \; M_2(x) = M\delta_0(x) \tag{53}$$

The response is shown on Figure 9, these results show that considering double-force point loads on the surface, the displacement field is **regular** and shows high gradients at the loaded point.

5 FAULT BEHAVIOUR AND FRICTION LAWS

In this section we would like to discuss some aspects of interface behaviour as encountered in seismology. Understanding fault friction represents a key question toward the comprehensive description of the seismic cycle (e.g. Scholz 1990). A fault is seen as a fracture of the earth crust but in most cases it is a layer of finite thickness of cataclastic rock (gouge material). This gouge is the result of the wear of sliding rock walls. Although gouge material is composed of brittle fragments, it behaves like a granular material when sheared, with cataclastic flow dominated by grain comminution and volume changes (Marone et al. 1990). Friction behaviour of gouge-rock interface is an important aspect for the understanding of the nucleation and propagation of seism. In particular, the transition between slow (aseismic) frictional creep to rapid (seismic) slip has been modelled mathematically by rate and state dependent friction (RSF) laws. This formalism has been successful in explaining the behaviour of simulated faults in the laboratory and is often used to model real fault zones (Dieterich 1979, Ruina 1983).

The Coulomb law with constant dynamic and static friction coefficients can be seen as a first order approximation for rock friction. The Dieterich-Ruina friction law is a widely used RSF law, which accounts for various effects such as (a) a logarithmic dependence of the static friction coefficient with the age of the contact, (b) a logarithmic dependence of the dynamic friction coefficient with the velocity, (c) an influence of the state of the contact (i.e. its history) on the dynamic friction coefficient (memory effect). It is written as follows:

$$\mu = \mu_0 + A \ln\left(\frac{V}{V_0}\right) + B \ln\left(\frac{V_0 \theta}{L}\right) \tag{54}$$

where μ is the friction coefficient, A and B are temperature dependent material parameters. Their values are very small, typically of the order of 0.01 as compared to the typical values of friction coefficients. V is the macroscopic frictional slip velocity, θ is the state variable of the contact with dimension of a time. It is interpreted as an average contact time for surface asperity. μ_0 is a reference friction coefficient, V_0 a reference velocity and L a characteristic distance for the evolution of θ. According to Dieterich (1979) it is a typical size of the asperity in contact with values of the order of 10^{-6} m.

The friction law (54) is completed by the evolution law for the state variable θ:

$$\frac{d\theta}{dt} = 1 - \frac{\theta V}{L} \tag{55}$$

For constant velocity V_1 the friction coefficient reaches a stationary value μ_s after a displacement of the order of L:

$$\mu_s = \mu_0 + (A - B) \ln\left(\frac{V_1}{V_0}\right) \tag{56}$$

Equation (56) expresses a logarithmic dependence of the friction coefficient at steady state with the velocity. If A-B < 0, the behaviour is velocity weakening, if greater than zero, velocity strengthening. For zero velocity, A = 0 and the evolution law (55) gives a linear dependence of θ with time. Equation (54) thus accounts also for a logarithmic increase of the static friction coefficient with time. It describes the re-strengthening effect in stationary contact.

For sudden changes of the sliding velocity, the Dieterich-Ruina friction law also describes correctly the experimental results, as schematically represented on Figure 10. Considering a jump in the velocity from V_1 to $V_2 > V_1$, there is first an immediate increase in friction, with magnitude determined by A, followed by a fall, of magnitude B, over the characteristic distance L (Fig. 10). Furthermore non-linear stability analyses show the general result that slip will be unstable for A-B < 0 and stable for A-B > 0 (Rice & Ruina 1983, Gu et al. 1984). L is thus interpreted as a critical slip distance over which strength breaks down during earthquake nucleation.

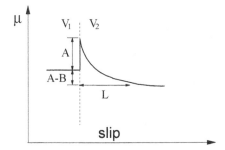

Figure 10. Illustration of slip-rate effects on friction

This distance has a key role for determining the rupture nucleation dimension, the amount of premonitory and post-seismic slip and the maximum seismic ground acceleration. However severe quantitative discrepancies appear between the micrometric length scale observed in laboratory friction experiments (Dieterich 1979) and the metric friction weakening distances typically derived from earthquake inversions (Ide & Takeo 1997). Some authors (Marone & Kilgore 1993) proposed that the critical slip distance is scale-dependent and is controlled by the thickness of the zone of localised shear strain (10^2 to 10^3 m thick for gouge zones in faults, 1 to 10 mm thick for gouge layer in the laboratory). The ring-shear apparatus presented in section 2 was used in recent experiments by Chambon (2000) to study the friction behaviour of granular assemblies, which simulate a gouge material. The experimental set-up allows for very large (several metres) shear displacements but is limited to 1 MPa of confining pressure. The results exhibited clearly two uncoupled frictional lengths. The smallest of about 100 μm is associated with changes of prescribed velocity as expected from the RSF law. The second length scale is associated with an overall slip-weakening of friction active over a characteristic distance of 60 cm. These data show a frictional mechanism compatible with field data and should be studied further in the near future.

6 CONCLUSION

In the various examples of interface behaviour discussed in this paper, we emphasised the fact that the two materials in contact interact at a scale where microstructure is important. Several length scales thus appear in the modelling of interface problems. One is related to the fabric length of the material (grain size), the other to the contact length of the interacting solids (size of asperity, roughness). These various lengths influence the extent the softening layer close to the contact surface (remoulded soil or degrading rock). Within the frame of continuum theories with microstructure, such as Cosserat and gradient theories, the constitutive equations contain additional parameters with the dimension of a length, and additional boundary conditions for the higher order static or kinematic quantities are introduced. These theories can thus account for the existence of both material length and contact length. As these parameters control the extent of the softening zone, they are of major importance for stability analyses. The experimental calibration of these parameters is a key question. In this paper, some examples of calibration through back analysis of interface thickness or scale effect in indentation tests have been presented. An important aspect of interface problems is related to evolution of the material microstructure when sheared due to the abrasion and the crushing of grains. This is still an open question when dealing with continuum theories, and the framework of modelling presented here should be enriched in the future by considering appropriate evolution laws for the material and contact lengths.

REFERENCES

Aifantis, E. C. 1978. A proposal for continuum with microstructure. *Mech. Res. Comm.* 5: 139-145.

Bauer, E. & Huang, W. 1999. Numerical study of polar effects in shear zone. In Pande, Pietruszczak & Schweiger (eds), *Proc. 7th Symp. Num. Mod. in Geom. - NUMOG VII*, 133-138. Rotterdam: Balkema.

Bencheikh, B. 1991. Interaction sol-structure: modélisation et résolution numérique. *Thèse de Doctorat de l'Université des Sciences et Techniques de Lille*, Lille, France.

Boulon, M. & Jarzebowski, A. 1991. Rate-type elastoplastic approaches for soil-structure interface behavior: a comparison. In: Beer, Booker & Carter (eds) *Proc. Comp. Methods and Adv. in Geomech.* 305-310. Rotterdam: Balkema.

Boulon, M. & Nova, R. 1990. Modeling of soil-structure interface behavior, a comparison between elasto-plastic and rate type laws. *Computers and Geotechnics* 9: 21-46.

Brumund, W.F. & Leonards, G.A. 1973. Experimental study of static and dynamic friction between sand and typical construction materials. *Journal of Testing and Evaluation* 1(2): 162-165.

Casal, P. 1961. La capillarité interne. *Cahier du groupe français de rhéologie* 6(3): 31-37.

Chambon, G. 2000. Etude expérimentale du comportement d'un matériau granulaire soumis à un cisaillement simple annulaire, *Res. rep.*, Ecole Normale Supérieure, Ecole Nat. des Ponts et Chaussées, 74 p.

Corfdir, A., Lerat, P. & Roux, J.-N. 2001. Translation and rotation of grains within an interface between granular media and structure. In: Kishino (ed.) *Proc. 4th Int. Conf. on Micromechanics of Granular Media (Powders & Grains 2001)*, Sendai, Japan: 315-318.

Daouadji, A, Hicher, P. Y., Rahma, A. 2001. An elastoplastic model for granular materials taking into account grain breakage. *Eur. J. Mech. A/Solids* 20: 113-137.

Day, R.A. & Potts, D.M. 1994. Zero thickness interface elements. Numerical stability and application. *Int. Journ. Num. Anal. Meth. Geomech.* 18: 689-708.

De Gennaro, V. 1999. Etude du comportement mécanique des interfaces sol-structure. Application à l'analyse du comportement des pieux. *Ph.D. thesis - Ecole Nat. des Ponts et Chaussées*, Paris, France.

De Gennaro, V. & Lerat, P. 2001. Modeling of simple shear tests on soil-structure interfaces. *Proc. 1st Int. Conf. Albert Caquot* Paris, France: in print.

Desai, C.S. & Fishman, K.L. 1991. Plasticity-based constitutive model with associated testing joints. *Int. J. Rock Mech. Min. Sci. and Geomech. Abstr.* 28(1): 15-26.

Desai, C.S. & Ma, Y. 1992. Modeling of joints and interfaces using the disturbed-state concept. *Int. J. Numer. Anal. Methods Geomech.* 16: 623-653.

Desai, C.S., Drumm, C. E. & Zaman, M.M. 1985. Cyclic testing and modeling of interfaces. *J. Geotech. Engrg. Div.* ASCE, 111 (GT6): 793-815.

Dieterich, J.H. 1979. Modelling of rock friction: 1. Experimental results and constitutive equations. *J. Geophys. Res.* 84: 2169-75.

Exadaktylos, G. 1998. Gradient elasticity with surface energy: Mode-I crack problem. *Int. J. Solids Structures* 35(5-6): 421-456.

Fakharian, K. & Evgin, E. 1996. An automated apparatus for three-dimensional monotonic and cyclic testing of interfaces. *Geot. Testing J.* ASTM 19(1): 22-31.

Gens, A., Carol, I. & Alonso, E. E. 1990. A constitutive model for rock joints: formulation and numerical implementation. *Computers and Geotechnics* 9: 3-20.

Germain, P. 1973a. La méthode des puissances virtuelles en mécanique des milieux continus. Part 2. *J. Mécanique* 12: 235-274.

Germain, P. 1973b. The method of virtual power in continuum mechanics. Part 2: Microstructure. SIAM, J. Appl. Math. 25: 556-575.

Gu, J. C., Rice, J. R., Ruina, A. L. & Tse, S. T. 1984. Slip motion and stability of a single degree of freedom elastic system with rate and state dependent friction. J. Mech. Phys. Sol. 32: 167-196.

Hoteit, N. 1990. Contribution à l'étude du comportement d'interface sable-inclusion et application au frottement apparent. Thèse de Doctorat de l'Institut Nat. Polytech. de Grenoble, Grenoble, France.

Huang, W. 2000. Hypoplastic modeling of shear localization in granular media. Thesis, Graz University of Technology, Austria, 107p.

Ide, S. & Takeo, M. 1997. Determination of constitutive relations of fault slip based on seismic wave analysis. J. Geophys. Res. 102: 27379-27391.

Johnson, K. L. 1985. Contact Mechanics. Cambridge: Cambridge University Press.

Lerat, P. 1996. Etude de l'interface sol-structure dans les milieux granulaires à l'aide d'un nouvel appareil de cisaillement annulaire. Ph.D. thesis - Ecole Nationale des Ponts et Chaussées, Paris, France.

Lerat, P., Schlosser, F. & Vardoulakis, I. 1997. Nouvel appareil de cisaillement annulaire pour l'étude des interfaces matériau granulaire-structure. In Publ. Comm. Of XIV ICSMFE (eds) Proc. 14th Int. Conf. Soils Mech. and Found. Eng., Hambourg, Vol. 2: 363-366.

Marone, C. & Kilgore, B. 1993. Scaling of the critical slip distance for seismic faulting with shear strain in fault zones. Nature 362: 618-621.

Marone, C., Raleigh, C.B. & Scholz, C. 1990. Frictional behavior and constitutive modeling of simulated fault gouge. J. Geophys. Res. 95: 7007-7026.

Mindlin, R.D. & Eshel, N.N. On first strain-gradient theories in linear elasticity. Int. J. Solids Structures 4: 109-124.

Mindlin, R.D. 1963. The influence of couple stresses on stress concentrations. Experimental Mech. 3: 1-7.

Mindlin, R.D. 1964. Microstructure in linear elasticity. Arch. Rat. Mech. Anal. 4: 50-78.

Pietruszczak, S. & Niu, X. 1993. On the description of localized deformation. Int. J. Numer. Anal. Meth. Geomech. 17: 791-805.

Plytas C. 1985. Contribution à l'étude expérimentale et numérique des interfaces sols granulaires-structures. Application à la prévision du frottement latéral des pieux. Thèse de Doctorat de l'Institut National Polytechnique de Grenoble, Grenoble, France.

Poorooshasb, H.B. & Pietruszczak, S. 1985. On yielding and flow of sand; a generalized two-surface model. Computers and Geotechnics 1: 33-58.

Potyondy, J.G. 1961. Skin friction between various soils and construction materials. Géotechnique 11(4): 339-353.

Rice, J. R. & Ruina, A. L. 1983. Stability of steady frictional slipping. J. Appl. Mech. 105: 343-349.

Ruina, A. L. 1983. Slip instability and state variable friction laws. J. Geophys. Res. 88: 10359-10370.

Sadrnejad, S. A. & Pande, G.N. 1989. A multilaminate model for sands. In Pande & Pietruszczak (eds) Proc. Int. Symp. Num. Models in Geomech. (NUMOG III), Niagara Falls, Canada: 17-27.

Scholz, C. H. 1990. The mechanics of earthquakes and faulting. Cambridge: Cambridge University Press.

Shahrour, I. & Rezaie, F. 1997. An elastoplastic constitutive relation for soil-structure interface under cyclic loading. Computers and Geotechnics 21: 21-39.

Sharma, K. G. & Desai, C. S. 1992. Analysis and implementation of thin-layer element for interfaces and joints. J. of Engin. Mechanics, ASCE 118(12): 2442-2462.

Sulem, J., Vardoulakis, I. & Exadaktylos, G. 2001. Microstructural effects in stress concentration and fracture problems in rock mechanics. In M. H. Aliabadi (ed), Nonlinear Fracture and Damage Mechanics 161-199, WITPress.

Tabucanon, J. T. & Airey, D. W. 1992. Interface tests to investigate pile skin friction in sands. Research Report No. R662, University of Sydney: pp. 14.

Tatsuoka, F. 1985. On the angle of interface friction for cohesionless soils. Soils and Foundations 25(4): 135-141.

Tejchman, J. & Wu, W. 1995. Experimental and numerical study of sand-steel interfaces. Int. Journ. Num. Anal. Meth. Geomech. 19: 513-536.

Uesugi, M. & Kishida, H. 1986. Influential factors of friction between steel and dry sands. Soils and Foundations 26(2): 33-46.

Unterreiner, P. 1994. Modélisation des interfaces en mécanique des sols: Application aux calculs en déformation des murs en sol cloué. Thèse de doctorat de l'Ecole Nat. des Ponts et Chaussées, Paris.

Vardoulakis, I., Shah, K.R. & Papanastasiou, P. 1992. Modeling of tool-rock shear interfaces using gradient-dependent flow theory of plasticity. Int. J. Rock Mech. Min. Sci. & Geom. Abstr. 29(6): 573-582.

Vardoulakis, I. & Sulem, J. 1995. Bifurcation analysis in geomechanics. Glasgow: Blackie Academic & Professional.

Yoshimi, Y. & Kishida, T. 1981. A ring torsion apparatus for evaluating friction between soil and metal surface. Geot. Testing J. ASTM 4(4): 145-152.

Zervos, A., Vardoulakis, I., Jean, M. & Lerat, P. 2000. Numerical investigation of granular interfaces kinematics. Mechanics of cohesive-frictional materials 5: 305-324.

Zong-Ze, Y., Hong, Z. & Guo-Hua, X. 1995. A study of deformation in the interface between soil and concrete. Computers and Geotechnics 17: 75-92.

Constitutive and Centrifuge Modelling: Two Extremes, Springman (ed.)
© *2002 Swets & Zeitlinger, Lisse, ISBN 90 5809 361 1*

Size effects in shear interfaces

J. Garnier
Laboratoire Central des Ponts et Chaussées, Nantes, France

ABSTRACT: In physical modelling (centrifuge tests, calibration chambers, laboratory tests), the size of the soil particles may not be negligible when compared to the dimensions of the models. Size effects may disturb the response of the models, and the experimental data obtained on these cannot be extended to true scale conditions. Different tests have been performed to study and quantify the size effects that may happen in shear interfaces between soils and structures: modified shear box tests, pull-out tests of plates and cylindrical rods in a centrifuge model or in a modified triaxial cell, torsional tests of piles in a centrifuge model. Results show that there are no significant size effects on the maximum mobilised shear stress at the interface if the model dimensions are larger than about 100 times the mean grain size. The effects on displacement at peak shear strength are less clear and more investigations are needed on this aspect of the grain size effects.

1 INTRODUCTION

This paper summarises the main results of several studies carried out at LCPC (Laboratoire Central des Ponts et Chaussées, Nantes) to try to quantify the size effects in shear interfaces. The final objective is to obtain rules to be applied to reduced scale models to minimise or to avoid the size effects that could be introduced by the relatively small dimensions of the model structures compared to the soil particles. The paper gathers works done in collaboration with several partners and already presented by different authors (Mainetti 1995, Dano 1996, Riot 1996, Hommers 1997, Le Collinet 1998, Garnier & König 1998, Dubreucq 1999). Data from recently published papers are also included (Alawneh et al. 1999, Dietz 2000). It is obvious that the size effects will only be a problem in sand samples, the dimension of clay particles is always very small compared to the size of the models.

2 SHEAR BOX TESTS ON SAND

In shear box tests on sand samples, two parameters which have dimension of length may generate size effects and must be studied (grain size and dimensions of the box). It is very difficult, if not impossible, to vary the grain size without modifying the rheological properties of the sand and it is the reason why the grain size effects were studied first using glass marbles. It is then assumed that all glass marbles are spherical and that the samples will only differ by the diameter of the grains.

2.1 *Grain size effects in shear box tests*

A square shear box is used (60 mm x 60 mm x 24.5 mm) and different classes of marbles have been tested with diameters ranging from 0.2 mm to 0.6 mm. Unfortunately, very strong slip-stick effects were observed on the shear stress vs. strain curves, that prevent accurate comparison of the results. A second attempt was to try to reconstitute the gradation curve of Fontainebleau sand ($d_{10} = 0.14$ mm, $d_{50} = 0.21$ mm and $d_{100} = 0.5$ mm), by mixing glass marbles of five different sizes. The shear tests results on this artificial Fontainebleau sand placed at a density of 15.6 kN/m^3 are shown in Figure 1 for three different normal stresses σ where F/S is the shear stress and U is the shear displacement.

Large slip-stick effects are observed on the shear stress vs. displacement curve. They seem to be almost periodic, and are perfectly synchronous with the vertical deformations of the sample during shearing, as seen in Figure 2. When the sample dilates, the shear stress increases and suddenly drops down when the sample contracts.

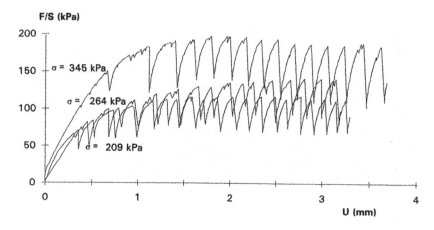

Figure 1. Shear box tests on glass beads of different diameters, with a grading close to Fontainebleau sand (Mainetti 1995)

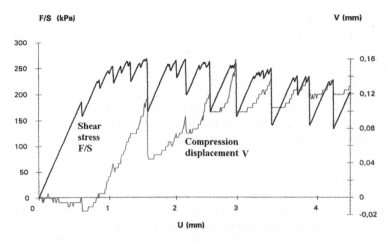

Figure 2. Synchronisation between mobilised shear stress F/S and vertical deformation V of the sample during slip-stick (Mainetti 1995)

These slip-stick effects are still too large and may hide the effects of the size of the particles. It was then decided to continue the work, but not by varying the particle size (true Fontainebleau sand was selected), but by using shear boxes of different sizes.

2.2 *Effect of the size of the shear box*

Three different square shear boxes (L = 30 mm, 45 mm and 60 mm) were designed and 30 tests were performed by varying the normal stress and the sample height (Riot 1996, Dubreucq 1999). An example of results is shown in Figure 3 for a normal stress σ = 345 kPa. The peak resistance decreases and the displacement at peak increases when the dimension of the box increases. The box size does not seem not to have a significant effect on the residual shear strength.

These first results have been confirmed by new series of tests on circular boxes with diameter 30 mm, 45 mm and 60 mm (Dubreucq 1999). In another series of tests, the height of the sample has also been varied, and some results obtained with the smallest square box (30 mm x 30 mm) are presented in Figure 4, indicating a good repeatability of the tests and no significant influence of the sample height in the tested range.

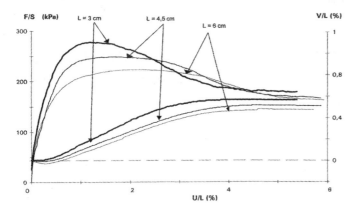

Figure 3. Effect of the dimension of the shear box on shear stress F/S vs. shear deformation U/L for σ = 345 kPa (Dubreucq 1999)

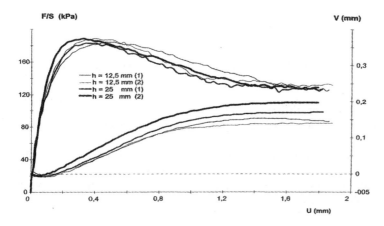

Figure 4. Effect of the sample height h on shear box tests results (Dubreucq 1999)

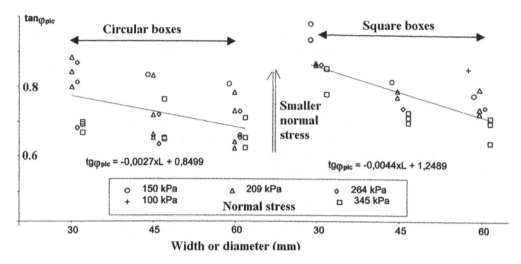

(a) Effect on peak shear resistance tan φ_{pic}

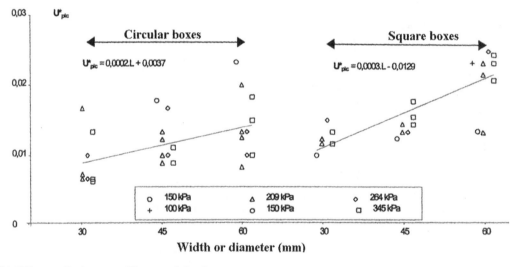

(b) Effect on displacement U_{pic} at peak load

Figure 5. Effect of the dimension of the box on shear interface tests results (from Dubreucq 1999)

The results from all shear tests under different normal stresses and in various box sizes are summarised in Figure 5a (effect on peak resistance φ_{pic}) and Figure 5b (effect on displacement U_{pic} at peak load).

The peak resistance always decreases when the size of the shear box increases. It is rather surprising to note that this clear size effect on shear strength measurements, that has been known about for a long time, has never been seriously taken into account in practice in geotechnical engineering. The effect on displacement at peak is more logical, as seen in Figure 5b.

3 INFLUENCE OF INTERFACE ROUGHNESS

3.1 *Interface shear box tests*

A shear box was modified to perform interface shear tests on Fontainebleau sand in contact with steel plates of different roughness. The 60 mm x 60 mm square plates have been machined using a numerical machining tool that created grooves perpendicular to the shear direction. Their surface roughnesses were measured and are summarised in Table 1. The interface shear tests results corresponding to a 345 kPa normal stress are plotted in Figures 6 and 7.

Table 1. Roughness of the steel plates used in shear interface tests (Mainetti 1995)

Plate n°	Mean roughness R (microns)	Maximum roughness R_{max} (microns)	Mean thread AR (microns)
1	3.1	12.3	60.6
2	23.8	32.3	195
3	90.5	100.8	379
4	210	228	583
5	400	400?	1200?

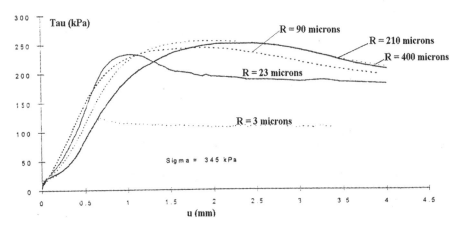

Figure 6. Shear stress Tau (kPa) vs. displacement u (mm) in shear box tests on plates of different roughness R (Mainetti 1995)

Tests on plates with roughness from 90 to 400 microns show very similar results. The much smoother plate (n°1), with R = 3 microns, gives a much smaller shear resistance and the results for plate n° 2, with a mean roughness R = 23 microns, is intermediate between the rough and smooth behaviour. A similar tendency is observed in Figure 7 showing the volume change (measured by the vertical deformation Wz of the sample) during shearing. The dilatancy is large for the rough plates n°4 and 5 (R = 210 and 400 microns) and is zero with the smooth one.

The roughness of the interface depends on the roughness of the surface of the structure as well as on the grain size of the soil. The normalised roughness R_n may be defined by R_{max}/d_{50}. The influence of R_n on the maximum shear strength of a soil-structure interface has been studied by Yoshima & Kischida (1981), Kishida & Uesugi (1987) and Paikowsky et al. (1995). Three zones are observed:

- For small values of $R_n \leq 0.01$, the interface is smooth. The ratio between maximum shear stress and normal stress is relatively low, and no dilatancy occurs.

- For the normalised roughness larger than $R_n \geq 0.1$ to 1, the surface is totally rough with high shear strength and dilatancy. For these rough interfaces, the maximum mobilised shear stress is not dependent on roughness.
- Between the smooth and the rough interfaces is a zone of intermediate interfaces. The ratio of shear stress to normal stress increases with increasing roughness.

The results of the tests performed at LCPC and those from Dietz (2000) are in good agreement with these findings (Fig. 8).

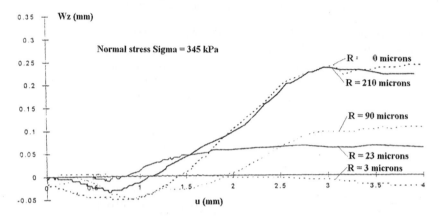

Figure 7. Vertical deformation Wz of the sample vs. displacement u in shear box tests on plates of different roughness (Mainetti 1995)

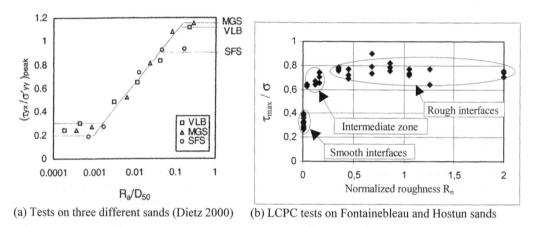

(a) Tests on three different sands (Dietz 2000)   (b) LCPC tests on Fontainebleau and Hostun sands

Figure 8. Influence of normalised roughness $R_n = R/d_{50}$ on peak interface resistance from shear box tests on steel-sand interfaces ($R_n = R_a/D_{50}$ with Dietz notation)

3.2 Centrifuge pull-out tests of piles

When a long inclusion (pile or nail) is axially loaded in a centrifuge model, size effects may influence both the average shear stress τ_p mobilised at peak load and the corresponding displacement u_p, as defined in Figure 9. To simplify the analysis, only rigid structures are considered here. The deformation of the tested models is negligible and the displacements of all points of the shear surface are assumed to be equal.

Model piles 12 mm in diameter and 300 mm long have been placed into Fontainebleau sand samples (sand pluviated around the piles at 16.7 kN/m^3 using the LCPC automatic hopper). Eight different values of roughness R of the pile wall have been obtained by machining the steel pipes, ranging from a few microns to 600 microns. Vertical pull-out loading tests have been carried out at 50g and all tests were duplicated. Results concerning the effect of normalised wall roughness on lateral friction resistance are shown in Figure 10 and are very similar to what has been observed in interface shear box tests and presented in Figure 8b (Garnier & König 1998).

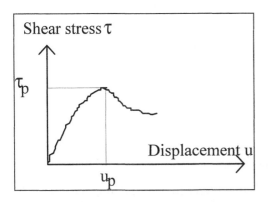

Figure 9. Typical shear stress-displacement curve and relevant parameters τ_p (shear strength) and u_p (displacement at peak)

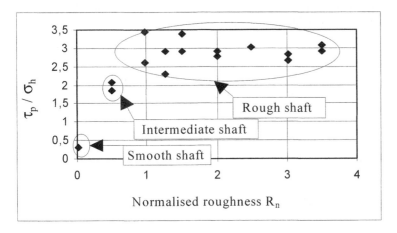

Figure 10. Influence of normalised roughness R_n on the ratio of maximum average shear stress to horizontal earth pressure in centrifuge pull out tests on piles (Fontainebleau sand)

4 INFLUENCE OF MODEL DIMENSIONS

To reduce the number of parameters involved, all tests described in the following sections have been carried out on totally rough interfaces (normalised roughness $R_n \geq 1$ in all cases). Interface shear strength will also be independent of roughness R_n. The appropriate roughness is obtained either by gluing sand or emery paper on the structures or by machining their surface.

4.1 *Pull-out tests of plates*

Modelling of models tests have been performed on rough plates under tension loads in the LCPC geotechnical centrifuge. The prototype depth of the plate is 3.2 m. This prototype has been modelled at several different g-levels from 10g to 40g. The dimensions of the different models are shown in Figure 11. The width and thickness of all models have been kept constant and are equal to 100 mm and 4 mm respectively.

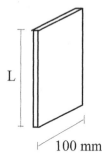

g-level	Length L (mm)
10	320
15	213
20	160
25	128
30	107
40	80

Figure 11. Dimensions of plate models used in modelling of models tension tests

An average shear stress is calculated from the pull out force measured during the test by dividing the measured load by the lateral area of the plate. The average maximum shear stress τ_p obtained at different g-levels is shown in Figure 12 and seems to be independent of the g-level. All values are covered by a band width between 22 kPa and 25 kPa. The corresponding displacement u_p (prototype scale) is plotted versus g-level in Figure 13. The results show a slight increase of the prototype displacements with increasing g-level.

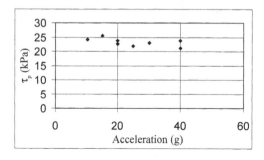

Figure 12. Average shear strength τ_p measured in modelling of models pull-out tests

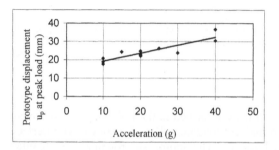

Figure 13. Displacements u_p (prototype scale) corresponding to the maximum shear stress mobilised in modelling of models pull-out tests on plates

4.2 Pull-out tests of piles

The diameter of the piles ranges from 2 mm to 36 mm and all piles are 300 mm long. The centrifuge pull-out tests are performed at 50 g and have all been duplicated. To obtain very well controlled rough interfaces, dowel steel screws of different diameters are used as model piles (Table 2). Fontainebleau sand is pluviated around the piles using the LCPC automatic hopper, at a density of 15.6 kN/m^3 (relative density $I_D = 0.7$). The maximum mobilised shear stress is calculated by dividing the peak load by the shaft area. Scale effect is determined by the ratio of the shear strength $\tau_p(D)$ observed on a given pile to the shear strength $\tau_p(Dmax)$ measured on the largest pile. Results have been presented in Garnier (1997) and are compared to previous tests performed in the LCPC centrifuge by Balachowski (1995).

Table 2. Roughness of the model piles (dowel screws)

Diameter (mm)	Thread (mm)	R_{max} (mm)	Normalised roughness R_n
2	0.4	0.43	2.2
4	0.7	0.76	3.8
6	1.0	1.08	5.4
12	1.75	1.89	9.5
26	3	3.24	16
36	4	4.32	22

4.3 Effect of pile diameter on the shear strength

Figure 14 shows that the two sets of results compare reasonably well although Balachowski's tests have been carried out on totally different materials than LCPC's ones (two different Hostun sands with $d_{50} = 0.32$ mm and $d_{50} = 0.7$ mm, pile diameters ranging from 16 mm to 55 mm).

No significant grain size effect can be seen if the ratio between the pile diameter D and the mean grain size d_{50} is larger than $D/d_{50} = 100$. For smaller piles, the maximum mobilised shear stress may be multiplied by a factor of up to 3.

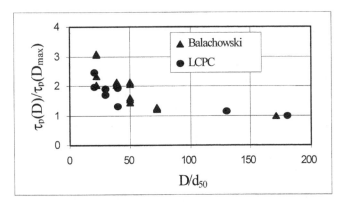

Figure 14. Effect of pile diameter D on shaft friction obtained from pull-out tests of pile of different diameters installed in different sands: $\tau_p(D)/\tau_p(D_{max})$ vs. D/d_{50}

4.4 Effect of pile diameter on displacement at peak load

Pile displacements at peak load versus model pile diameter are plotted in Figures 15a, b. Figure 15a shows that the model pile displacement u_p varies only from 0.6 mm to 1.2 mm when the pile diameter increases from 2 mm to 36 mm. The ratio u_p/D is presented in Figure 15b versus pile diameter D. For piles with diameters larger than about 20 mm (D/d_{50} higher than 100), this ratio is nearly constant and is about 0.04. For smaller piles, the ratio u_p/D increases drastically and may be multiplied by about 10 for the smallest piles with a diameter of 2 mm ($D/d_{50} = 10$).

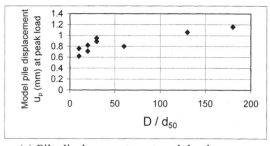 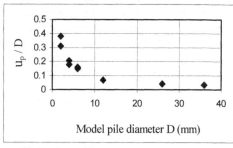

(a) Pile displacement u_p at peak load (b) Ratio u_p/D

Figure 15. Model displacements u_p corresponding to the maximum shear stress measured in pile pull-out tests

5 CONCLUSIONS

All friction tests (modified shear box, centrifuge pull-out tests) confirm that the interface roughness R may be normalised by the grain size $R_n = R/d_{50}$ and that there are three different types of interface behaviour:

- R_n higher than 0.5 to 1 (totally rough interface, large dilatancy, shear bands into the soil, no influence of R_n)
- R_n less than about 0.02 (totally smooth interface, no dilatancy, shear at the contact soil-structure, no influence of R_n)
- R_n between 0.02 and 0.5 (intermediate roughness with significant effect of R_n on the interface behaviour).

Both totally rough and smooth interfaces may be properly simulated in centrifuge models, but modelling intermediate roughness is much more complex.

Regarding the effect of size, results of all tests (pull-out tests on piles or plates) indicate that there is no significant scale effect on the maximum shear stress mobilised at the interface if the model dimensions are large enough compared to the grain diameter (D higher than about 100 d_{50}). For very small piles, with a diameter much smaller than 100 d_{50}, the mobilised shear strength may be two to three times the value observed for large piles.

The scaling laws concerning the displacement at peak resistance are not so clear. Centrifuge pull-out tests of plates and piles seems to show that prototype displacements are dependent on the size and scale of the models and on the g-level. However, the results of these tests concerning displacement measurements may be questionable due to negative friction that develops along the shaft during the increase in acceleration (Rezende at al. 1998, Garnier & König 1998). Part of the shear strength may indeed be mobilised before starting the pull-out loading test. New investigations are needed to clarify this important point for modelling soil-structure interactions that are driven by displacement.

REFERENCES

Alawneh, A.S., Malkawi A.H. & Al-Deeky H. 1999. Tension tests on smooth and rough model piles in dry sand. *Can. Geotechn. J.* 36: 746-753.

Balachowski, L. 1995. Modélisation physique du comportement des pieux en chambre d'étalonnage et en centrifugeuse. *Thèse INPG* Grenoble, 360 p.

Dano, C. 1996. Etude expérimentale des lois d'interface. *Mémoire de fin d'études*, INSA Rennes-LCPC, 166 p.

Dietz, M. S. 2000. Developing an holistic understanding of interface friction using sand within the direct shear apparatus. *Degree of Doctor of Philosophy*, University of Bristol, 282 p.

Dubreucq, T. 1999. Renforcement des fondations superficielles par inclusions planes horizontales extensibles – Quelques effets de taille dans le frottement sable-inclusions. *Thèse de doctorat ENPC*, 340 p.

Garnier, J. 1997. Validation of numerical and physical models: Problem of scale effects. In W. Wittke (ed.) *14th ICSMFE, Hamburg*: 659-662. Rotterdam: Balkema.

Garnier, J. & König, D. 1998. Scale effects in piles and nails loading tests in sand. In T. Kimura et al. (eds), *Int. Conf. Centrifuge 98, Tokyo*: 205-210. Rotterdam: Balkema.

Hommers, S. 1997. Zur physikalischen Modellierung des Tragverhaltens von Pfahlgründungen. *Diplomarbeit,* Ruhr-Universität Bochum, 114 p.

Kishida, H. & Uesugi, M. 1987. Tests of the interface between sand and steel in simple shear apparatus. *Géotechnique* 37(1): 45-52.

Le Collinet, J. 1998. Effet de cisaillement et d'accélération dans le cisaillement des interfaces sol-pieu. *Mémoire de stage de DEA*, Université de Nantes, LCPC, 80 p.

Mainetti, D. 1995. Etude du frottement sable sur plaque-acier. *Mémoire de stage IUT Génie mécanique et productique*, LCPC.

Paikowsky, S.G., Player, C.M. & Connors, P.J. 1995. A dual interface apparatus for testing unrestricted friction of soil along solid surfaces. *Geot. Test. J.* 18(2): 168-193.

Rezende, M.E., Garnier, J. & Cintra, J. 1998. Effect of shaft friction developed by centrifuge acceleration on pile loading tests results. In T. Kimura et al. (eds), *Int. Conf. Centrifuge 98, Tokyo*: 501-506. Rotterdam: Balkema.

Riot, A. 1996. Etude des interfaces sol/structure - Effets d'échelle. *Mémoire de stage IUT Génie mécanique et productique*, LCPC, 34 p.

Yoshimi, Y. & Kishida, T. 1981. A ring torsion apparatus for evaluating friction between soil and metal surfaces. *Geot. Test. J.* 4(1): 145-152.

Constitutive and Centrifuge Modelling: Two Extremes, Springman (ed.)
© *2002 Swets & Zeitlinger, Lisse, ISBN 90 5809 361 1*

Observing friction fatigue on a jacked pile

D.J. White & M.D. Bolton
Cambridge University Engineering Department, United Kingdom

ABSTRACT: Calibration chamber testing combined with a new technique of displacement measurement using image analysis has allowed the penetration mechanism of a jacked pile to be quantified. Observation of soil adjacent to an advancing pile revealed movement towards the shaft. This movement is linked to 'friction fatigue'. The mechanism of this process is observed to be volume reduction in the boundary layer at the pile-soil interface combined with horizontal unloading in the far field. The possibility of replicating this behaviour in a constitutive or centrifuge model is discussed.

1 INTRODUCTION

1.1 *Jacked piles*

The axial capacity of displacement piles in sand is arguably the subject of greatest uncertainty in geotechnical engineering (Randolph et al. 1994). There is no accepted consensus of the mechanism by which a pile penetrates soil. As a consequence, the most widely used design methods (eg. API 1993, Fleming et al. 1992, Eslami & Fellenius 1997) remain highly empirical, and must be accompanied by large safety factors.

Most design methods have evolved from experience of dynamically installed piles. Since pile hammers and vibrators are rarely permitted in urban areas, displacement piles are increasingly being installed by jacking. Press-in pile drivers can install pre-formed piles up to a maximum jacking force of 400 tonnes (4 MN) without exceeding urban noise and vibration limits (White et al. 2002). For this installation method to be utilized safely in axially loaded applications, the differences in the behaviour of jacked and driven piles must be established.

1.2 *'Friction fatigue'*

Recent field and laboratory testing (notably Lehane 1992, Chow 1997, De Nicola 1996, Bruno 1999) has revealed some aspects of pile behaviour which are not captured in conventional design methods. One key observation is the 'friction fatigue' or 'h/R' effect, where h/R represents the vertical distance above the pile tip normalised by pile radius. This is the phenomenon by which the horizontal effective stress, σ'_h (and hence local shaft friction, τ_s) acting on the pile shaft at a given soil horizon decreases as the pile tip penetrates deeper (Fig. 1).

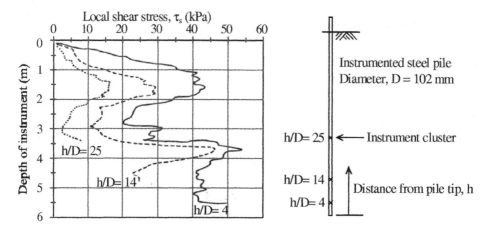

Figure 1. Local shear stresses during installation of an instrumented pile (after Lehane 1992)

The design framework described by Randolph et al. (1994) captures the 'h/R' effect by predicting the horizontal earth pressure coefficient (K) to decay exponentially with distance from the pile tip (Equation 1). The variables in this formulation are the maximum and minimum values (K_{max}, K_{min}) and the decay rate (μ). Design values for μ have been deduced from four different sources (Table 1). Figure 2a shows normalized curves of local shaft friction for each decay rate shown in Table 1, assuming $K_{min} = 0.2$, $K_{max} = 1$ and L/D = 20. These curves predict significantly different profiles of local shaft friction. This raises the question; why does each dataset reported in Table 1 display a different decay rate?

$$K(h) = K_{min} + (K_{max} - K_{min})\, e^{-(\mu h/D)} \tag{1}$$

Table 1. Friction fatigue parameters deduced from field and laboratory testing

Author	Description	Pile size	Decay rate, μ	Distance from pile tip to $\tau_s = \tau_{max}/2$ ($h_{50\%}$)			
				Actual $h_{50\%}$	Prototype $h_{50\%}$	$h_{50\%}/D$	$h_{50\%}/D_{50}$
Randolph et al. (1994)	Best fit to database of field tests.	Database of various. L/D=15–60	0.05	20 m*	20 m*	13*	≈200000* (Silty sand $D_{50} \approx 0.1$mm)
De Nicola (1996)	Best fit to database of centrifuge tests.	L= 150 mm D= 16 mm	0.25–0.35	66 mm	6.6 m	3.7	1300 ($D_{50} = 45$ μm)
Bruno (1999)	Best fit to database of centrifuge tests.	L= 200 mm D=11.5 mm	0.65	24 mm	2.4 m	2.1	500 ($D_{50} = 45$ μm)
Bruno (1999)	Best fit to field test.	L= 45 m D= 0.76 m	0.2	4 m	4 m	5.2	≈40000 (Fine sand & silt $D_{50} \approx 0.1$ mm)

* Data from worked example in original reference for L= 50m, D= 1.5m.

It might be expected that the different scales and embedment ratios of the pile tests shown in Table 1 would offer some insight into the origin of μ. If the initial ('unfatigued') local shaft friction is defined as the shear stress acting one diameter behind the pile tip, a reference distance over which the shearing process reduces local shaft friction by a factor of 2 can be defined as $h_{50\%}$ (Fig. 2b). This definition neglects any change in τ_{max} over the short distance $h_{50\%}$. This reference distance has been extracted from the original data reported in Table 1, and has been non-

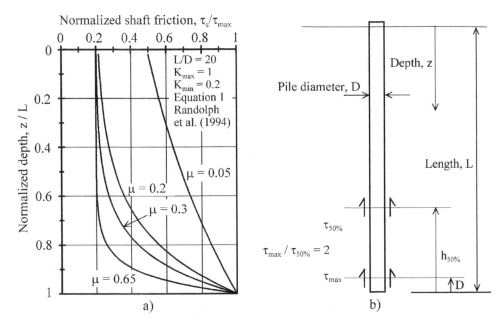

Figure 2. The decay of local shaft friction

dimensionalized by normalizing with two local length scales; pile diameter and original D_{50} grain size. No clear trend is evident. The decay in normal effective stress is not a direct function of absolute distance sheared (prototype or centrifuge scale) or distance sheared normalized by pile diameter or original D_{50} grain size.

The pile tests listed in Table 1 all involved dynamic installation. Would a different rate of decay have been observed if the piles were installed by jacking? De Nicola & Randolph (1997) examined the distribution of horizontal earth pressure coefficient within a tubular model pile. It was found that the internal shaft friction in a driven (hammered) pile was best predicted by assuming a degradation of K typically from $K_{max} = 1$ to $K_{min} = 0.4$ along the internal soil column. In contrast, for a jacked model pile, a best match was achieved if K was assumed to remain constant at K_{max} along the internal soil column. This suggests that friction fatigue predictions obtained from dynamically installed piles may under-predict the capacity of jacked piles.

2 EXPERIMENTAL METHODOLOGY

2.1 *Calibration chamber*

Calibration chambers are widely used to study penetration resistance (e.g. Houlsby & Hitchman 1988, Salgado et al. 1997, Yasufuku & Hyde 1995). The stresses and deformations around the tip of an advancing CPT or pile can be replicated correctly by applying a surcharge pressure. In order to observe the deformation around an advancing pile, a plane strain chamber with observation windows has been constructed (Fig. 3). Sheets of glass are placed on the inner faces of the box to reduce side friction. A surcharge pressure is applied through a rubber bag. The model pile is jacked into the chamber at a rate of 1 mm/minute. Digital cameras are used to record images of the soil and pile at regular intervals. In this paper, some results from a test on dry medium dense (relative density 44%, voids ratio 1.48) Dog's Bay carbonate sand are presented. The mechanical behaviour of this soil is described by Coop (1990).

349

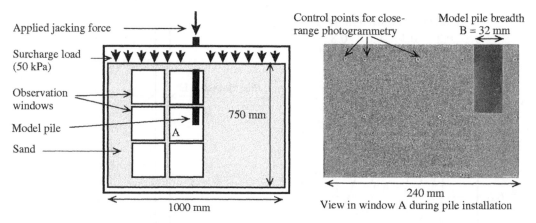

Figure 3. Plane strain calibration chamber

2.2 *Displacement measurement using PIV image analysis and close-range photogrammetry*

A novel technique for non-contact measurement of soil deformation in physical models has been developed. This system combines digital photography, close-range photogrammetry and image analysis using Particle Image Velocimetry (PIV). Soil displacements are measured to a high precision without requiring intrusive target markers to be installed in the soil.

Image processing algorithms based on PIV have been written to track the movement of small patches of soil (typically 2 - 4 mm in size) through a series of digital images to a precision of $1/15^{th}$ of a pixel (White et al. 2001a). The images presented in this paper were captured using a Kodak DC280 digital camera, with a pixel resolution of 1760 x 1168. Having measured the image-space coordinates of the deforming soil by PIV, these must be converted into model-space coordinates. This process is known as camera calibration. The calibration routine developed in this research has 18 parameters to describe the model-space to image-space transformation and has a measured precision of 1/18500 of the field of view (White et al. 2001b).

3 RESULTS

The phenomenon of friction fatigue has been examined by measuring the movement of a horizon of soil adjacent to the pile shaft as the pile penetrates beyond this horizon. Image 1 (Fig. 4a) shows the tip of the pile entering the field of view, with a mesh of PIV patches established adjacent to the pile shaft (Fig. 4c). Comparison with a subsequent image (Figs. 4b, d) taken after 80 mm (2.5 pile widths) of further penetration allows the intervening soil movement to be measured. The PIV analysis reveals that the soil is moving towards the pile shaft (Fig. 4e), with the greatest vector (250 μm, 1.9 pixels) being measured in the patch closest to the pile shaft. Similar vectors were obtained from analysis of the opposite face of the pile.

Differentiation of the horizontal component of displacement with respect to the gauge length between adjacent PIV patches allows horizontal strain to be plotted (Fig. 4f). This reveals that the soil is unloading in horizontal extension, with the greatest strain (0.6%) occurring close to the pile shaft. After completion of the test, the chamber was disassembled, and the sand adjacent to the shaft was photographed (Fig. 5a). The measured displacements are superimposed on this image.

Particle size analysis of sand taken from zone B (within 3 mm of the pile shaft) was carried out using a single particle optical sensing method (AccuSizer 780/DPS, http://www.christison .com). This device measures the mean diameter of individual particles flowing past a laserdiode sensor. A significant shift in the grading curve is evident, with the d_{10} size being reduced from 330 µm to 150 µm (Fig. 5b).

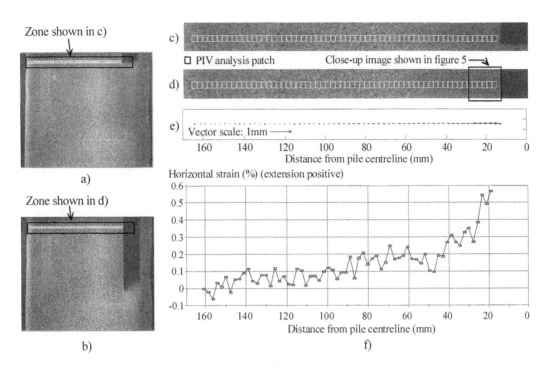

Figure 4. PIV analysis of sand adjacent to pile shaft

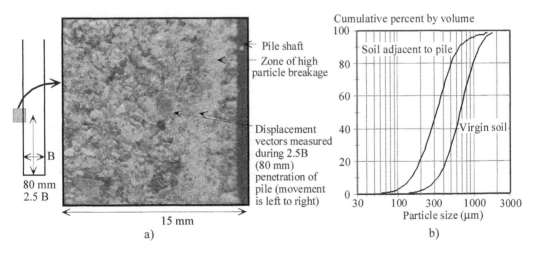

Figure 5. Post-mortem analysis of sand adjacent to pile shaft

4 DISCUSSION

The measured displacements and strains, combined with the insight offered by the close-up photography, allow the kinematics of friction fatigue to be deduced (Fig. 6a). The 3 mm thick zone of sand closer to the pile than the closest PIV patch undergoes volume reduction, due to continued shearing at the pile-soil interface. This volume reduction permits the soil further from the shaft to relax inwards.

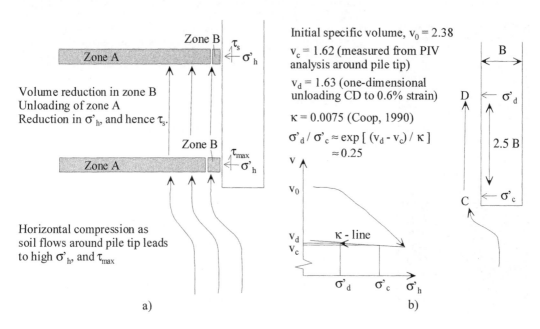

Figure 6. Friction fatigue mechanism

Figure 6b shows a simple calculation in which the measured volume changes are used to predict the change in horizontal stress adjacent to the pile shaft, with the unloading stiffness in $\ln \sigma'_h - v$ space being approximated as κ. This framework predicts a four-fold decay in horizontal stress after 80 mm of shearing, which is comparable to the decay rates presented in Table 1 for reduced-scale modelling (Bruno 1999, De Nicola 1996).

The friction fatigue mechanism hypothesized in Figure 6 suggests that the reduction of horizontal effective stress acting on the pile shaft is governed by two processes:

Process 1: Volume reduction in zone B due to continued shearing at the pile-soil interface.

It is hypothesized that this process is associated with two mechanisms of volume reduction. Firstly, rearrangement and repacking of the sand grains is caused by the agitative action of the rough pile surface. Secondly, further repacking in the boundary layer is permitted by diffusion of the fine broken particles away from the pile-soil interface into the more open matrix of uncrushed soil in the far field. The one-way shearing created by jacking is likely to create less rearrangement than two-way cycling during dynamic pile installation.

Process 2: Horizontal unloading in zone A.

This is a continuum unloading process. The governing stiffness will depend not only on the *in situ* soil properties, but also on the installation-induced stress level at that soil horizon (which will have occurred as the pile tip passed) and also the installation-induced strain level (high in the near field, low in the far field).

Since the focus of this workshop is the contrast between centrifuge and constitutive modelling, a key question is whether this mechanism of friction fatigue can be correctly replicated using either of these modelling techniques.

Process 1 is not continuum behaviour and hence is not captured by conventional constitutive models. This process *will* occur in centrifuge models, but may not scale correctly. Conventional centrifuge laws of scaling particle size and surface roughness are unlikely to create self-similar diffusion of fine particles. Also, attempts to model shaft friction by maintaining the ratio of pile roughness to particle size fail when breakage occurs since particle strength (and hence change in grading curve for a given stress history) is a function of particle size (McDowell & Bolton 1998).

Process 2 will occur in a centrifuge model, and the stress and strain-histories will be correctly replicated. However, if process 2 is to be modelled correctly in a numerical simulation using a conventional constitutive model, the entire installation process must be simulated to capture the stress- and strain-histories of the soil around the pile.

The difference in particle size between the virgin soil and the soil around the pile shaft (Fig. 6b) has an influence not just on friction fatigue and the horizontal stress on the pile shaft, but also affects the coefficient of pile-soil friction. Correlations between mean particle size and coefficient of friction derived from interface shear box testing (e.g. Kishida & Uesugi 1987, Jardine et al. 1993) do not acknowledge that the soil adjacent to the pile shaft has a significantly different grading curve.

5 CONCLUSIONS

Calibration chamber testing combined with a new technique of displacement measurement using image analysis has allowed the penetration mechanism of a jacked pile to be quantified. Observation of soil adjacent to an advancing pile revealed movement towards the shaft. This movement is linked to the decrease in horizontal stress acting on the pile shaft; a process known as 'friction fatigue'.

A mechanism for this process consists of i) volume reduction in the boundary layer at the pile-soil interface combined with ii) horizontal unloading in the far field. This process involves physical behaviour that cannot be captured by a continuum constitutive model nor replicated at small scale using conventional scaling laws.

REFERENCES

American Petroleum Institute (API). 1993. RP2A: Recommended practice of planning, designing and constructing fixed offshore platforms- Working stress design, 20th edition. Washington: 59-61.
Bruno, D. 1999. Dynamic and static load testing of driven piles in sand. PhD thesis, University of Western Australia.
Chow, F.C. 1997. Investigations into the behaviour of displacement piles for offshore foundations. PhD thesis, University of London (Imperial College).
Coop, M.R. 1990. The mechanics of uncemented carbonate sands. *Géotechnique* 40(4): 607-626.
De Nicola, A. 1996. The performance of pipe piles in sand. PhD thesis, University of Western Australia.
De Nicola, A. & Randolph, M.F. 1997. The plugging behaviour of driven and jacked piles in sand. *Géotechnique* 47(4): 841-856.
Eslami, A. & Fellenius, B.H. 1997. Pile capacity by direct CPT and CPTu methods applied to 102 case histories. *Canadian Geotechnical Journal* 34(6): 886-904.

Fleming, W.G.K., Weltman, A.J., Randolph, M.F. & Elson, W.K. 1992. *Piling Engineering*. Glasgow: Blackie (Halsted Press).

Houlsby, G.T. & Hitchman, R. 1988. Calibration chamber tests of a cone penetrometer in sand. *Géotechnique* 38(1): 39-44.

Jardine, R.J., Lehane, B.M. & Everton, S.J. 1993. Friction coefficients for piles in sands and silts. *Offshore Site Investigation and Foundation Behaviour, Soc. for Underwater Technology* 28: 661-677.

Kishida, H. & Uesugi, M. 1987. Tests of the interface between sand and steel in the simple shear apparatus. *Géotechnique* 37(1): 45-52.

Lehane, B.M. 1992. Experimental investigations of pile behaviour using instrumented field piles. PhD thesis, University of London (Imperial College).

McDowell, G.R. & Bolton, M.D. 1998. On the micromechanics of crushable aggregates. *Géotechnique* 48 (5): 667-679.

Randolph, M.F., Dolwin, J. & Beck, R. 1994. Design of driven piles in sand. *Géotechnique* 44(3): 427-448.

Salgado, R., Mitchell, J.K. & Jamiolkowski, M. 1997. Cavity expansion and penetration resistance in sand. *ASCE Journal of Geotechnical and Geoenvironmental Engineering* 123(4): 344-354.

White, D.J., Take, W.A. & Bolton, M.D. 2001a. Measuring soil deformation in geotechnical models using digital images and PIV analysis. *Proc. 10th Int. Conf. on Computer Methods and Advances in Geomechanics. Tucson, Arizona:* 997-1002. Rotterdam: Balkema.

White, D.J., Take, W.A, Bolton, M.D. & Munachen, S.E. 2001b. A deformation measuring system for geotechnical testing based on digital imaging, close-range photogrammetry, and PIV image analysis. *Proc. 15th Int. Conf. on Soil Mechanics and Geotechnical Engineering. Istanbul, Turkey.* 1: 539-542, Rotterdam: Balkema.

White, D.J., Finlay T.C.R., Bolton, M.D. & Bearss, G. 2002. Press-in piling: Ground vibration and noise during pile installation. *Proc. International Deep Foundations Congress. Orlando, USA. ASCE Special Publication* 116: 363-371.

Yasufuku, N. & Hyde, A.F.L. 1995. Pile end-bearing capacity in crushable sands. *Géotechnique* 45(4): 663-676.

Constitutive and Centrifuge Modelling: Two Extremes, Springman (ed.)
© *2002 Swets & Zeitlinger, Lisse, ISBN 90 5809 361 1*

Experimental observations of interface friction

M.S. Dietz
University of Bristol, United Kingdom

ABSTRACT: A modified direct shear apparatus has been used to investigate interface friction using surfaces of wide-ranging roughness. The modifications have allowed the role of interface dilation to be clarified and prompted a classification of interfaces based on stress-dilatancy considerations. Multi-reverse interface tests indicate the run-in angle of friction, mobilised after interface abrasion, is independent of initial surface roughness. It may represent the appropriate operative angle of interface friction occurring in the field.

1 INTRODUCTION

Load transfer between a granular material and a solid surface occurs across an interface. Such systems have received a good deal of experimental investigation. Concurrence between the failure characteristics of interfaces and the kinematics of the Direct Shear Apparatus (DSA) has led to the pervasion of this apparatus in this field, its output often condensed into peak angles of interface friction for publication. However, the literature is full of criticism of the DSA and details of a number of purportedly rectifying adaptations have been published. Here, the modifications necessary for a working apparatus have pre-empted a study of sand-steel interface friction. Escaping from convention, an attempt has been made to view interface response in the light of *all* the recorded data, and thus provide a better understanding of interfacial processes.

2 EXPERIMENT

An experimental investigation of interface behaviour was undertaken using a modified 100mm DSA. Figure 1 contrasts the essential features of the archetype with the new Winged Direct Shear Apparatus (WDSA), as used for interface testing.

The modified articulation of the WDSA simultaneously produces reliable estimates of vertical stress and permits unimpeded dilation. With the load pad secured within the upper frame (Jewell 1989), the test method is simplified and rotation of the apparatus' upper components is significantly reduced. Whilst the archetype's simplicity is preserved, high quality data are generated.

Interface behaviour was investigated by replacing the lower half of the apparatus with a series of solid steel blocks, the uppermost surface of each having a distinct topography: the roughest

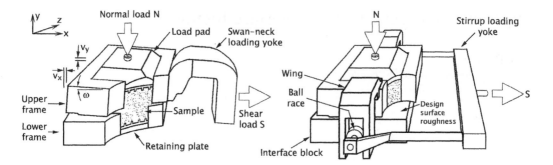

Figure 1. The archetypal DSA contrasted with the WDSA as used for interface tests

with a layer of sand grains attached and the smoothest finely polished to a mirror finish. Topographies were digitized using a Talysurf stylus profilometer and statistical descriptors of roughness derived from the resulting set of ordinates or "profile".

Three sands were investigated: fine (D_{50} = 0.2 mm), medium (D_{50} = 0.4mm) and coarse (D_{50} = 0.8 mm, Leighton Buzzard 14-25), each with rounded to sub-rounded particles. Samples were pluviated into the WDSA. The stress levels spanned the limits of the apparatus and produced vertical stresses on the apparatus' horizontal mid-plane, σ'_{yy}, between 25 and 250 kPa.

3 INTERPRETATION

The WDSA, when properly optimised, produces reliable plane strain strength parameters. The appropriate test interpretation is based on four assumptions. Firstly, that there is sufficient uniformity that the deforming sand can be described in terms of a single state of stress and incremental strain; secondly, that the elastic stiffness is sufficiently high that the elastic strains are negligible when compared to the plastic strains; thirdly, that the principal axes of stress and strain rate coincide; and lastly, that the horizontal is a direction of zero linear incremental strain. Ratios of the WDSA's boundary measurements (Fig. 1) can give firstly the direct shear angle of friction (ϕ'_{ds}), and secondly the angle of dilation (ψ), which, by employing the Mohr's circles of Figure 2, can be combined to produce the plane strain angle of friction (ϕ'_{ps}):

$$\tan\phi'_{ds} = \tau_{yx} / \sigma'_{yy} = S/N \tag{1}$$

$$\tan\psi = -d\varepsilon_{yy} / d\gamma_{yx} = dv_y / dv_x \tag{2}$$

$$\sin\phi'_{ps} = \frac{\tan\phi'_{ds}}{\cos\psi + \sin\psi \tan\phi'_{ps}} \tag{3}$$

At the critical state, when the dilation rate is zero, Equation 3 reduces to:

$$\sin\phi'_{crit} = \tan\phi'_{ds} \tag{4}$$

During interface tests, it is the forces mobilised parallel to the solid surface which are of interest. Thus, independent of dilation rate, the angle of interface friction (δ') is given by:

$$\tan\delta' = \tau_{yx} / \sigma'_{yy} \tag{5}$$

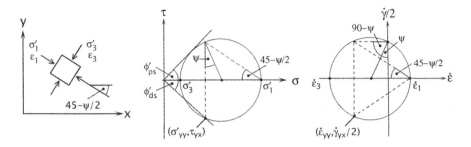

Figure 2. Mohr's circles of stress and incremental strain for dense sand in the WDSA

4 EXPERIMENTAL FINDINGS

Data from six interface tests on surfaces of increasing roughness are displayed alongside those of an equivalent direct shear test in Figure 3. Block arrows indicate the trends for data as the interface's constituent surface roughens. Increased interface strength is accompanied by increases both in dilatancy rate (dv_y/dv_x) and vertical displacement (v_y) and rotation (ω) of the apparatus' upper half. The interface able to offer the highest resistance, i.e. that featuring the roughest surface, behaves similarly to the same sand in the equivalent state when tested in direct shear.

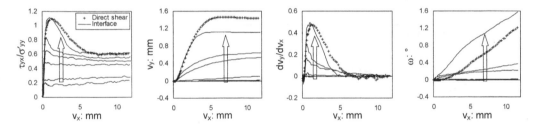

Figure 3. The response of interfaces featuring dense Leighton Buzzard 14-25 at low stress levels (25 kPa)

4.1 Peak interfacial state

Within the literature it has become the norm to display interface response on axes of peak resistance versus some measure of surface roughness. Uesugi & Kishida (1986a), the proponents of this model, employed the Japanese Standard R_t (the vertical height between the highest peak and lowest valley on a surface profile) to quantify roughness, although this parameter is not exceptional. At least eight other conventional roughness parameters along with a newer fractal parameter are equally adept at correlating experimental data (Dietz 2000). Here, and as indicated in Figure 4a, the British Standard (BS 1134 1961) roughness parameter R_a (the arithmetical mean absolute deviation a profile from its centre line) is employed.

Uesugi & Kishida (1986b) unified peak resistance data using D_{50} to normalise the ordinate, as demonstrated in Figure 4b. Paikowsky et al. (1995) explained that smaller particles found a given surface rougher than larger particles, as indicated in Figure 5 by an increase in the angle α between the plane of contact and the plane of movement. Using figures similar to Figure 4b, Paikowsky et al. (1995) distinguished between "smooth", "intermediate" and "rough" interfaces.

The WDSA's unrestrained upper half allows the peak dilatancy response of interfaces to be analysed in a similar fashion. Figure 4c and d show peak dilatancy behaviour to fit into the framework suggested by Paikowsky et al. (1995) for peak resistance.

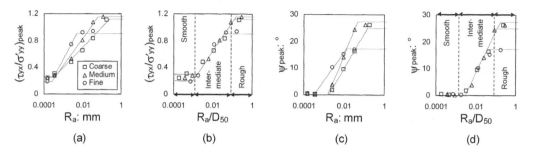

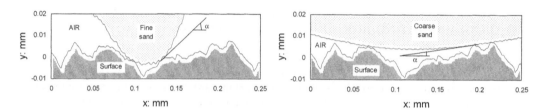

Figure 4. Peak state interface data for coarse, medium and fine dense sand at low stress levels (25 kPa)

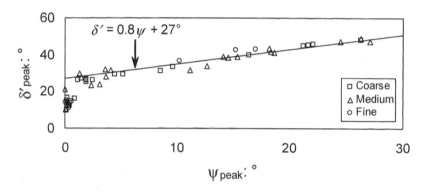

Figure 5. Sand grain profiles superimposed above that of a solid surface

Figure 6. Peak stress-dilatancy response of interfaces comprising different sands, surfaces, stress levels and densities

An increase in strength coinciding with an increase in dilatancy response is a familiar concept. Bolton's (1986) comprehensive review of published friction and dilation angles produced a simple empirical fit to experimental data, while expressing this idea succinctly:

$$\phi'_{ps} = 0.8\psi + \phi'_{crit} \tag{6}$$

By plotting the peak stress-dilatancy response, a similar relationship is revealed for dilative interfaces. A two-tiered model of interface behaviour may thus be more appropriate than the three-tiered characterisation of Paikowsky et al. (1995). On the axes of Figure 6, non-dilative interfaces lie close to abscissa, somewhere between the origin and a transitional value of δ'. Dilative interfaces mobilise a δ'_{peak} above this transitional value and in accordance with their dilational characteristics. The transitional value of δ' that provides a reasonable fit to the test data of Figure 6 is 27°.

4.2 Beyond peak state

Within a few millimetres of shear displacement from peak, the rate of change of interface data becomes small. Jardine et al. (1993) proclaimed this to be indicative of an interfacial critical state activated once sand grains become "unlocked" from the interface's surface. Such a definition is only appropriate if the paths followed by all tests arrive at a unique end state. Contradictory to this, Figure 3 shows the post-peak resistance to be roughness dependent and, at least for smooth interfaces, to have a tendency to increase with v_x.

Reverse sand-steel interface tests were performed to investigate this behaviour more fully. During these tests, the cumulative shear displacement v_x was increased by displacing the interface block to and fro. The response of the "intermediate" interface of Figure 7a suggests a critical state has been mobilised where deformation can continue without changes in stress or, were it not for the reversals in shear direction, volume.

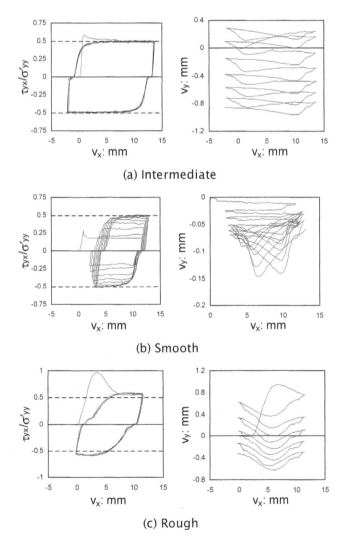

(a) Intermediate

(b) Smooth

(c) Rough

Figure 7. Reverse interface data for dense Leighton Buzzard 14-25 at high stress (250 kPa)

In contrast, the "smooth" interface of Figure 7b exhibits large increases in resistance during early traverses. A clearer picture is presented in Figure 8 wherein absolute data is plotted excluding the valleys associated with shear reversal. The strain hardening of smooth interfaces ends once dilatancy is activated, and only then does behaviour indicate the existence of a critical state.

The "run-in" δ' values mobilised after reverse shear are indicated by the subscript r. The smooth and intermediate sand-steel interfaces of Figure 7 mobilise $(\tau_{yx}/\sigma'_{yy})_r$ values of similar magnitude (0.50), producing δ'_r values of 27°. As indicated in Figure 7c, the rough interface behaves differently. Reverse shear stress ratios are appreciably higher (0.57), corresponding to a δ'_r value of 30°.

Dismantling the WDSA post-test revealed a powder to have collected across the interface block. Judging by its colour, the dust comprised primarily abraded sand particles. Swathes of colour of differing size and hue remained attached to the interface block even after thorough scrubbing and these were only removed by surface re-preparation. Changes in surface topography were also apparent. Ploughing of the relatively hard sand grains caused anisotropic roughening of initially smooth steel surfaces. In contrast, rougher steel surfaces had their asperities flattened, appearing smoother at the end of the test.

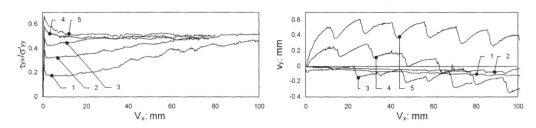

Figure 8. Absolute reverse interface data for dense Leighton Buzzard 14-25; increases in initial roughness indicated by increase in line number

5 CONCLUDING REMARKS

5.1 Summary and implications

Considering first the peak state, quantitative dilatancy analysis of interfaces has revealed a dilatancy response that fits the Paikowsky et al. (1995) framework for characterising resistance. To avoid masking the dilatancy behaviour, it is suggested that a two-tiered representation of interface response on axes of δ' versus ψ provides a more realistic view of interface behaviour. Post-peak behaviour is roughness dependent and, although states resembling critical are eventually attained by all of the interfaces investigated, Jardine et al. (1993) assert that a critical state for interfaces is mobilised within a few millimetres of shear displacement: this is misleading.

The roughest interface, comprising sand grains adhered to the interfacial block, did not reduce to a δ'_r value of similar magnitude to that recorded on the metallic interfaces, instead remaining at a value corresponding to $\sin\phi'_{crit}$. Thus, either a fundamental difference exists between sand-sand friction and sand-steel friction or the measured run-in data, which seems to indicate a unique end state, is actually transitory and slow increases in the resistance are to be expected. The observations of interface abrasion and degradation suggest that the first hypothesis is true, the hardness gradient between sand and steel playing a role in metallic friction absent when particles displace over themselves. Judging from this abrasion evidence, it seems likely that the Paikowsky et al. (1995) conceptual model purporting the inclination of grain/steel sliding planes to determine interface resistance is overly simplistic.

5.2 The choice of angle of interface friction

In the absence of experimental data, BS 8002 (1994) recommends that the "representative strength" of interfaces "should not exceed" one of two values: $\delta' = 20°$, for "smooth surfaces with a texture finer than that of the median particle size"; or $\delta' = \phi'_{crit}$, for "rough surfaces with a texture coarser than that of the median particle size". Following from this work, it is suggested that the choice of a representative angle of sand-steel interface friction should be based on the following reasoning: firstly, that only small shear displacements are needed to overcome peak response; secondly, that the relative security of the critical state, with its avoidance of the uncertainties of dilatancy, provides a better springboard for predictive estimates; and lastly, that even on initially "mirror-finish" surfaces, only moderate shear displacements are required to mobilise the run-in resistance.

With this in mind, and extrapolating from the limited data presented herein, it is tentatively suggested that δ'_r best represents the operative angle of interface friction, and reasonable estimates of its magnitude on both standard and uncharacteristically rough surfaces are given by:

$$\delta'_r = 3\phi'_{crit}/4 \quad \text{and} \quad \tan\delta'_r = \sin\phi'_{crit} \qquad \text{(7) and (8)}$$

Thus, the representative interface strengths suggested by BS 8002 (1994) for smooth interfaces are woefully conservative and, for rough interfaces, unsafe.

REFERENCES

Bolton, M.D. 1986. The strength and dilatancy of sands. *Géotechnique* 36(1): 65-78.
BS 1134. 1961. Surface texture. London, UK: British Standards Institute.
BS 8002. 1994. Earth retaining structures. Milton Keynes, UK: British Standards Institute.
Dietz, M.S. 2000. Developing an holistic understanding of interface friction using sand within the direct shear apparatus. PhD Thesis, University of Bristol.
Jardine, R. J., Lehane, B. M. & Everton, S. J. 1993. Friction coefficients for piles in sands and silts. Soc. Underwater Tech. 28: Offshore site investigation and foundation behaviour: 661-677.
Jewell, R.A. 1989. Direct shear tests on sand. *Géotechnique* 39(2): 309-322.
Paikowsky, S.G., Player, C. M. & Connors, P.J. 1995. A dual interface apparatus for testing unrestricted friction of a soil along solid surfaces. *Geotech. Test. Jrnl, GTJODJ* (18)2: 168-193.
Uesugi, M. & Kishida, H. 1986a. Influential factors of friction between steel and dry sands. *Soils & Foundations* 26(2): 33-46.
Uesugi, M. & Kishida, H. 1986b. Frictional resistance at yield between dry sand and mild steel. *Soils & Foundations* 26(4): 139-149.

Discussion on problems governed by INTERFACES

S.M. Springman & T. Weber
Swiss Federal Institute of Technology, Zurich, Switzerland

ABSTRACT: In this series on problems governed by interfaces, a general discussion follows a summary of the short research contributions.

1 OBSERVING FRICTION FATIGUE ON A JACKED PILE (D. WHITE)

The speaker presented his research into penetration mechanisms at the base of a jacked pile in carbonate sand, as seen through a perspex window in a calibration chamber. He was interested in the phenomenon of friction fatigue, for which the local shear stress acting on the shaft of a jacked pile was shown to decrease with the distance that the pile tip had travelled past this location. He mentioned an instrumented pile test published by Lehane (1992), from which surface shear stress transducers had shown that local shear stress depended on the depth and the distance that the pile tip had progressed past a given soil horizon. This had led to degradation in the maximum shear stress that could be mobilised, and this was significant for pile design.

He acknowledged the importance of this data but noted that a field test was unable to deliver insight into the mechanisms. He had been able to determine these from tests in a calibration chamber, in terms of the soil flow around a pile during jacking, by using the Particle Imaging Velocimetry (PIV) technique. Details of pattern matching of texture using PIV, as defined on a grid of patches, were described and the patches were tracked in image space during a further 80 mm of pile penetration. It was noted that it was now possible to measure displacements of only one third of a mean grain size so that errors from short focal length cameras might now be significant, and that these included fisheye effects and barrelling of images. He also commented that other errors might include non-perpendicularity of the camera to the plane and distortion if the pixels were not square.

The close range photogrammetry allowed a robust reconstruction of image space data (with displacement in pixels) to model space data (in terms of microns) and showed that soil moved firstly away from the pile tip as the tip passed the particles concerned, and then moved back towards the shaft. This meant that as the pile continued to shear soil at the interface, the soil in the far field was relaxing in horizontal extension, which was only permissible if the soil in the interface zone reduced in volume as shearing continued. This led to very stiff unloading and hence a large reduction in the normal stress acting on the pile.

Rigid body rotations, shear and volumetric strains were deduced from differentiation of displacement vectors from the patches and pile installation caused up to 100% shear strain and volumetric strains of 30% close to the shaft. The speaker suggested that centrifuge modellers should look carefully at the two zones of soil adjacent to the shaft, with the virgin soil in the far field whereas fine crushed sand could be found in the 3 mm adjacent to the pile. The grading curve of these crushed particles showed that the mean grain size had been reduced by a factor of two (determined by a single particle optical particle sizing machine since the sample was too small to sieve). He referred to Jacques Garnier's presentation and wondered what normalised roughness should be used to reduce scale effects since a jacked or driven pile might now exhibit a smooth interface now that the particles were so much smaller!

Having observed these complicated aspects, he also questioned the level of confidence of replicating friction fatigue for driven piles in numerical or centrifuge modelling and asked whether field testing was really the only way of obtaining appropriate results.

2 EXPERIMENTAL OBSERVATIONS OF INTERFACE FRICTION (M. DIETZ)

The speaker described an experimental investigation of interface friction featuring sand in a new form of winged direct shear apparatus and focused on issues relating to interface degradation or abrasion rather than on peak states. He confirmed that a fine layer of crushed particles was found adjacent to the interface, and quoted geomechanics literature suggesting that small particles would find a given surface rougher than larger particles, so that the geometry of interface interaction would change as crushing progressed. On the other hand, some tribology literature indicated that small particles would find any given steel surface harder than large particles.

In short, both the geometry and the material properties would be altered as interface displacement progressed. He also noted that the surface topography would change as well and that a fractal characterisation produced an effective means of visualisation via a relative length defined as the apparent length normalised by the projected profile length. He explained that a typical steel surface had four regions varying from smooth, with a relative length of approximately 1, via transitional regions on the smooth-rough crossover to those exhibiting fractal characteristics, representative of the complexity of the surface roughness of interfaces. Analyses on surfaces before and after interface tests showed that rough surfaces became smoother and initially smooth surfaces became rougher.

Physical attributes relating to the Vickers hardness were discussed, with values for mild steel lying between 121-149 and for silica between 750-1200, which produced a hardness ratio of at least 5. Sand particles would plough wear scars into smooth surfaces, anisotropically roughening the surface texture, whereas asperities on rough surfaces would be flattened by sand particles, and changes in surface texture and sand particle grading would be brought about.

Consequently, reverse shear tests were carried out with a solid surface displaced to and fro underneath the upper section of the shear box containing the soil, thus increasing cumulative shear displacement. Rough surfaces gave dilation from the outset, and an absolute steady state stress ratio (in terms of shear strain normalised by normal effective stress) of $\tau_{yx}/\sigma'_{yy} \sim 0.5$ was soon mobilised on the interface, whereas smooth surfaces showed hardening associated with the traverse of apparatus until the surface had degraded. Dilatancy was mobilised slowly with cumulative relative displacement to the same level as for the rough surface, with approximately the same stress ratio of 0.5, which was identified as the run-in angle of interface friction of $\delta' \sim 27°$.

The speaker posed a question to the physical modellers about whether it was a feasible objective to replicate all aspects of interface behaviour at reduced scales when interface geometry and material attributes were not only dependent on scale but also changed as interface displacement progressed. If not, he asked whether it was worthwhile carrying out reduced scale model tests of scenarios when interface friction was dominant.

3 STEEL-GRANULAR MATERIAL INTERACTION FROM SMALL TO LARGE DEFORMATIONS (N. YASUFUKU)

The speaker introduced his work on steel-granular material interaction behaviour under a constant normal stress of 100 kPa in a ring shear apparatus with an outer and inner diameter of 300 mm and 200 mm respectively, and a sample height of 40 mm (Fig. 1). He noted that many factors influenced the shear stress - shear displacement response and that these included the roughness of the steel, soil crushability, particle angularity, soil grading (coefficient of uniformity U_c), soil relative density (D_r) etc. He declared that the focus of this research was on soil crushability and the relative roughness between the mean grain size and steel surface, where relative roughness (R_{max}/D_{50}) was defined as the change in the height of asperities on the surface of the steel (R_{max}), normalised by the mean grain size (D_{50}).

He showed a typical result of the stress ratio mobilised in terms of the shear stress normalised by the normal effective stress (τ/σ'_n) against the horizontal displacement (H). Two points, A and B, have been defined, where A represents the peak stress ratio and B is the critical or residual state (Fig. 2). The interaction behaviour observed when relative roughness or crushability increases is shown in matrix form in Figure 3.

Shear test   **Frictional test**

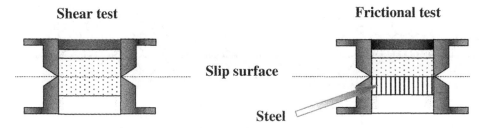

Figure 1. Test set up for soil-soil shear and soil-steel frictional tests (Yasufuku 2001)

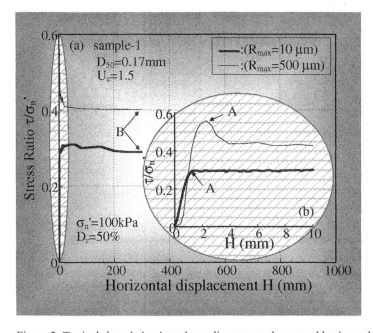

Figure 2. Typical shear behaviour depending on roughness and horizontal displacement (Yasufuku 2001)

Tests for smooth surfaces with low crushability delivered no significant peak stress (Fig. 3, dotted line) and mobilised more or less a constant angle of friction (solid line), whereas very rough surfaces with low crushability show that a higher peak strength can be mobilised and that this was associated with dilatancy at low displacements. A completely different response is found for very smooth surfaces for particles with high crushability in that a small peak value of shear strength is mobilised before particle degradation becomes significant and this is accompanied by a gradual increase of friction angle (ϕ'_δ) with horizontal displacement.

The mobilised interface friction angle ϕ'_δ was found to increase as the surface (Fig. 4a) or relative roughness (Fig. 5) increased for both peak and residual states, and these converged towards a steady state value (Fig. 5). The friction angle decreased at both peak and residual states from more or less the same steady state values for fine sands as the mean grain size increased (Fig. 4b).

The speaker felt that centrifuge modelling delivered 'lucky harmony' in that even if the g-value was increased, the relative roughness would be constant, if R_{max} and D_{50} remained in the same ratio, and that if this parameter was controlled, the mobilised angle of friction should not be affected.

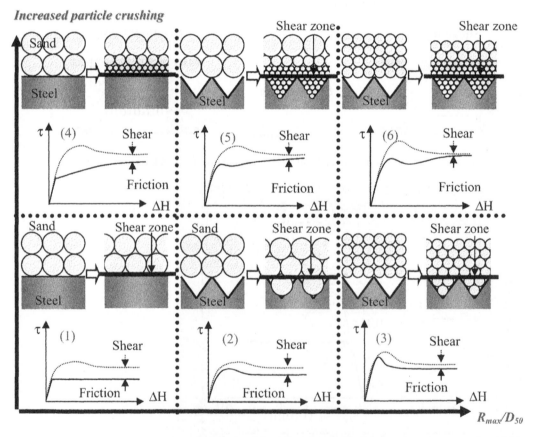

Figure 3. Surface friction behaviour with increased particle crushing and relative roughness (Yasufuku 2001)

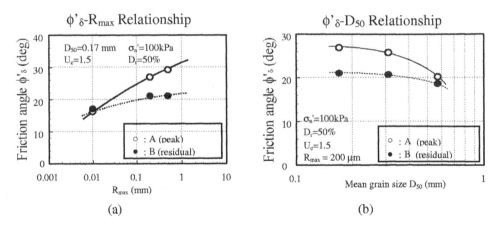

Figure 4. Relationship between friction angle, roughness and mean grain size (Yasufuku 2001)

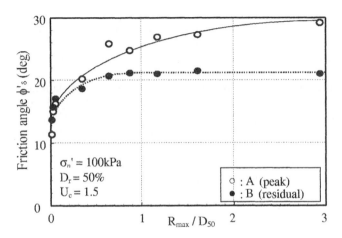

Figure 5. R_{max} / D_{50} relationship at peak and residual state (Yasufuku 2001)

4 CONSTITUTIVE MODELLING OF SOIL AND ROCK INTERACTING WITH INTERFACES (V. DE GENNARO)

The speaker noted that the states of deformation and stress were not homogeneous inside a shear box, so what was usually considered as an element test on an interface, did not fulfil the essential conditions of homogeneity of stresses and strains. He said that there were enormous difficulties in isolating the interface, the thickness of the shear zone was unknown and that an experimental visualisation of this layer would be helpful to estimate and correlate the thickness to an internal dimension of the material, which was often selected as the mean grain size D_{50} in granular soils.

He referred to the model presented by Jean Sulem (Sulem & De Gennaro 2002) in the context of elastoplasticity and the simulation of interface shear tests. He recalled that there were 13 parameters: two were elastic and defined the stiffness in the tangential and vertical directions to the interface; four referred to frictional parameters that gave the limit of the elastic domain, the coefficient of friction at peak (or failure states) and the coefficients of friction at the phase transformation state and ultimate state. One parameter described the evolution of the hardening law, and the remaining four parameters controlled the rate of mobilisation of dilatancy and friction, and how the softening region would be modelled.

He presented an example based on the comparison between experimental results and numerical simulation of a direct shear test at constant normal stress (σ_n) (Fig. 6, experimental data after Fakharian & Eving 2000). Tests were conducted on a Cyclic 3-Dimensional Simple Shear Interface (C3DSSI) apparatus between samples of dense Silica sand (with a D_{50} of 0.6 mm and a relative density, $I_D = 0.88$) and a rough steel plate. He showed the evolution of shear stress and normal displacement versus tangential displacement and observed the initial contractancy and subsequent development of dilatancy as well as the stabilisation of dilatancy at large tangential displacements corresponding to the ultimate state of the material.

In order to highlight the intrinsic nature of the constitutive parameters of the model and the shortcomings associated with the experiments performed, he described subsequent simulations using the same set of parameters for tests following different stress paths. He compared results of tests carried out at constant normal stiffness (i.e. with a constant ratio K_n between normal displacement and normal stress, where σ_{ni} was the initial normal stress) with the numerical simulations (Fig. 7, experimental data after Fakharian & Eving 2000) and showed overestimation of the dilatancy of the interface (see dotted line). He noted that if the parameter C, which controls the mobilisation of dilatancy at the interface, was increased by about 30%, from an initial value of 0.005 to a final value of 0.0065, a much better fit to the numerical results was achieved. It was emphasised that behaviour was dominated by the response in the normal direction.

He commented that since the parameters were based on results obtained from a constant normal stress test, it was likely that the behaviour in the normal direction reflected purely the behaviour of the interface layer, and ignored the part played by the soil above this layer. He noted, however, that this layer would have a finite stiffness, and would reduce the dilatancy at the interface. He concluded that the determination of the parameters was a key point in a constitutive model based on elastoplasticity and confirmed that the nature of the stress path followed had a significant effect on the determination of the constitutive parameters and important consequences for the modelling.

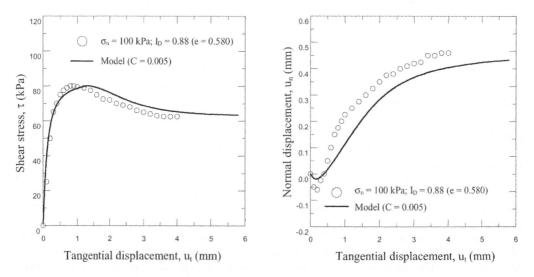

Figure 6. Simulation of a direct shear test on an interface at constant normal stress (De Gennaro 2001, after Fakharian & Eving 2000)

368

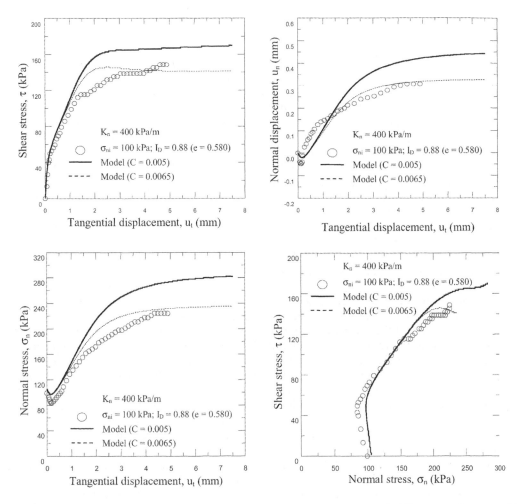

Figure 7. Simulation of a direct shear test on an interface at constant normal stiffness (De Gennaro 2001, after Fakharian & Eving 2000)

5 GENERAL DISCUSSION

Michael Davies noted that two experiments had been described, which treated both sides of the problem: in one there was crushing of the soil grains, and in the other, the interface was wearing away. The question was, could friction fatigue be modelled in the centrifuge? Malcolm Bolton mentioned that Jacques Garnier had showed that rough piles were able to mobilise significant shaft friction in the centrifuge, so the engineer should imagine how piles could be created with a wavy surface for which there were a number of satisfactory practical solutions.

He thought that the real phenomena of friction and wear were being exposed whereas mathematical approaches, that ignored concepts of wear by failing to introduce crushing of particles and the diffusion of fine particles to form a thin layer, were unable to reproduce friction fatigue. Vincenzo De Gennaro agreed but thought that two mechanisms were competing during installation of a pile, and these were very difficult to separate. He felt that there was too much focus on the degradation of the material, whilst forgetting the displacement of a significant volume of material, which led to compaction due to grain crushing. He noted that this produced a reduction in the normal stress and probably a reduction of mobilised shaft friction and suspected that the strong compacting influence was also the origin of the diffusion of the particles inside the intact material. He mentioned that cavity expansion could be used to model jacking or driving of a pile, although the effect of crushing would not be taken into account, so this aspect should be developed further. He postulated that grain crushing was a key point and that this produced the finest particles along the shaft of the pile, with an increasing and not a decreasing roughness, so that proper considerations of the tribology could be represented by accepting that there was another material forming along the interface.

Bolton approved of this approach and concluded that this meant that there was no such thing as an (effective) element test. He pointed out that the stress (i.e. the force divided by the projected area of a particle) necessary to break a silica sand grain was about 100 MPa, whereas the average lateral pressure on the side of Jacques Garnier's pile was about 100 kPa. He concluded that any physical evidence of crushing of particles meant that one particle in a thousand was taking all the force, so this would discourage the use of continuum mechanics in favour of looking at the breakage of individual particles and the consequences of that breakage. He proposed that future research should focus on studies of the mechanics of particulate media and not on continuum mechanics.

Jean Sulem commented that breakage tended to occur due to normal force, whereas in this interface and other shear band problems, intense grain crushing was likely to develop. He was convinced that the rôle of the shear was important and that this was borne out from practical application when trying to break granular material, that it was easier when shear was applied as well as pressure. He thought that important progress had been made in understanding this phenomenon in that an engineering solution had been found to a difficult engineering problem. He described another example of a similar engineering problem based on damage of particles but unfortunately there was presently no effective solution. He described observations from shear bands obtained from triaxial tests on Fontainebleau sandstone at various confining pressures, in which there was high grain comminution or breakage as well as significant crushing. He had analysed the grain size with various methods, starting with the initial size of the particles in the Fontainebleau sandstone matrix at 200 microns, reduced to small particles of maybe one or two microns in the shear band. He mooted that this was the same important phenomenon of grain breakage, as seen by geologists at large scale in faulted ground, and noted that a fluid was circulated, which would (in rock mechanics terms) be expected to augment the permeability following damage. He drew a parallel with petroleum engineering, in which a hole was drilled and followed by some perforation to fragment the rock in order to augment the permeability, which would increase the porosity and improve the circulation. But he pointed out the difference between crushing and breakage, since small particles would fill the pore space and this zone would not become a *channel for* but a *barrier to* the fluid flow, and said that sometimes this had happened following perforation and had reduced fluid migration. He did not feel that there was a good solution at present and that understanding of these phenomena were needed first, as well as good models.

Ivo Herle returned to the effects of grain crushing and what was happening at the interfaces, because the numerical modellers had been asked how to model the effects of grain crushing of the interfaces. He felt that this question could not be answered numerically but should be investigated in experiments first. He noted that the literature focused on the effect of the mean grain size only, whereas he didn't believe that this was the crucial factor and that there was no uniformity of the grain size distribution. He also took issue with the supposition that compaction always occurred due to crushing because of the change in the grain distribution, which might also affect the non-uniformity, and which could change the upper and lower limits of the density dramatically. He explained that this could lead to compression of the sample but that the relative density could reduce at the same time! He thought that this behaviour could be replicated by traditional continuum models if the change in void ratio (e) and the upper (e_{max}) and lower (e_{min}) void ratios were taken into account.

Bolton thought that if the critical state was defined as the stress level at which there was no more damage, then a stress level below this would probably lead to net dilation. But he added that the details of what was happening between the grains would show that the asperities were being removed from the matrix of rather large particles that were still dilating, and that fewer and fewer asperities would be removed at lower stress levels. In this case, he supposed that the angle of dilatancy would increase due to the interlocking of the rough grains at low stress levels, but at a stress level greater than the critical state stress level required to split particles, there would be continuous plastic compaction of volume. He thought that stress states on the dry side of critical would be the most difficult phase, where fine particles were being knocked off the asperities of the large particles and were falling into larger voids, even though the matrix of large sand particles were still dilating. Bolton agreed that it was possible to have damage, contact crushing and dilatancy all at the same time, as was found by Jacques Garnier in the triaxial tests on sand under 100 kPa confining pressure.

Gerd Festag asked about whether it was appropriate to try to capture post-peak behaviour of triaxial tests or shear tests by 'playing with parameters', when the material had been changed in terms of shape, texture and orientation of the grains. He wondered whether adapting one parameter, for example C (which controlled the mobilisation of dilatancy along a shear interface, De Gennaro 2001), was acceptable or should there be a more sophisticated model. De Gennaro noted that elastoplastic modelling of the material did not include the effects of evolution, i.e. did not allow for the properties of the material along the interface to change. He acknowledged that there was a difference in the response between an element test and the behaviour both inside and outside a soil-interface shearing zone. He thought that the material adjacent to the interface would be enormously damaged during the test, and although nothing had been done yet in the context of modelling degradation, the near future should deliver some theoretical micromechanics to take the evolution of the material properties into account.

David White mentioned ways of modelling this problem by assuming that the interface material was different to the one in the far field by assuming other values of the constitutive parameters in the shear zone, but thought that this could be problematic. He noted that the interface material was reducing in volume, even after it had been subjected to very high shear strains and unloaded again, so that dilation would be expected, but he confirmed that this did not eventuate and volume reduction continued. As for the material in the far field, which was modelled as virgin soil, he thought that conventional modelling was difficult because the soil around the pile had already gone through a very complicated stress and strain path. He explained that any rigid body rotation was important in a carbonate soil because the soil was heavily anisotropic, and this, as well as the stress and strain paths, had a major influence on the stiffness, for which it was very hard to deduce parameters. He said that this was crucial to the amount of stress change experienced by the pile shaft during this fatigue process and wondered how these effects would be modelled constitutively in a convincing way.

Andrew Schofield recalled the work of Amontons (1699), who observed that the coefficient of friction between sliding plates was independent of the area and proportional to the pressure, and the work of Belidor (1737) who suggested that roughness could be represented by hemispheres on one surface sitting in pockets between hemispheres on the other surface. Belidor had calculated the angle at which a hemisphere must climb out of a pocket and explained the coefficient of friction as being derived from geometry of roughness and linked to the angle of dilation. Coulomb (1773) had learned that friction was due to dilation, as was thought in France throughout the whole of the 18th century, but only showed a tangential displacement along a shear surface and no coefficient of dilation normal to this surface. Schofield presumed that Coulomb (1785) was troubled by the lack of dissipation energy in this mechanism in his paper on friction. His 1785 prize paper speculated on the mechanisms by which energy could be dissipated with elastic energy dissipation in brushes of deflecting fibres along rubbing wooden surfaces. In post-revolutionary France, Navier (1819) discussed the work of Belidor. He said that the facts relating to friction had been known since Amontons, that friction was independent of the area and was dependent on the pressure. For wood or iron and copper or lead, the coefficient of friction of a surface well lubricated with oil was approximately a third, and so a geometrical theory about the nature of friction was attractive. But Navier's dilative roughness on sliding surfaces was not an acceptable way of characterising friction, as it did not dissipate energy.

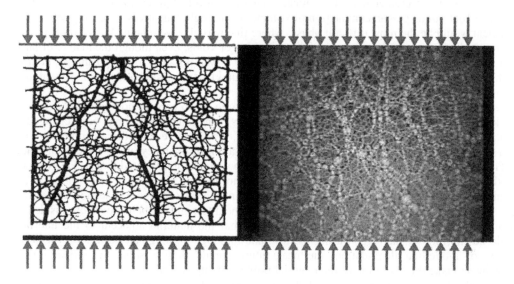

Figure 8. Load paths in a granular material shown by means of photo-elastic particles (Allersma 2001)

Schofield drew a picture of a framework of highly stressed grains in an aggregate, thinking of photo-elastic pictures created by Allersma (2001) in Delft (e.g. Fig. 8). He noted that high loads were carried within an aggregate in a framework of highly stressed grains. Most other grains were not stressed but there was continuing reforming and buckling of the highly stressed members of the frameworks, seen in flickering of the polarised light due to changes in the columns of stressed grains. He concluded that if dissipation of energy in internal friction was attributable to losses of elastic energy, and if this explained the angle of a slope at repose (the fundamental parameter defining soil strength), then our physical model tests may have a fundamental ability to show us mechanisms of failure that triaxial test data and computations can never reveal.

Garnier attempted to answer questions posed by David White and Matthew Dietz about the possibility of simulating interface behaviour in centrifuge models and related this to the important domain of pile problems in geotechnical engineering. He thought that a distinction should be made between displacement piles and non-displacement piles, as adopted by some current codes of practice. He thought that non-displacement piles could be constructed at 1g and then spun up in the centrifuge and loaded in-flight. But he was convinced that displacement piles should be penetrated into the soil in-flight, either by a robot or a driving system, since past work and current research had already demonstrated the complexity of the interface behaviour, particularly due to crushing or wear of the soil grains.

Garnier did not think that the question 'were centrifuge models of interest or not?' was valid because the problem was strongly stress dependent, so that the stresses should be simulated properly. He was convinced that centrifuge technology offered one way of simulating the stresses properly, as did a calibration chamber. He suggested that 'where were the limits of these techniques?' was a better question and noted that these technologies could not be avoided. He concluded that many people had focused on these grain size effects to determine the limits so that experimenters could stay within and not outside the limits.

6 CONCLUSIONS

Andrew House and his group (Jolanda Trausch Giudici, Andy Take, Ravi Chikatamarla) agreed that any physical geotechnical modelling did indeed involve the issue of interfaces, so House declined to summarise what had been discussed primarily the day before, preferring to focus on what was not discussed, and in particular on the issue of interface friction between a foundation and a clay soil. He was not sure why this had not been considered seriously at the workshop until now, and wondered whether it was because it was thought that it was fully understood or perhaps that it was very difficult to model, for example in a centrifuge. He mentioned the very simple example of the penetration resistance of T-bar penetrometer, for which there was an exact numerical solution, but which could not be validated in the centrifuge because the surface smoothness of the T-bar in relation to a clay soil could not be modelled exactly.

He created a theoretical scenario with an offshore foundation designed for installation in soft clay, which might bear a purely coincidental resemblance to a recent experience. He postulated that the offshore foundation had required significantly less installation pressure than expected, which had caused the designers to accept that they had a problem. He imagined the designers wondering about whether the soil had not been characterised correctly and if the foundation would also have a significantly lower uplift capacity, which was the main design requirement. If, on removal of the foundation, clay was not found underneath it, the problem would have turned into an interface one. He suggested that closer inspection might show that the surface had been painted, which would be expected intuitively to mobilise lower frictional resistance than a fully rough surface, such as a rusted steel foundation.

House noted that this scenario exhibited more of a communication than an engineering problem in that interface interaction may be understood but that this would be irrelevant if the design assumptions were not communicated properly. He wondered who might have been responsible for having the surface painted: was it the engineer, who specified it should be painted or was it an insurer, who was concerned about the risks of a rusty foundation? He thought that design engineers needed to focus more on communicating their assumptions to those responsible for construction, or perhaps offshore engineers needed to be more vigilant when foundations were installed. He concluded that it was dangerous to get too carried away with modelling when communication was just as important.

ACKNOWLEDGEMENTS

We are grateful to the Chair, Michael Davies, and the discussion leader, Nori Yasufuku, for their clear leadership and guidance of this session, and Andrew House and his team for their summary. We also thank all of the authors and contributors for their patience with our interpretation of their statements, and for their help in ensuring that this record of the discussions is as accurate as possible.

REFERENCES

Allersma, H.G.B. 2001. http://dutcgeo.ct.tudelft.nl/allersma/hgball.htm.
Amontons, G. 1699. De la résistance causée dans les machines, tant la frottements des parties qui les composent, que le roideur des Corde qu'on employe, et la maniere de calculer l'un et l'autre. Histoire de l'Académie Royale des Sciences. 206. Paris 1702.
Belidor B. F. de. 1737. Architecture Hydraulique. Paris.
Coulomb, C.A. 1785 Théories des machines simples en ayant égard au frottement de leur parties. *Académie Royale des Sciences. Mémoire de mathématique et de physique* 10.
Coulomb, C.A. 1773. Essai sur une application des regeles de maximis & minimis a quelques problemes de statique relatifs a l'architecture. *Mem. de Math. et de Phys., presentes a l'Acad. Roy. des Sci. Paris* 7: 343-382.
De Gennaro, V. 2001. Discussion on "Constitutive modelling of soil and rock interacting with interfaces". *Workshop on Constitutive and Centrifuge Modelling: Two Extremes, Monte Verità, Ticino, July 2001.*
Fakharian, K. & Evgin, E. 2000. Elasto-plastic modelling of stress-path-dependent behaviour of interfaces. *Int. Journal Num. Anal. Meth. Geomech.* 24: 183-199.
Yasufuku, N. 2001. Discussion on "Effects of Steel Granular Material Interaction". *Workshop on Constitutive and Centrifuge Modelling: Two Extremes, Monte Verità, Ticino, July 2001.*
Lehane, B.M. 1992. Experimental investigations of pile behaviour using instrumented field piles. PhD thesis, University of London (Imperial College).
Navier C-L.M.H. 1819. New edition of Belidor. Paris.

Closure

Constitutive and Centrifuge Modelling: Two Extremes, Springman (ed.)
© *2002 Swets & Zeitlinger, Lisse, ISBN 90 5809 361 1*

Closure

David Muir Wood
Department of Civil Engineering, University of Bristol, United Kingdom

I have a colleague David Blockley who believes that, whenever one organises any activity, you should have in advance some idea about your criteria for success. In fact what he usually does at the start of the activity - whether it is a class or a project meeting – is to brainstorm these criteria for success. At the end of the meeting you can decide – again collectively - whether you actually met these criteria.

One of the criteria for success in this workshop, which we did not actually share collectively at the start, is 'communication', mentioned just now by Andrew Schofield. Communication has occurred between very different groups of researchers who are clearly all working in the general area of geomechanics and geotechnical engineering. There is clearly a need for all of us to recognise that there is much common ground in the problems on which we are working, whether from a numerical point of view, or a constitutive point of view, or a physical modelling point of view. I hope that we can feel that there has been some success in achieving some level of communication across these various different groups. I am impressed that there have been people here from various different constituencies, with many of which I have not interacted in the past, but others I know have not interacted with each other either. That intermeshing is a key result that we had hoped to achieve through this workshop this week – aided by the nice surroundings and comfortable accommodation. So, if we have helped to set up some lines of communication which did not previously exist, then I think we may judge this workshop to have been a success.

I should like to thank all the speakers, all the debaters, all the chair people, all the discussers, all the young researchers and everybody in any other category who has taken part in this meeting this week. I would like to add my especial appreciation and gratitude for Sarah's immense energy and drive and insight and inspiration in setting up this meeting and choosing this wonderful place to hold this workshop.

Constitutive and Centrifuge Modelling: Two Extremes, Springman (ed.)
© *2002 Swets & Zeitlinger, Lisse, ISBN 90 5809 361 1*

Author index

T - #0037 - 101024 - C16 - 254/178/21 [23] - CB - 9789058093615 - Gloss Lamination